# Six Sigma for
# Electronics Design
# and Manufacturing

# Six Sigma for Electronics Design and Manufacturing

### Sammy G. Shina
**University of Massachusetts, Lowell**

## McGraw-Hill

New York   Chicago   San Francisco   Lisbon   London   Madrid
Mexico City   Milan   New Delhi   San Juan   Seoul
Singapore   Sydney   Toronto

**Library of Congress Cataloging-in-Publication Data:**

Shina, Sammy G.
    Six Sigma for electronics design and manufacturing / Sammy G. Shina.
      p. cm.
    ISBN 0-07-139511-3
    1. Electronic apparatus and appliances—Design and construction—Quality
control—Standards.   2. Six Sigma (Quality control standard).   I. Title.

TK7836 .S4824 2002
621.381—dc21                                    2002021283

## McGraw-Hill

*A Division of The **McGraw·Hill** Companies*

1 2 3 4 5 7 8 9 0   DOC/DOC   0 9 8 7 6 5 4 3 2

ISBN 0-07-139511-3

The sponsoring editor for this book was Steve Chapman and the production supervisor
was Pamela Pelton. It was set in Century Schoolbook by Ampersand Graphics, Ltd.

*To my wife, love, friend, and companion Jackie,
and our children and grandchildren*

# Contents

# Illustrations and Tables

## Illustrations

## Tables

# Abbreviations

| | |
|---|---|
| AQAP | Advance product quality planning and control plan |
| ANOVA | Analysis of variance |
| AV | Appraiser variation |
| BIST | Built in self-test |
| BOM | Bill of materials |
| CAD | Computer-aided design |
| CAE | Computer-aided engineering |
| CAM | Computer-aided manufacturing |
| CEM | Contract electronic manufacturers |
| CLT | Central limit theory |
| CIM | Computer-integrated manufacturing |
| CPI | Continuous process improvement |
| Cp | Capability of the process |
| Cpk | Capability of the process, with average shift |
| CR | Criteria rating |
| DA | Decision analysis |
| DFD | Data flow diagrams |
| DFM | Design for manufacture |
| DFT | Design for testability |
| DoE | Design of experiments |
| DOF | Degrees of freedom |
| DPMO | Defect per million opportunities |
| DPU | Defects per unit |
| ECO | Engineering change orders |
| ERP | Enterprise requirements planning |
| ESI | Early supplier involvement |
| EV | Equipment variation |
| IPC | Institute for Interconnecting and Packaging of Electronic Circuits |
| FMEA | Failure mode effect analysis |
| FT | Functional test |
| FTY | First-time yield |
| GMP | Good manufacturing practices |
| GR&R | Gauge repeatability and reproducibility |
| Hipot | High potential |
| IC | Integrated circuit |
| ICT | In-circuit test |
| JIT | Just in time |
| MR | Moving range |
| MTBF | Mean time between failure |
| NIH | Not invented here |

| NS | Normal (probability) score |
| NTF | No trouble found |
| OA | Orthogonal arrays |
| OEM | Original equipment manufacturers |
| PCB | Printed circuit board |
| PPM | Parts per million |
| PTF | Polymer thick film |
| QA | Quality assurance |
| QFD | Quality function deployment |
| QLF | Quality loss function |
| RFI | Radio frequency interference |
| ROI | Return on investment |
| RPN | Risk priority number |
| RSS | Root sum of the squares |
| SA | Structure analysis |
| SMT | Surface mount technology |
| SOW | Same old way |
| SL | Specification limits |
| SS | Sum of the squares |
| TH | Through-hole (technology) |
| TQC | Total quality control |
| TQM | Total quality management |

# Preface

Six sigma is becoming more important as companies compete in a worldwide market for high-quality, low-cost products. Successful implementations of six sigma in different companies, large and small, do not follow identical scripts. The tools and methodologies of six sigma are fused with the company's culture to create a unique and successful blend in each instance.

This book is intended to introduce and familiarize design, production, quality, and process engineers and their managers with many of the issues regarding the use of six sigma quality in design and manufacturing of electronic products, and how to resolve them. It is based on my experience in practicing, consulting, and teaching six sigma and its techniques over the last 15 years. During that time, I confronted many engineers' natural reservation about six sigma: its assumptions are too arbitrary, it is too difficult to achieve, it works only for large companies, it is too expensive to implement, it works only for manufacturing, not for design, and so on. They continuously challenged me to apply it in their own areas of interest, presenting me with many difficult design and manufacturing six sigma application problems to solve. At the same time, I was involved with many companies and organizations whose engineers and managers were using original and ingenious applications of six sigma in traditional design and manufacturing. Out of these experiences came many of the examples and case studies in this book.

I observed and helped train many engineers in companies using tools and methodologies of six sigma. The companies vary in size, scope, product type, and strategy, yet they are similar in their approach to successfully implementing six sigma through an interdisciplinary team environment and using the tools and methods mentioned in this book effectively by altering them to meet their particular needs.

I believe the most important impact of six sigma is its use in the design of new products, starting with making it one of the goals of the new product creation process. It makes the design engineers extremely cognizant of the importance of designing and specifying products that can be manufactured with six sigma quality at low cost. Too many times, a company introduces six sigma by having manufacturing adopt it as its goal, a very daunting task, especially if current products were not designed with six sigma in mind.

The approach I use in this book is not to be rigid about six sigma. I have attempted to present many of the options available to measure and implement six sigma, and not to specifically recommend a course of action in each instance. Engineers are very creative people, and they will always try to meld new concepts into ones familiar to them. Many will put their own stamp on its methodology or add their own way of doing things to the six sigma techniques. The one sure way to make them resist a new concept is to force it down their throats. I believe these individual engineers' efforts should be encouraged, as long as they do not detract from the overall goal of achieving six sigma.

I hope that this book will be of value to the neophyte as well as the experienced practitioners of Six Sigma. In particular, it will benefit the small to medium size companies that do not have the support staff and the resources necessary to try out some of the six sigma ideas and techniques and meld them into the company culture. The experiences documented here should be helpful to encourage many companies to venture out and develop new world-class products through six sigma that can help them grow and prosper for the future.

## Acknowledgments

The principals of six sigma discussed in this book were learned, collected and practiced through 14 years on the faculty of the University of Massachusetts, Lowell, where working as a teacher, researcher, and consultant to different companies increased my personal knowledge and experience in the fields of design, manufacturing, quality, and six sigma.

I am indebted to several organizations for supporting and encouraging me during the lengthy time needed to collect my materials, write the chapters, and edit the book. I thank The University of Massachusetts, Lowell for its continuing support for product design and manufacturing, especially Chancellor Bill Hogan; the Dean of the James B. Francis College of Engineering, Krishna Vedula; and the chairman of the Department of Mechanical Engineering, John McKelliget. The Reed Exhibition Companies and SMTA, through their NEPCON and SMTI conferences in Anaheim and Chicago, encourage and nurture the design and manufacturing of electronic products.

In addition, I offer my thanks to Mr. Steve Chapman of McGraw-Hill, Inc., who was my editor for this book, as well as my previous two books on concurrent engineering. He always believed in me and encouraged and guided me through three books, and for that I am very grateful.

Finally, many thanks to my family for emotional support during the writing, editing, and production of the book, including my wife Jackie and our children, Mike, Gail, Nancy, and Jon, as well as my grandchildren, who brought me great joy between the many days of writing and editing. I also thank the many attendees of my seminars on six sigma and quality methods, including the in-company presentations, who kept alive my interest and faith in six sigma. I wish them success in implementing six sigma tools and methods in their companies.

Sammy G. Shina
January, 2002

# The Nature of Six Sigma and Its Connectivity to Other Quality Tools

## 1.1 Historical Perspective

The modern attention to the use of statistical tools for the manufacture of products and processes originated prior to and during World War II, when the United States of America geared up to a massive buildup of machinery and arms to successfully conclude the war. The need to manage the myriad of complex weapon systems and their varied and distributed defense contractors led to the evolution of the system of Statistical Quality Control (SQC), a set of tools that culminated in the military standards for subcontracting, such as MIL-Std 105. The term "government inspector" became synonymous with those individuals who were trained to use the tables that controlled the amount of sampling inspection between the different suppliers of parts used by the main weapons manufacturers. The basis of the SQC process was the use of 3 sigma limits, which yields a rate of 2700 defective parts per million (PPM).

Prior to that period, large U.S. companies established a quality strategy of vertical integration. In order to maintain and manage quality, companies had to control all of the resources used in the product. Thus, the Ford Motor Company in the early part of the 20th century purchased coal and iron mines for making steel for car bodies and forests in Brazil to ensure a quality supply of tires. This strategy was shelved during the rapid buildup for the war because of the use of coproducers as well as subcontractors.

The war was won and U.S. companies returned to their original strategy while the defeated countries were rebuilding their industries. In order to revive the Japanese economy, General McArthur, who was the governor general of Japan at that time, imported some of the U.S. pioneers of SQC to help train their counterparts in Japan. These efforts were largely successful in transforming Japanese industry from a low-technology producer of low-quality, low-cost products such as toys to the other side of the spectrum. By the 1970s and 1980s Japanese products were renowned for their quality and durability. Consumers and companies flocked to buy Japanese electronics, cars, and computer chips, willing to pay a premium for their high quality. In recognition of this effort, Japan established the Deming prize for quality, which was later emulated in the United States, with the Baldrige award.

U.S. companies' response to their loss of market share to Japanese companies was to investigate the Japanese companies' secrets of success. Many U.S. companies organized trips in the 1980s to Japanese companies or branches of U.S. companies in Japan. Initial findings were mostly unsuccessful. Japanese concepts such as "quality circles" or "zero defects" did not translate well into the U.S. companies' culture. Quality circles, which were mostly ad hoc committees of engineers, workers, and their managers, were created to investigate quality problems. In many cases, they were not well organized, and after many months of meetings and discussions, resulted in frivolous solutions. It was also difficult to implement quality circles in unionized shops. The term zero defects was also ambiguous, because it was hard to define: Does the fact that a production line produces a million parts and only one is found to be defective constitute a failure to reach the zero defects goal?

The industrial and business press in the 1980s was filled with articles comparing Japanese and U.S. quality. The pressures mounted to close the quality gap. U.S. Companies slowly realized that quality improvements depended on the realization of two major elements—they have to be quantifiable and measurable, and all elements that make the company successful must be implemented: superior pricing, delivery, performance, reliability, and customer satisfaction. All of the company's elements, not just manufacturing, have to participate in this effort, including management, marketing, design, and external (subcontractors) as well as internal suppliers (in-house manufacturing). The six sigma concept satisfies these two key requirements, which has led to its wide use in U.S. industry today.

The Motorola Company pioneered the use of six sigma. Bill Smith, Motorola Vice President and Senior Quality Assurance Manager, is widely regarded as the father of six sigma. He wrote in the *Journal of Machine Design* issue of February 12, 1993:

For a company aiming to design products with the lowest possible number of defects, traditional three-sigma designs are completely inadequate. Accordingly in 1987, Motorola engineers were required to create all new designs with plus or minus six sigma tolerance limits, given that the sigma is that of a world-class part or process in the first place. This marked the start of Motorola's Six Sigma process and its adoption of robust design as one capable of withstanding twice the normal variation of a process.

Early in 1987, Bob Galvin, the CEO of Motorola and head of its Operating/Policy Committee, committed the corporation to a plan that would determine quality goals of 10 times improvement by 1989, 100 times improvement by 1991, and six sigma capability by 1992. At that time, no one in the company knew how to achieve the six sigma goal, but, in their drive for quality, they committed the company to reach the six sigma defect rate of just 3.4 defective parts per million (PPM) in each step of their processes. By 1992, they met these goals for the most part. At several Motorola facilities, they even exceeded six sigma capability in some products and processes. On average, however, their manufacturing operations by 1992 were at about 5.4 sigma capability, or 40 defective PPM—somewhat short of their original goal.

The six sigma effort at Motorola has led to a reduction of in-process defects in manufacturing by 150 times from 1987 to 1992. This amounts to total savings of $2.2 billion since the beginning of the six sigma program. Richard Buetow, Motorola's Director of Quality, commented that six sigma reduced defects by 99.7% and had saved the company $11 billion for the nine-year period from 1987 to 1996.

Today, Motorola has reached its goal of six sigma. The complexity of new technology has resulted in a continued pressure to maintain this high level of quality. As product complexity continues to increase—such as semiconductor chips with billions of devices and trillions of instructions per second—it will be essential that Motorola master the process of producing quality at a parts-per-billion level. That is quite a challenge. One part per billion is equivalent to one second in 31 years!

Therefore, Motorola expanded the six sigma program in 1992 and beyond to achieve the following:

1. Continue their efforts to achieve six sigma results, and beyond, in everything they do
2. Change metrics from parts per million to parts per billion (PPB)
3. Go forward with a goal of 10 times reduction in defects every 2 years

Many other companies have also adopted these high levels of quality, as well as cost reduction, responsiveness, flexibility, and inventory turnover. One of the most notable is the General Electric Company (GE). Several GE executives commented on the six sigma program in an article by Rachel Lane, a reporter for Bloomberg news, in 1997 and in the GE annual report for the same year. James McNerney, CEO of GE Aircraft Engines said:

> Foremost among our initiatives, Six Sigma Quality is driving cultural change throughout our entire operation and accelerating our business results. Six Sigma tools allow us to improve results dramatically by enhancing the value we provide to our customers. Almost one third of our employees have been trained to lead projects and spread Six Sigma tools to co-workers, resulting in more than $70 million in productivity gains in 1997.

The same year, GE Appliance Director/CEO David Cote said: "This is a leap of faith, when people see the actual results that come from this and make money, you think, 'Son of a gun, this thing really does work!'"

Jeffery Immelt, CEO of GE Medical Systems said in 1997: "If you want to change the way you do things, you have to have people who are in the game." To that end, GE created a class of six sigma practitioners that take their titles from the martial arts. Extensive Training was provided to all employees. Those at the top were called "black belts" and "master black belts." They work on six sigma full time and assist in training and leading six sigma projects. Regular employees who receive abridged training are called "green belts."

## 1.2   Why Six Sigma?

During the last few decades, advances in the high-technology and electronics industries have accelerated. The price/performance ratios continue to follow the industry idioms of more performance for lower price. Intel's Gordon Moore first proposed the law that bears his name in the late 1960s: chip complexity (as defined by the number of active elements on a single semiconductor chip) will double about every device generation, usually about 18 calendar months. This law has now been valid for more than three decades, and it appears likely to be valid for several more device generations. The capacity of today's hard drives is doubling every nine months; and the average price per megabit have declined from $11.54 in 1988 to an estimated $0.02 in 1999.

Great expansion has also been occurring in the field of communication, both in the speed and the availability of the Internet. It is estimated that that global access to the Internet has increased from 171 million people in March 1999 to 304 million in March 2000, an increase of 78%.

In quality, similar improvements have been made, as shown by some of the numbers quoted above. These improvements have led to an increase in customer expectations of quality. Companies have responded to this increase by continuously measuring themselves and their competition in several areas of capabilities and performance. This concept, also known as benchmarking, is a favorite tool of managers to set goals for the enterprise that are commensurate with their competition. They can also gauge the progress of enterprises toward achieving their goals in quality, as well as cost, responsiveness, flexibility, and inventory turnover. Figure 1.1 is a spider diagram of U.S. versus world class benchmarks outlining annual improvements generated by Motorola in 1988, showing the range of capabilities and their annual percentage improvements over a 4 year average period. At that time, it was estimated that the average business in the United States is somewhat profitable, with market prices declining and new competitors entering the marketplace. These companies were spending 10–25% of sales dollars on reworking defects. Concurrently, 5–10% of their customers were dissatisfied and would not recommend that others purchase their products. These companies believed that typical six sigma quality is neither realistic nor achievable, and were unaware that the "best in class" companies are 100 times better in quality.

The inner closed segment in Figure 1.1 represents an average U.S. company in 1988, profiled above. The middle segment represents a world class company, and the outer segment represents the best in class companies. World class is the level of improvements that is

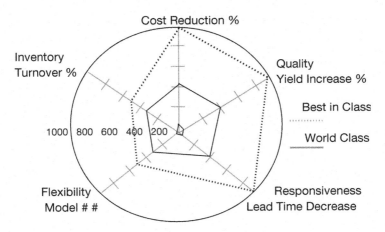

**Figure 1.1**   World-class benchmarks, percentage improvements per year.

needed to compete globally. Best in class represents the best achievable annual improvements recorded anywhere, and not necessarily in the business segment that the company competes in. It is the benchmark of what is achievable in any measure of performance.

It is apparent that an accurate method for developing and improving quality systems in design and manufacturing as well as customer satisfaction is needed to achieve these high quality and capability results, and to compete with products that can be designed, manufactured, and sold anywhere. Six sigma is an excellent tool to achieve world class status as well as best in class results in quality, especially given the increased complexity of designs and products.

At the same time, the requirements for developing new products in high-technology industries have followed these increases in complexity and improvements in quality, necessitating faster product development processes and shorter product lifecycles. Many of the leading technology companies have created "virtual enterprises," aligning themselves with design and manufacturing outsourcing partners to carry out services that can be performed more efficiently outside the boundaries of the organization. These partnerships enabled a company to focus on its core competencies, its own product brand, its customers, and its particular competency in design or manufacturing.

These newly formed outsourcing companies are providing cost-effective and timely services. In manufacturing, they provide multidisciplinary production; test and support services, including printed circuit board (PCB) assembly and testing and packaging technology such as sheet metal and plastic injection molding; and software configuration and support services such as repair depot and warranty exchanges. They also offer lower cost, higher flexibility, and excellent quality, eliminating the need to spend money on capital equipment for internal capacity. This new outsourcing model allows all links in the supply chain to focus on their own core competencies while still reducing overall cycle times.

In design outsourcing, the supply chain offers the flexibility of single or multiple competencies, including specialized engineering analysis and design validation, testing, and conformance to design standards for multiple countries or codes. In addition, suppliers can offer their own supply chain of strategic alliances in tooling and manufacturing services worldwide. Most of these outsourcing companies offer design feedback in terms of design for manufacture (DFM) through early supplier involvement (ESI). These design service providers have reduced the need for high-technology companies to purchase or maintain expensive engineering and design competencies, such as specific design analysis, some of which are used infrequently in project design cycles.

Several industries, especially the auto industry, have worked to standardize their relationship with their suppliers. They created the Advance Product Quality Planning (APQP) Task Force. Its purpose was to standardize the manuals, procedures, reporting format, and technical nomenclature used by Daimler-Chrysler, Ford, and General Motors in their respective supplier quality systems for their design and manufacturing. The APQP also issued a reference manual developed by the Measurement Systems Analysis (MSA) Group for insuring supplier compliance with their standards, especially QS9000. These standards contain many of the principles of six sigma and associated quality tools, such as Cpk requirements. These manuals were published in the mid-1990s and are available from the Automotive Industry Action Group (AIAG) in Southfield Michigan.

Six sigma can be used as a standard for design and manufacturing, as well as a communication method between design and manufacturing groups, especially when part of the design or manufacturing is outsourced. This is important for companies in meeting shorter product lifecycles and speeding up product development through faster access to design and manufacturing information and the use of global supply chains.

## 1.3  Defending Six Sigma

Six sigma, like many new trends or initiatives, is not without its critics and detractors. The author has run into several issues brought up by engineers and managers struggling with six sigma concepts, and has attempted to address these concerns by writing this book. Some of the most frequent critiques of six sigma, and the author's approach to addressing these problems are listed below.

1. The goal of six sigma defects, at 3.4 PPM, and some of its principles, such as the ±1.5 sigma shift of the average manufactured part from specification nominal, sound arbitrary. In addition, there is no solid evidence as to why these numbers have been chosen.

These are reasonable assumptions that were made to implement six sigma. There are other comparable systems, such as Cpk targets used in the auto industry, that could substitute for some of these assumptions. Discussions of these concepts are in Chapters 2 and 3.

2. The cost of achieving six sigma might result in a negative return on investment. Conventional wisdom once held that higher quality costs more, or that there is an optimum point at which cost and quality balance each other, and any further investment in quality will result in negative returns (see the discussion of the quality loss function in Chapter 6).

These beliefs are based on the misconceptions that more tests and inspections are needed in the factory prior to delivery to the customer, in order to deliver higher quality. Six sigma advocates the identification of these costs during the design stage, prior to the manufacturing release of the product, so that these costs are well understood. In addition, it has been demonstrated in six sigma programs that the cost of changing the product in the design stage to achieve higher quality, whether through design changes, different specifications, better manufacturing methods, or alternate suppliers, are much lower than subsequent testing and inspection in manufacturing. These issues are discussed in the chapters on product testing (Chapter 4) and cost (Chapter 6).

3. Many companies feel that the six sigma programs only work well for large-volume, well-established, and consumer-oriented companies such as Motorola and GE, but do not work for other industries such as aerospace, defense, or medical, since their volumes are small or they are more focused on maximizing the performance of products or reducing the time of development projects.

There are many statistical methods that can be used to supplant the sampling and analysis required for six sigma, allowing smaller companies the full benefits of six sigma in product design and manufacturing. Six sigma methods can be used successfully to introduce new low-volume products as well as quantifying marginal designs. These methods will be discussed in the chapters on high and low volume (Chapter 5) and six sigma current and new products (Chapter 8).

4. Many engineers feel that six sigma is for manufacturing only, not for product design, and that it is very difficult to accomplish and cannot be achieved in a timely manner.

In this book, there will be many examples of using six sigma and its associated tools, such as design of experiments (DoE), in product design. These methods can help in realizing the six sigma goals and targets in a timely and organized manner in design and manufacturing. In addition, there are many examples where design engineers were surprised to find out that they are already achieving six sigma in current designs. Six sigma can also be used to flush out "gold plated" designs: designs that are overly robust, beyond the six sigma limits, and therefore costing more than required. These issues are discussed in Chapter 7 on DoE and Chapter 8 on designing current and new products.

## 1.4   The Definitions of Six Sigma

Six sigma integrates well with all of the quality programs and trends of the last few decades. The purpose of this section is to outline conceptually where the six sigma program connects in the quality hierarchy and some of the quality tools that are in common use today. Specific mathematical background and formulations are discussed in detail in later chapters.

Six sigma is a condition of the generalized formula for process capability, which is defined as the ability of a process to turn out a good product. It is a relationship of product specifications to manufacturing variability, measured in terms of Cp or Cpk, or expressed as a numerical index. Six sigma is equivalent to Cp = 2 or Cpk = 1.5 (more on that in the next chapter). The classical definition of the capability of the process or Cp is:

$$Cp = \frac{\text{specification width (or design tolerance)}}{\text{process capability (or total process variation)}} \tag{1.1}$$

Specifically,

$$Cp = \frac{USL - LSL}{6\sigma \text{ (total process range from } -3\sigma \text{ to } +3\sigma)} \tag{1.2}$$

This formula can be expressed conceptually as

$$Cp = \frac{\text{product specifications}}{\text{manufacturing variability}} \tag{1.3}$$

Six sigma is achieved when the product specifications are at ±6σ (σ is the symbol for standard deviation) of the manufacturing process corresponding to Cp = 2 (or Cpk = 1.5, discussed in Chapter 2)

Six sigma or Cp is an excellent indicator of the capability of a process, which can be expressed numerically. This numerical expression can be translated into a defect level using normal distribution statistical assumptions. It is a useful tool for manufacturing process comparisons, as well as a common language of design and manufacturing personnel during the development phase of a product. The design project team and their managers can use it to set new product quality goals. It can be used to assess the quality of internal manufacturing plants anywhere in the world or to measure the capability of a supplier. Companies can use it to communicate a particular contractual level of quality for their supply chain.

## 1.5  Increasing the Cp Level to Reach Six Sigma

The quality tools in wide use today can easily be integrated within the six sigma definitions. The object of six sigma is to steadily increase the process capability index until it reaches the desired level: the specification limits of the design are equal to six sigma of the manufacturing variability.

Design engineers normally set the product specifications, whereas manufacturing engineers are responsible for production variability. The object of increasing the process capability to six sigma or $Cp = 2$ is twofold: increase the product specifications, either by widening them or reducing the manufacturing variability. Either effort can have a positive effect on reaching six sigma.

The design specifications for any part or process are related to the top published product specifications. Ultimately, it is the customer that determines the relative importance of each specification and the desired level of performance. Good market research and project management for new products can determine the best level of specification. This level can be set to balance the wishes of the customer, tempered by what the competition is offering and considering inputs from design and manufacturing engineers as to the difficulty of meeting that specification level.

The quality of supplied parts and the efforts of the manufacturing engineers in production solely determine the denominator of the six sigma equation, or the manufacturing variability. Implementing the traditional quality tools of manufacturing, such as statistical quality control (SQC) and associated quality tools, can reduce the manufacturing variability. The tools of SQC and their relationship to process capability are discussed in Chapter 3.

The Cp formula can then be rewritten as

$$Cp = \frac{\text{specifications}}{\text{variability}} = \frac{\text{design engineering}}{\text{manufacturing engineering}} = \frac{\text{customer}}{\text{supplier}} \quad (1.4)$$

## 1.6   Definitions of Major Quality Tools and How They Affect Six Sigma

Before a six sigma effort is launched, it is mandatory to have a well-defined and successfully managed total quality management (TQM) program. The tools of TQM encourage the use of well-established methodologies for quantifying, analyzing, and resolving quality problems. A brief description of the TQM tools and examples of each will be given in Chapter 2.

## 1.7   Mandatory Quality Tools

It is widely recognized that TQM tools and techniques should be in full utilization before the launch of any six sigma program. SQC should also be well implemented in the organization, with wide use of control charts in manufacturing and the supply chain. Both of these tools will be discussed in Chapter 3 regarding process control.

The major tools of quality can be arrayed as to their use in achieving six sigma. Some tools can effect the numerator, denominator, or both elements in Equation 1.4. However, a definition of each major tool is given below, in order to examine its relationship with six sigma.

## 1.8   Quality Function Deployment (QFD)

QFD is a structured process that provides a means for identifying and carrying the customer's voice through each stage of product development and implementation. QFD is achieved by cross-functional teams that collect, interpret, document, and prioritize customer requirements to identify bottlenecks and breakthrough opportunities.

QFD is a market-driven design and development process resulting in products and services that meet or exceed customer needs and expectations. It is achieved by hearing the voice of the customer, directly stated in their own words, as well as analyzing the competitive position of the company's products and services. Usually, a QFD team is formed, consisting of marketing, design, and manufacturing engineers, to help in designing new products, using customer inputs and current product capabilities as well as competitive analysis of the marketplace. QFD can be used alternately for new product design as well as focusing the efforts of the QFD team on improving existing products and processes. QFD combines tools from many traditional disciplines, including engineering, management, and marketing.

### 1.8.1   Engineering

Tools such as structured analysis or process mapping, which is a top-down division of requirements into multiple elements in several charts, each related to a requirement in the higher chart, are employed. An example of two tiers of structured analysis is given in Figures 1.6 and 1.7 and will be discussed later in this chapter.

### 1.8.2   Management

Tools such as decision analysis (DA) or criteria rating (CR) are employed. This technique consists of breaking a complex decision into distinct criteria, ranking each alternative decision versus each criterion, then adding the total weighted criteria to determine the most effective overall decision. An example of criteria rating is the decision on a soldering material for PCB assembly given in Table 1.1. There are four alternatives being considered by the selection team, and the criteria for the decision are listed on the left side of Table 1.1, each

**Table 1.1**   Criteria rating (CR) to select a solder system for PCB assembly

| Criterion | Weight | A | B | C | D |
|---|---|---|---|---|---|
| Resistance | 10 | 70/7 | 50/5 | 70/7 | 60/6 |
| Quality | 10 | 10/1 | 80/8 | 10/1 | 80/8 |
| Foaming | 3 | 30/10 | 21/7 | 21/7 | 30/10 |
| TLV | 4 | 40/10 | 32/8 | 24/6 | 32/8 |
| History | 4 | 40/10 | 24/6 | 40/10 | 40/10 |
| Supplier | 3 | 30/10 | 30/10 | 30/10 | 12/4 |
| Total | | 220 | 237 | 195 | 254 |
| Rank | | 3 | 2 | 4 | 1 |

with its own weight or rank of importance. The selection team decides on the criteria topics and their relative weight based on team discussions and members' individual experiences. Each alternative soldering material is then rated against each criterion, and a relative score is given. In this example, both the criteria weight and the alternative score were recorded with a maximum value of 10. This choice is arbitrary and smaller numbers can be used for the maximum, such as 3, 5, or 9. Each alternative score is multiplied by the criteria weight and recorded in the table. The total weighted score for each alternative is then calculated, and the final decision is selected based on the highest score. In the case shown in Table 1.1, alternative D has the highest score and should be selected.

### 1.8.3  Marketing

Tools such as customer surveys and competitive analyses are employed. These are traditional elements used by marketing to determine customer needs and perceptions about the company's products versus their competition.

In its simplest form, QFD could be used as a relationship matrix whose input is the customer requirements or needs, and outputs are the product specifications. The QFD process is an interaction between the customer needs and the product characteristics, tempered by a competitive analysis and a ranking of the importance of the different customer needs. The QFD matrix is commonly known as the "house of quality," or QFD chart. A simplified approach to the general QFD chart is shown in Figure 1.2. The "hat" on top of the matrix is used to indicate the presence, if any, of interaction(s) between the various product design characteristics. This interaction should be considered when setting the final product specifications. For example, in a disc design, changing the disc characteristic or storage capacity might influence other characteristics such as the data access time for the disc.

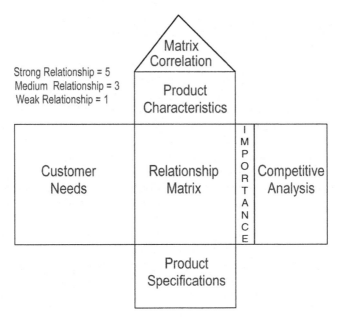

**Figure 1.2**  QFD product planning matrix.

The relation between the customer needs and the product charac-teristics can be considered by the QFD team as having one of four states: strong, medium, weak, or none. Each customer need is given an importance ranking, and that ranking is multiplied by the rela-tionship to generate a total score for each of the product design char-acteristics. In some cases, the importance ranking could further be modified by the marketing emphasis on that customer need. For ex-ample, lighter weight of a product might be considered an important customer need, and customer feedback indicated it should be ranked as medium in importance, for a value of 5. Marketing managers might decide that the new product could compete better if they could empha-size this attribute as a sales point. The importance level could be mul-tiplied by a factor of two, increasing its value to 10. In this manner, the design of the product is forced to be of lighter weight than would otherwise be indicated by the customer's wishes.

A high characteristic total score indicates that the design character-istic is important and the related specification should be enhanced, ei-ther in the positive or negative direction, depending on the direction of "goodness" of the specification. A low score indicates that the speci-fication of the current product design is adequate, and should be left alone or even widened or decreased in value. In this manner, QFD acts as a guide to the design team on what areas of design or specifications to improve, and which others could be left alone.

QFD could be used as a design tool to generate appropriate specifications for new product designs based on customer expectation and competitive analysis as well as marketing inputs. An example is given by Figure 1.3, a modified QFD matrix for the design of a new cable TV connector by Raychem Corporation. This case study was authored by Marylin Liner and published in a book edited by the author (Shina, 1994). Only a portion of the matrix is shown for brevity. Several customer needs obtained from a survey of cable installers are shown, with each having an importance rating (not shown). The relationships are outlined in the top left-hand part of the matrix, with a strong relationship given a value of 9 instead of the commonly used 5, to emphasize the strong customer input contribution in the design of the product. Some of the product characteristics are shown at the top of the matrix. The QFD relationship matrix output is the target value of the design characteristics, where symbolic numbers are shown. The arrows at the bottom show the direction of the enhanced specifications. One of the product characteristics, the number of installation modes, scored the highest total for weighted requirements. This indicated a need for the most important specification change, to a smaller number indicated by the arrow. Hence, the target value or specification for the number of installation modes was assigned a 1. Another product characteristic, the force on equipment panel, which also scored high on weighted requirements, should have its target value or specification

| Relationships | Target Values | ... | 1 mode | x lbs | xx dB | n steps | |
|---|---|---|---|---|---|---|---|
| ■ Strong = 9 pts<br>● Medium = 3 pts<br>▲ Weak = 1 pt | Product Characteristics | ... | Installation modes | Force on Equipment Panel | RF Shielding | Installation steps | ... |
| **Customer Needs** | Importance | | | | | | |
| Clear picture | ... | | ● | | | | ▲ |
| Easy to tell when installed | ... | | ■ | ■ | ■ | | |
| Long lifetime | ... | ■ | ■ | ■ | ▲ | | |
| Simple to install | ... | | ■ | ● | ● | | ■ |
| Weighted Requirements | ... | | 327 | 314 | 322 | 234 | ... |
| Enhanced Specifications Direction | | | ↓ | ↑ | ↑ | ↓ | |

**Figure 1.3**   Raychem CATV new connector QFD matrix.

raised to a higher value, noted symbolically in the figure in units of weight (lbs.).

A QFD example for improving the quality of a manufacturing process is shown in Figure 1.4. In this case, the PCB assembly, consisting of surface mount technology (SMT) solder processes, was analyzed. The QFD team used the QFD process to identify customer needs for quality and delivery of PCBs and rank their importance, as well as the process characteristics of various elements in SMT manufacturing, such as process steps and suppliers of PCBs. The customers of PCB assembly were the personnel in the next stage of production: final product assembly and test technicians. The output of the QFD chart indicates which process element was the most important in meeting customer needs. This is the element that the team should focus on to reduce process defects or manufacturing variability. In the example given in Figure 1.4, the relationship matrix and their calculations for the weighted requirements are outlined. It shows that the team should work most effectively on improving the quality of the screening process before all others, to increase internal customer satisfaction. Indeed, the team decided to run a DoE to optimize the process, similar to the DoE example 8.2.4, given in Chapter 8.

The customer needs were identified in a survey of the appropriate customers that use the PCBs, which are the output of the manufacturing process, divided into primary and secondary needs. The customers also indicated their ratings of importance for each need. This rating is qualitative and is ranked by the team using a scale of 1 to 5, with the larger number being the most important. The process engineers also identified the PCB assembly process characteristics. The team then generated the relationship matrix by matching the customer needs to the process characteristics, in terms of four levels (strong, medium, weak, and none). There should at least one match for each item in the matrix. If an item from the customer needs is not matched by at least one item in the quality characteristics, then the team has to reevaluate the QFD analysis. This is true of the opposite case of a process characteristic not matched by a least one customer need.

The results of the analysis, or the weighted requirements, are determined by multiplying the importance factor by the relationship strength. The screening operation achieved the highest score, indicating that customer needs are best satisfied when that process is improved before the others. This chart represents the analysis by the QFD team at that moment in time, and their collective findings; it does not necessarily reflect a universal solution to improving an SMT process.

**Relationship Legend**

- ⊙ Strong Relationship = 5
- ○ Medium Relationship = 3
- △ Weak Relationship = 1

| Customer Needs (Primary) | Process Characteristics (Secondary) | Screen | place | Load | Reflow | Vision | Supply | Importance | Competitive Analysis (Worse \| Better) |
|---|---|---|---|---|---|---|---|---|---|
| Quality | No solder shorts | ⊙ | | | ⊙ | | | 9 | |
| | No opens | ⊙ | | | ⊙ | | | 10 | |
| | No loading errors | | ⊙ | ○ | | ⊙ | | 8 | |
| | Paste height | ⊙ | | | | | △ | 7 | |
| Delivery | Low downtime | | △ | | | | △ | 6 | |
| | Consistent output | ⊙ | ⊙ | | | ○ | | 5 | |
| **Weighted Requirements** | | 155 | 71 | 24 | 95 | 55 | 13 | | |
| **Target Specification** | | | | | | | | | |
| **Technical Analysis** | | | | | | | | | |

Figure 1.4  SMT process QFD matrix.

The competitive analysis portion of the QFD chart is used mostly for product design. It outlines the team's evaluation of the position of the company's current products against the competition as perceived by the customers. The team could decide to counteract a particular deficiency of the current design in meeting one of the customer needs, and therefore add a multiplier to the importance factor. This multiplier forces the design team to focus on reversing this deficiency in the new product. This occurs when the deficient customer need generates a higher score when multiplied by the importance factor.

As was shown by both design and manufacturing examples, QFD can be an excellent tool to improve the design quality and to attain six sigma levels through focusing on customer needs. In the design example, it can be used to show which specifications should be widened and which can be left alone or even reduced. Widened specifications would affect the numerator of the six sigma equation, making the goal of six sigma easier to achieve. In the manufacturing example, it was used as a defect reduction tool by the manufacturing quality team to identify which process should be investigated to reduce defects and hence manufacturing variability. Such processes could undergo a design of experiments (DoE) project to reduce variability, which is the denominator of the six sigma equation.

It is important to note that QFD is a process designed to solicit customer needs from experienced users of established products or processes. In both examples, those directly involved in the use of the product, such as cable installers or the recipients of PCBs, were part of the customer needs assessment. Products and processes using new technology would benefit less from QFD. For example, it would not be beneficial for slide rule users to quantify their experience into customer needs for calculators. In this case, more traditional marketing research methods could substitute for QFD.

## 1.9   Design for Manufacture (DFM)

The principles of design for manufacture and design for electronic assembly have been widely been used in industry through design guidelines and DFM systems for effectively measuring the efficiency of designs for manufacture and cost. The most important guidelines for DFM design for parts are:

1. Use minimum parts types
2. Use standard components
3. Use parts that fit or snap together with no fasteners
4. Tools are not required for product assembly

DFM analysis results in reduced production time and need for operator skills. The DFM design guidelines, such as the ones mentioned above, are based on common lessons learned while developing electronic products. Prior to formal DFM systems, checklists were being used by major electronic companies as a repository for the collective wisdom of their successful design engineers.

DFM design guidelines emphasize the design of electronic products using self-locating and self-aligning parts, built on a suitable base part. The number of parts should be minimized by using standard parts and integrating functionality and utility. Several cost saving techniques should be used, such as standard and automatic labeling, self-diagnosis capability at the lowest level, and using symmetrical and tangle-free part designs.

In the formal methodology of DFM, a scoring system is used to measure the design efficiency, based on the performance objective and the manufacturing capability. Several alternate designs can be created using the principles of DFM, and the best design can then be chosen based on the scoring system. A conceptual view of a DFM scoring system is shown in Figure 1.5. A typical output of well-designed DFM products is shown in Table 1.2, which compares the design of a new product to older non-DFM designs. Such a product is the Hewlett Packard (now Agilent) 34401A Multi-meter. This case study was authored by Robert Williams and published in a book edited by the author (Shina, 1994). The product was designed using six sigma and QFD. It can be seen that the number of parts and assemblies have been reduced significantly over previous generations of multi-meters through the application of DFM as well as QFD princi-

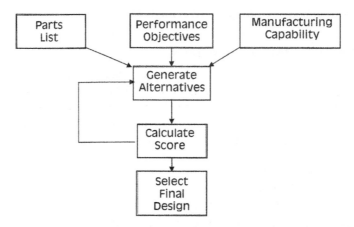

**Figure 1.5**  Use of a DFM scoring system.

Table 1.2   HP 34401A multimeter DFM results

| DFM metric | HP 34401A % | Previous generation | Previous similar generation |
|---|---|---|---|
| Material $ | 80 | 100 | 200 |
| Nonmaterial $ | 55 | 100 | 250 |
| Assembly time | 37 | 100 | 210 |
| Average repair time | 33 | 100 | 400 |
| Number of mechanical parts | 30 | 100 | 190 |
| Number of fasteners | 31 | 100 | 172 |
| Number of fastener types | 8 | 100 | 8 |
| Number of connects/disconnects/adjusts | 36 | 100 | 120 |
| Final assembly part count | 40 | 100 | 153 |
| Total parts | 68 | 100 | 190 |
| Total part numbers | 77 | 100 | 150 |
| Number of suppliers | 70 | 100 | N/A |
| Inventory days | 4 | 100 | 100 |
| Throughput | 100 | 100 | 100 |
| First year engineering change orders (ECOs) | 0 | 100 | 58 |

ples during the design stage. In addition, the new product was introduced to manufacturing without any engineering change orders (ECOs) in the first year of production. A typical successful new development project for a new product using DFM could include the following activities:

- Score product and part designs in breadboard or early prototype stage, prior to initiating CAD drawings. This is important, since once the drawings are completed, it is difficult for design engineers who invested valuable time in the current drawings to redraw them based on DFM evaluations.
- Identify difficult assembly steps and determine if part design changes can make them easier to assemble.
- Test for redundant parts and review the use of nonstandard parts.
- Based on the DFM review, simplify and redesign the parts or final product, using competitive benchmarks, especially if the competition is successfully applying DFM. This design review may include changing process plans or assumptions. Generate a new design that is more efficient by eliminating redundant parts, making parts symmetrical and minimizing assembly motions.
- Rescore the new design and weigh benefits of redesign versus cost and quality adverse consequences, if any. Consider the impact on schedule, tooling, production, and part cost.
- Pursue chosen design approach.

The objectives of DFM are more focused on design for low cost. This is accomplished through fewer parts, parts that are standardized, or parts that are easier for operators or production machines to assemble, hence requiring lower operator skills. The result of DFM analysis could be very beneficial toward achieving the goal of six sigma. A well-designed DFM part or assembly can have a much wider tolerance, or it can be easier to manufacture, resulting in reduced assembly defects. In addition, the design team can focus better on a smaller number of parts.

An interesting consequence of applying DFM to new designs, which will be discussed in the next chapter, results from the reduction in the number of parts. Each additional part carries with it a potential for more defects. A smaller number of parts reduces the opportunities to generate defects, hence making the part design more robust and closer to the six sigma goal.

## 1.10    Design of Experiments (DoE)

Though this quality tool will be discussed in detail in Chapter 7, a quick review is given in this chapter in order to round out the quality tools integration with the six sigma principles. Much like QFD, design of experiments (DoE) can be used in both design and manufacturing, and hence can influence both parts of the six sigma equation: design specifications and manufacturing variability.

DoE can be used in order to focus the new product development project not only on cost, as in DFM, but on several other areas such as quality, variability reduction, and specification selection. The same set of experiments can be used to optimize any of the parameters mentioned above: product cost, quality, or specifications. DoE has been widely used in manufacturing, but not in design, much like six sigma. It is the intent of the author to demonstrate the successful use of DoE in design as well as manufacturing, especially in case studies where it was used to enhance the attainment of the six sigma goals.

## 1.11    Other Quality Tools

There is a wide range of tools necessary for the planning, maintenance, and troubleshooting of quality problems and defects. These tools include quality planning tools that are helpful in estimating and planning for contingencies when a new product is launched, or when a production process is being upgraded or improved. They include the tools described in the following subsections.

### 1.11.1  Process mapping

Process mapping is a structured approach focused on improving processes to deliver the highest quality and value of products and services to the customer. It is based on structured analysis (SA) and structured designs, which were tools that were developed for the software industry as a means toward hierarchical decomposition and description of software modules. Structured analysis and design were developed to replace the traditional tools of flowcharting as software projects and programming complexities increased.

The advantage of process mapping is the presentation of information flows between different systems and departments in a graphical manner. Using a hierarchical approach, process mapping allows for easy understanding of a complex system or process. Process mapping has been used successfully in management information systems to design the information and data flows for manufacturing operations. It could also be used to describe the complex marketing, sales, manufacturing, and quality systems that are used to develop and introduce new products to manufacturing and the marketplace.

Structured analysis uses few symbols and techniques to present a complex system or operation. The top-level boundary of the system being described is called the context diagram, and the decomposition of the system into smaller, more detailed units is called data flow diagraming. This process, known as "top-down partitioning," occurs when data flow diagrams are decomposed from a very high level and general view of the system, to a very detailed view of specific operations.

A data flow diagram may be defined as a network of related functions showing all data interfaces between its elements. These elements are:

- The data source or destination, represented by a rectangular box. A source is defined as an originator of data and a destination is defined as the target for data receipt. Sources and destinations are used to determine the domain of the study of the system, such as departments, suppliers, and customers.

- The data store is represented by two parallel lines or an open box. It represents a repository of information. Data can be stored in electronic files or in physical locations such as file drawers. The name of the file or the storage system should be written alongside the symbols. In complex diagrams, the same data stores might be drawn at different locations to minimize nesting of the lines. In these cases, another vertical line is added to the symbol to indicate that it is repeated elsewhere in the diagram.

- The data flow, represented by an arrow, symbolizes the information being carried from different parts of the systems in order to be transformed or recorded. The name of the data flow should be written alongside the arrow.

Every data flow and data store should have an associated data dictionary, which provides a single document to record information necessary to the understanding of the data flow diagrams. The information can take the form of what records are kept for each data item and the associated information for each record. The definition of each element of process mapping is as follows:

- Process—activities to satisfy customers' requirements
- Inputs—the material or data that is changed by the process
- Outputs—the results of the operations of the process

The basic elements of structured analysis or process mapping are shown in Figure 1.6. The process mapping procedures consist of the following steps:

1. Establish process boundaries ("as is" flow), including discussions among the team members regarding the basic elements of the

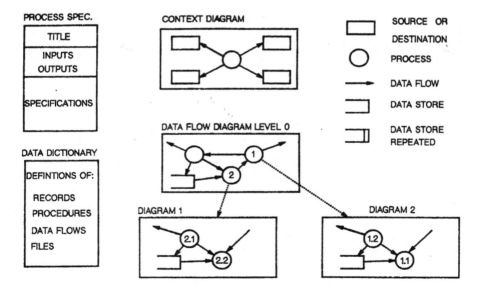

**Figure 1.6**  Structured analysis.

process to be examined, their inputs, and outputs. This would constitute the context diagram of the process and the data flows of current information originating from sources, transformed by processes and arriving at destinations and data stores. The current process operation is recorded, including the relationship of the various processes and the types of data stores and how the data is manipulated.

2. For each process, establish definitions of inputs, outputs, customers, and key requirements. Document process specifications and data dictionary for each process.

3. Analyze the current "as is" process map, then create a more efficient process map, called "should be."

4. Reestablish the process definitions and data flows for the "should be" process map.

These elements of structured analysis, shown in Figure 1.6, are very useful in documenting and explaining to the enterprise how the methods, techniques, responsibilities, and operations of the different parts of the organization interact with one another. It serves as a powerful documenting tool for current processes. In addition, the inherent inefficiencies of the process can be visualized easily, and can be optimized by eliminating excess loops and data transcriptions.

Each department should record its procedures, responsibilities, and functions in its own data flow diagram. This serves as an excellent documentation tool for the total process and its interactions. The visual presentation of the diagrams is much easier to comprehend than written procedures and documentation. For example, design engineers can quickly grasp the interconnection of the different parts of the organization in such cases as design implementation and production of prototypes.

**1.11.1.1  Case study: Using process mapping to schedule a production system.** A team was formed, comprised of associates from different shifts as well as the shop scheduling personnel, to analyze and recommend a new operational strategy for a whiteboard communication system between the different shifts of an electronics factory using process mapping. Team members were quickly able to establish how the different shifts and scheduling departments in their plant carry out their tasks, and interact with other departments. The team elected to formulate the challenge of improving the system in the following three steps:

1. Problem statement: Establish a dispatching system for shop floor scheduling using whiteboards.

2. Establish a set of rules and guidelines.
   - Work the plan: Do not expedite from the next production period and do not start more parts than scheduled.
   - Do not start a job before materials are scheduled or physically in-house.
   - Identify and follow schedule control points or whiteboards.
   - Reduce inventory by developing flexible catch-up plans.
   - Schedule all whiteboards on the floor at the same time.
3. Goals of the scheduling system using whiteboards:
   - Visibility and communication of the plan
   - Track performance to plan
   - Prioritize jobs
   - Recovery plan from problems
   - Improve work flow
   - Communications with upstream and downstream processes
   - Production associates assume responsibility to execute the plan

Using the tools of process and data flow diagrams, the team members collectively produced the context diagram and the top-level data flow diagram, as shown in Figure 1.7. The charts were helpful for

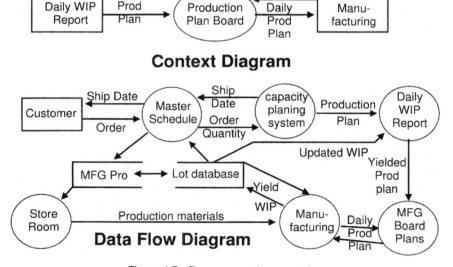

**Figure 1.7**  Process mapping example.

team members to understand the overall manufacturing processes and their interactions, and were used as the basis for formulating a new strategy for the production function of the company.

The DFD diagram in Figure 1.7 contains data stores, which are named by acronyms particular to this manufacturing operation. Their intent was to document the manufacturing process flows in general, and not to specifically detail every existing operation and process. Although no data dictionaries or process specifications were provided for the current process, the reader can follow the information and data flows through the different departments, and understand the complexity and interconnection of the different systems involved in scheduling and manufacturing the product. When designing new manufacturing processes, it is advisable to create the data dictionaries and process specifications to identify each procedure in as detailed a manner as possible.

The data flow diagrams can be used as a quick reference to understand and follow the manufacturing system procedures and requirements. They can lead to better management of the manufacturing function and the data structure needed to support it. They provide a visual representation of the connectivity of the different departments, databases, and functions to be performed. The results of using process mapping are well-managed and efficient operations made possible by:

- Eliminating redundant operations, which will become apparent once the total process is visualized.
- Improving the efficiency of existing operations by clearly identifying the responsibilities of each and its relationship to other operations, as well as by providing the information necessary for correctly performing its functions.
- Better integration with outside activities and sharing of existing resources rather than developing new ones, based on the description of the procedures and documentation of the current process.
- Increasing data integrity by eliminating excess operations. More accuracy will result when databases are well connected, consulted more frequently, and used in more applications. With more focused attention, data has a greater chance of being maintained correctly.

Process mapping methodologies could be very useful when new processes and products are designed or improved to six sigma levels. A good understanding of the system components and their interactions is very beneficial in successfully achieving the goal of six sigma quality for the entire enterprise.

### 1.11.2   Failure modes and effects analysis (FMEA)

FEMA provides a formal mechanism for resolving potential problems in a product, system, or manufacturing process. It is a structured approach to identifying the ways in which products, systems, or processes can fail to meet customers' requirements by:

1. Estimating the risk of specific causes of the failures
2. Evaluating the control plan for preventing the failures from occurring
3. Prioritizing the actions taken to improve the product or process

FMEAs can be performed by teams focused on solving problems in systems, designs, or manufacturing processes. The teams should also perform process mapping of the product, process, or system to be analyzed. The types of FMEAs that can be performed are:

1. System FMEA: Performed in order to analyze systems and their subfunctions in the early concept and design stages. It should be started after systems functions are completed but before detailed design is initiated.
2. Product Design FMEA: Performed on products before they are released to manufacturing. It should be started after product functionality is defined and completed prior to release to manufacturing.
3. Manufacturing FMEA: Performed to analyze manufacturing, assembly, and transaction processes started when preliminary drawings are released. This activity should be ongoing, completed only when the product is obsolete.

### 1.11.1.2   FMEA process.

The FMEA methodology begins with identifying each element, assembly, or part of the process, and listing the potential failure modes, potential causes, and effects of each failure. A risk priority number (RPN) is calculated for each failure mode. It is an index used to measure the rank importance of the items listed in the FMEA chart. These conditions include the probability that the failure takes place (occurrence), the damage resulting from the failure (severity), and the probability of detecting the failure in-house (detection). High RPN items should be targeted for improvement first. The FMEA analysis suggests a recommended action to eliminate the failure condition by assigning a responsible person or department to resolve the failure by redesigning the system, design, or process and recalculating the RPN.

In summary, the FMEA process is comprised of the following steps:

1. FMEA preparation
   - Select FMEA process team and complete a process map. Identify all process steps.
   - List process outputs to satisfy internal and external customers.
   - List process inputs for each process step and rank them.
   - Develop a relationship matrix, relating product to process steps.
2. FMEA process
   - List ways process inputs can vary and identify failure modes.
   - List other causes and sources of variability.
   - Assign severity, occurrence, and detection rating for each cause.
   - Calculate risk priority number (RPN) for each failure.
3. FMEA improvements
   - Determine recommended actions with time needed to reduce RPN.
   - Forecast risk reduction and take appropriate action to reduce failure risk.
   - Recalculate RPN and put controls in place to ensure that failure is completely eliminated from the system or process. An example of an FMEA chart is given in Figure 1.8.

### 1.11.2.2  FMEA definitions

*Failure mode:* A statement of fact describing what would happen when a system, a part, or a process has potentially failed to meet the designer specification intent or performance requirements. The cause might be a design flaw or a change in the product that prevents it from functioning properly.

*Effect:* A description of what the end user will experience or notice. The users might be line operators, the next department to receive the parts, or the customers.

*Cause:* The reason why a failure occurred.

*Severity(SEV):* How significant is the impact of the effects to the customers (internal or external)?

*Occurrence (OCC):* How likely is the cause of the effect to occur?

*Detection (DET):* How likely will the current system detect the cause of the failure mode?

*Risk priority number (RPN):* A numerical calculation of the relative risk of a particular failure mode, obtained by multiplying the severity, occurrence and detection numbers of each failure listed in the FMEA chart.

Product _____
Primary Responsibility _____
Outside Suppliers _____

Model Affected _____
Schedule Start date _____
Project Engineer _____

Page ___ of ___
FMEA Date ___
Approved By ___

| Part name Part # | Process Function | Potential Failure Mode | Potential Effect(s) Failure | Potential Cause(s) Failure | Current Status | | | | | | | | Resulting Status | | |
|---|---|---|---|---|---|---|---|---|---|---|---|---|---|---|---|
| | | | | | Current Controls | occurrence | severity | detection | Risk Priority # (RPN) | Recommend Action /Status | | Action Taken | RPN | Person responsible |

**Figure 1.8** Failure mode and effect analysis (FMEA) chart.

$$RPN = SEC \times OCC \times DET \qquad (1.5)$$

All items with an RPN that exceeds 120 should be investigated first. An item that could cause a safety-related failure, a field recall, or one with a high customer requirement should be considered critical and dealt with promptly.

**1.11.2.3 FMEA results.** FMEA is an excellent tool for investigating potential failures in products or processes. It could lead directly to improving the design or manufacturing quality, especially when prioritizing which parts or processes to work on first. Ideally, it should be used for all parts of the process, product, or system. In practice, a methodology such as QFD should be established to prioritize which elements are to be analyzed using FMEA.

FMEA is a good example of using tools to identify and prioritize quality problems in design and manufacturing. It is another tool to guide the enterprise on where to start quality improvements on the road to six sigma. Some of the benefits of FMEA projects are:

- Establish priorities as to which of the failure items should be improved first
- Identify potential failure modes for each item
- List the types, risks, and causes of failures, and the effects these failures might have
- Calculate a risk priority number, and then use the same number to benchmark improvement in design or manufacturing
- Encourage the planning of a proposed corrective action
- Establish an ordered list of current controls
- List completed quality actions and who performed them
- Document improvements to the process or design

**1.12   Gauge Repeatability and Reproducibility (GR&R)**

The use of six sigma to communicate quality issues between the company and its supply chain is increasing, especially in cases where industries have adopted these techniques as standards for operations, such as the auto industry. Given that the six sigma or Cpk requirements are spelled out in contractual agreements, it is imperative that there be mutual agreement on the measurements of the specifications or manufacturing variability, the two major constituents of six sigma. Differences in measurements due to operator or equipment variability must be accounted for within the six sigma calculations. Gauge repeatability and reproducibility (GR&R) is an excellent tool to quantify these variations in measurements. A more detailed analysis of GR&R is given in Chapter 5.

## 1.13   Conclusions

Six sigma encompasses all the elements necessary for ensuring high-quality design and manufacturing. It draws from the historical perspective of quality, starting with three sigma design and statistical quality control, and moves forward to doubling the quality to six sigma and bringing these quality methods into systems for product design as well as manufacturing.

This chapter presented a historical perspective on quality tools and techniques and how they relate to six sigma. These methods have been used by world class companies to produce new products, aiming at the greatest customer satisfaction, with high quality and low cost. Six sigma is a requisite for companies developing new products, and must be used to develop aggressive but achievable goals of improving new product quality at lower costs, and with high serviceability and customer satisfaction. Examples of executive comments on six sigma were quoted from two companies that pioneered six sigma: Motorola and General Electric.

The techniques presented in this chapter included tools that one might think are independent of each other and six sigma, such as quality function deployment (QFD), design for manufacture (DFM) and design of experiments (DoE). This chapter showed how they are an integral part of the six sigma efforts. The use of these tools is indispensable in reaching six sigma quality, and will be discussed in greater detail in subsequent chapters.

Other quality planning techniques such as process mapping and failure mode and effects analysis (FMEA) were discussed. They are used for documenting and studying the potential defects of a system, process, or new product design and manufacturing. They can help in creating the environment in which quality is more proactive, allowing engineers and designers to search for ways to reduce defects by creating a methodology to prioritize potential problems and then resolve them. Finally, gauge repeatability and reproducibility (GR&R) was introduced in terms of its use in determining sources of measurement variability due to operators or measuring equipment. This is an important tool in communication between the company and its manufacturing and supply chain to resolve measurement problems.

The techniques, tools, and methodologies of six sigma are meant to augment the traditional R&D development and manufacturing process in terms of making it more responsive to customer needs.

## 1.14   References and Bibliography

*Advanced Product Quality Planing and Control Plan (APQP).* Automotive Industries Action Group (AIAG), Southfield, MI, 1995.

Ahmed, M. and Sullivan W. *Factory Automation and Information Management.* CRC Press, Boca Raton, FL, 1991.

Akao, Yoji (ed.). *Quality Function Deployment: Integrating Customer Requirements into Product Design.* Productivity Press, Cambridge MA, 1991.

Bajaria, H. "Six Sigma—Moving Beyond the Hype." In *Annual Quality Congress Proceedings,* ASQ, Milwaukee, WI, 1999.

Boothroyd, G. and Dewhurst D. *Product Design for Assembly.* Wakefield, RI, Boothroyd & Dewhurst Inc., 1997.

Clausing, D. and Simpson, H. "Quality By Design." *Quality Progress,* January, 41–44, 1990.

Cutts, L. *Structured Analysis and Design Methods.* Van Nostrand Reinhold, New York, 1990.

Galvin, R. Personal communication, June 1996.

General Electric Corporation. Annual Report, 1997.

Gryna, Frank. *Quality Planning and Analysis, from Product Development through Use,* Fourth Edition. New York, McGraw-Hill, 2001.

Harry, M. "The Nature of Six Sigma Quality." Motorola University, Schaumburg, IL, 1990.

Harry, M. and Schroeder R. *Six Sigma.* Doubleday, New York, 2000.

Hahn, G. et al. "The Evolution of Six Sigma." *Quality Engineering, 12,* 3, 317–326, 1992.

Hauser, J. and Clausing, D. "The House of Quality." *Harvard Business Review, 3,* May–June, 63–73, 1988.

Iversen, W. "The Six Sigma Shootout." *Assembly Magazine,* June 1993, 20–24.

Kemtovicz, R. *New Product Development: Design and Analysis.* Wiley, New York, 1992.

Koch, R. *The 80/20 Principle.* Doubleday, New York, 1998.

*Measurement Systems Analysis (ASA).* Automotive Industries Action Group (AIAG), Southfield, MI 1995.

Motorola Corporation. Annual Report, 1992.

"Six Sigma Technical Institute Level II Notes." Motorola University Conference, Dallas, TX, October 1993.

Otto, K. and Wood, K. *Product Design: Techniques in Reverse Engineering and New Product Development."* Upper Saddle River, NJ, Prentice Hall, 2001.

*Production Part Approval Process (PPAP).* Automotive Industries Action Group (AIAG), Southfield, MI, 1995.

Ross, Phillip J. "The Role of Taguchi Methods and Design of Experiments in QFD." *Quality Progress, XXI,* 6, 41–47, June 1988.

Shina, S. (ed.). *Successful Implementation of Concurrent Engineering Products and Processes.* Wiley, New York, 1994.

Shina, S. *Concurrent Engineering and Design for Manufacture of Electronic Products.* Kluwer Academic Publishers, Norwell MA, 1991.

Smith, B. "Six Sigma Quality, a Must Not a Myth." *Machine Design,* February 12, 1993, 13–15.

Srihari, K. "A DFM Framework for Surface Mount PCB Assembly." In *International Conference on Technology Management, Design for Competitiveness.* Colorado State University, Denver, CO, 1993, pp. 207–217.

Sullivan, L.P. "Quality Function Deployment." *Quality Progress, XIX,* 6, 39–50, June 1986.

Ulrich, K. and Eppinger S. *Product Design and Development,* 2nd ed., Irwin McGraw-Hill, New York, 2000.

Wolf, D., "Design for Manufacturability—A Printed Wiring Board Fabricator's Perspective." In *Proceedings of the Surface Mount International Conference.* Surface Mount Technology Associates, San Jose, CA, 1991, pp. 953–971.

# The Elements of Six Sigma and Their Determination

In this chapter, the concepts needed to define six sigma quality in design and manufacturing are differentiated from each other. Several techniques are developed for analyzing individual parts, as well as higher orders of complexity such as assemblies, modules, systems, and product designs. In addition, techniques for measuring manufacturing line performance are also developed for use in the six sigma concept. The following topics are discussed in this chapter:

1. The quality measurement techniques: SQC, six sigma, Cp and Cpk. This section is a review of the different methods used to design for quality as well as to control quality. Several techniques are outlined and the differences between the methods are contrasted.
2. The Cpk approach versus six sigma. In this section, the concept of Cpk is analyzed and compared to six sigma. The Cpk approach reduces some of the ambiguities of the 1.5 σ shift of the process average used in the traditional Six Sigma calculations. Cpk calculations, including negative Cpk, are analyzed, and the effects of average shifts on Cpk are also shown.
3. Calculating defects using normal distribution. In this section, defect calculations are shown for variable and attribute processes and designs. Many examples are shown for different conditions of average shift and process variability.
4. Are manufacturing processes and supply parts always normally distributed? Assuming normality of manufacturing process distri-

bution is an important part of calculating defects, yields, and performing other statistical analyses of six sigma. In this section, the requirements for assuming normal distribution of manufacturing processes are examined, as well as tests that can be made to review normality of data. In addition, methods for handling nonnormal distribution of data for six sigma analysis are also shown.

## 2.1   The Quality Measurement Techniques: SQC, Six Sigma, Cp, and Cpk

These quality techniques were developed originally for manufacturing quality and then used for determining product design quality. Six sigma has been used alternately with various assumptions of the manufacturing process average shift from the design specifications to set the defect rate due to design specifications and manufacturing variability.

### 2.1.1   The statistical quality control (SQC) methods

Control charts have been traditionally used as the method of determining the performance of manufacturing processes over time by the statistical characterization of a measured parameter that is dependent on the process. They have been used effectively to determine if manufacturing is in statistical control. Control exists when the occurrence of events (failures) follows the statistical properties of the distribution of production samples.

Control charts are run charts with a centerline drawn at the manufacturing process average and lines drawn at the tail of the distribution at the 3 σ points. If the manufacturing process is under statistical control, 99.73% of all observations are within the limits of the process. Control charts by themselves do not improve quality. They merely indicate that the quality is in statistical "synchronization" or "in control" with the quality level at the time when the charts were created.

A conceptual view of control charts is given in Figure 2.1. The out-of-control conditions indicate that the process is varying with respect to the original period of time when the process was characterized through the control chart, as shown in the bottom two cases. In the bottom case, the process average is shifted to the right, whereas in the next higher case, the process average is shifted to the left. For the two processes shown in control, the current average of the process is equal to the historical one that was determined when the chart was created. The top chart shows a process that is centered with the historical average, and with a small amount of variability, indicating that the

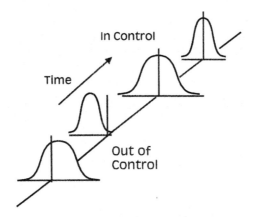

**Figure 2.1**   Conceptual view of control charts.

standard deviation ($\sigma$) is small. It is important to note here that the control charts do not reflect the relation of the process to the specification limit, only the performance of the process to historical standards. Six sigma gives that additional dimension of relating the process performance to the specification tolerance.

### 2.1.2   The relationship of control charts and six sigma

There are two major types of control charts: variable charts, which plot continuous data from the observed parameters, and attribute charts, which are discrete and plot accept or reject data. Variable charts are known as $\overline{X}$ and $R$ charts. They can be directly related to the six sigma calculations through the product specification. Attribute charts are measures of good or bad parts, and therefore are indirectly related to specifications. The relationship of attribute charts to six sigma is that of an assumed set of specifications that produces the particular defect rate plotted in the charts. More on these charts in the next chapter.

The selection of the parameters to be control charted is an important part of the six sigma process. Too many parameters plotted tend to adversely confuse the beneficial effect of the control charts, since they will move together in the same direction when the process is out of control. It is very important to note that the parameters selected for control charting are independent from each other, and are directly related to the overall performance of the product. When a chart shows an out-of-control condition, the process should be investigated and the cause of the problem identified on the chart.

When introducing control charts to a manufacturing operation, it is preferred to use parameters that are universally recognized and with simplified data collection, such as temperature and relative humidity, or take readings from a process display monitor, such as the temperature indicator in a soldering system. These initial control charts can be used to introduce and train the operators in data collection and plotting of parameters. The same principles in selecting these elements also apply to six sigma parameter selections.

### 2.1.3   The process capability index (Cp)

Electronic products are manufactured using materials and processes that are inherently variable. Design engineers specify materials and process characteristics to a nominal value, which is the ideal level for use in the product. The maximum range of variation of the product characteristic, when products are in working order (as defined by customer needs), determines the tolerance of that nominal value. This range is expressed as upper and lower specifications limits (USL and LSL), as shown in Figure 2.2.

The manufacturing process variability is usually approximated by a normal probability distribution, with an average of $\mu$ and a standard deviation of $\sigma$. The process capability is defined as the full range of normal manufacturing process variation measured for a chosen characteristic. Assuming normal distribution, 99.74% of the process output lies between $\mu - 3\sigma$ and $\mu + 3\sigma$.

A properly controlled manufacturing process should make products whose average output characteristic or target is set to the nominal value of the specifications. This is easily achieved through control charts. If the process average is not equal to the product specification nominal value, corrective actions could be taken, such as recalibrating production machinery, retraining the operators, or inspecting incoming raw material characteristics to fix this problem.

The variation of the manufacturing processes (process capability)

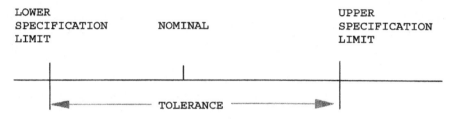

Figure 2.2   Specification and tolerance of a typical product.

should be well within the product tolerance limits. Process capability is commonly depicted by a standard normal distribution. The intersection of the process capability and the specification limits determines the defect level, as shown in Figure 2.3

Process capability could be monitored using control charts. The manufacturing process variability can be reduced by increased operator training, using optimized equipment calibration and maintenance schedules, increased material inspection and testing, and by using design of experiments (DoE) techniques to determine the best set of process parameters to reduce variability.

The classical design for manufacturing (DFM) conflict of interests between design and manufacturing engineers is usually about controlling product quality and cost. The design engineers would prefer the narrowest possible process capability, so they can specify the minimum tolerance specifications to ensure the proper functioning of their designs. The manufacturing and process engineers would prefer the widest possible tolerance specification, so that production can continue to operate at the largest possible manufacturing variability with a reduced amount of defects. The process capability index and six sigma are good arbiters of the two groups' interests.

A good conceptual view of this argument is the use of the term "capability." A process could be either "in control," or "capable," or both. Obviously, the desired condition is both in control and capable, as shown in Figure 2.4. Six sigma assures that the desired outcomes are processes that are highly capable and always in control. If there is a short-term out-of-control condition in manufacturing, then the robustness of the process, which is its capability versus its specifications, is

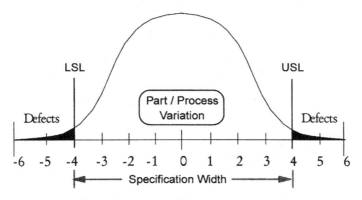

**Figure 2.3** Intersection of process capability and specification limits to determine the defect level.

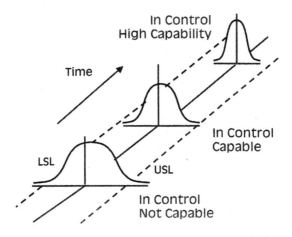

**Figure 2.4**   Conceptual view of control and capability concepts.

good enough to withstand that deviation and continue to produce parts with low defects.

There are two methods used to increase the quality level and hence approach six sigma for new product designs: either increase the product specification limits to allow manufacturing variability to remain the same, or keep product specifications limits constant and reduce manufacturing variability by improving the quality level of materials and processes. The latter can be achieved through inspection, increased maintenance, and performing design of experiments (DoE) to determine variability sources and counteract them. The ratio of the interaction of two sources of defect is the measure of design for quality, called the process capability index or Cp. Six sigma is a special condition in which Cp is equal to 2:

$$Cp = \frac{\text{specification width (or design tolerance)}}{\text{process capability (or total process variation)}} \qquad (2.1)$$

$$Cp = \frac{USL - LSL}{6\sigma \text{ (total range from } -3\sigma \text{ to } +3\sigma)} \qquad (2.2)$$

where
USL = upper specification limit
LSL = lower specification limit
    $\sigma$ = manufacturing process standard deviation

The Cp value can predict the reject rate of new products by using normal probability distribution curves. A high Cp index indicates that the process is capable of replicating faithfully the product characteristics, and therefore will produce products of high quality.

The utility of the Cp index is that it shows the balance of the quality responsibility between the design and manufacturing engineers. The quality level is set by the ratio of the efforts of both. The design engineers should increase the allowable tolerance to the maximum value that still permits the successful functioning of the product. The manufacturing engineers should minimize the variability of the manufacturing process by proper material and process selection, equipment calibration and control, operator training, and by performing design of experiments (DoE).

An example of design and manufacturing process interaction in the electronics industry is the physical implementation of electronic designs in printed circuit board (PCB) layout. The design engineer might select a higher number of layers in a multilayer PCB, which will speed up the layout process because each additional layer increases the PCB surface available for making electrical connections. Speedier layout time could result in a faster new product introduction, bringing in new revenues into the company faster. Minimizing the number of layers requires more layout time, but would produce lower-cost PCB's and fewer defects, because there are fewer process steps. This is a classical case of the balance between new product design and development expediency and manufacturing cost and quality. Six sigma helps focus all engineers toward making the proper decision in these cases by quantifying the quality and cost benefits of the alternatives. A case study of resolving this problem is given in Chapter 6, Section 6.3.4.

### 2.1.4  Six sigma approach

The six sigma concept requires that each process element and each part necessary for the product have a defect rate of no more than 3.4 PPM (parts per million). The underlying assumption is that the variations occurring in all the parameters associated with these process elements and parts follow a normal statistical distribution function and that the specification limits are situated six sigma away from the process average. A further assumption is made that the average value of a parameter can shift from the specification nominal by as much as $\pm 1.5\ \sigma$. With this shift, one of the specification limits is at 4.5 $\sigma$ away from the process average, instead of 6 $\sigma$, while the other specification limit is at 7.5 $\sigma$, where defects can be ignored. This will result in a defect rate, based on one side of the normal distribution, of 3.4 PPM. This defect rate results from the interaction of the normal distribution of parts versus the 4.5 $\sigma$ limit.

It has been a historical practice, based on the control charts methodology, to use a natural tolerance of ±3 σ of the manufacturing processes as design specification limit criteria. This would result in a defect rate on both sides of the normal distribution representing the manufacturing process of 2700 PPM (2 × 1350 PPM for each side) for processes whose average is equal to the specification nominal. At six sigma, the result is a 0.002 PPM defect rate. In Figure 2.5, a normal distribution with 4 σ specification limits is shown with process average shifted by 2.5 σ to either side of the distribution. If the average shift is to the left, the specifications are at 1.5 σ on the LSL and at 6.5 σ on the USL. The defect rate at the LSL can be calculated at 66,810 PPM, and is practically zero at the USL. For specification limits of ±4 σ and an average shift of ±1.5 σ, the specification limits will occur at 2.5 σ and 5.5 σ. The defect rates are 6210 and 0.02 PPM, respectively, for a total defect rate of 6210 PPM.

The defect rates resulting from combinations of different quality levels and process distribution average shifts are shown in Table 2.1. The strong effect of the distribution shift on the resulting failure rate is clearly evident. A reduction in distribution average shift from ±1.5 σ to ±1 σ, with a design specification limit of ±5 σ, allows the defects to be reduced from 230 to 32 PPM.

Achieving the six sigma defect rates of less than 3.4 PPM depends on the manufacturing processes distribution averages and standard deviations, and the product design nominal values and its specification limits. The manufacturing process distribution can be centered or shifted with respect to the nominal value, and it can be tight or broad relative to the specification limits. Setting the specification limits significantly tighter than functionally required could result in an unnec-

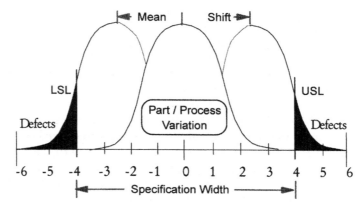

**Figure 2.5**  Normal distribution with mean shifted by 2.5 σ.

**Table 2.1**  Defect rates in PPM for different quality levels and distribution shifts

| Cp | ±SL | 0 Shift | ±1 σ Shift | ±1.5 σ Shift | |
|------|--------|------------|------------|------------|--------|
| 1.0 | ±3 σ | 2700.0 | 22782.0 | 66803.0 | PPM |
| | | 99.73 | 97.72 | 93.32 | % FTY |
| 1.33 | ±4 σ | 64.0 | 1350.0 | 6210.0 | PPM |
| | | 99.9936 | 99.87 | 99.38 | % FTY |
| 1.67 | ±5 σ | 0.6 | 32 | 233 | PPM |
| | | 99.99994 | 99.997 | 99.977 | % FTY |
| 2.0 | ±6 σ | 0.002 | 0.3 | 3.4 | PPM |
| | | 99.9999998 | 99.99997 | 99.99966 | % FTY |

essary increase of defects. Calculations of defect rates are shown later in this chapter.

### 2.1.5   Six sigma and the 1.5 σ shift

An advantage of six sigma is that design quality can be described in a single number equal to Cp = 2. Its disadvantage is when the process average does not equal the specification nominal. In that case, the defect rate is not well defined, and is dependent on the average shift, as shown in Table 2.1. The six sigma concept, as prescribed by most companies, assumes that the average quality characteristic of parts being produced can vary as much as ±1.5 σ from the specification nominal. According to Bill Smith, Vice President and Senior Quality Assurance Manager at Motorola, and the recognized "father of six sigma," this ±1.5 σ shift of the average was developed from the history of process shifts from Motorola's own supply chain. This makes six sigma defect calculations inclusive of normal changes in the manufacturing process. A possible cause of this shift in Motorola's supply chain average is that control charts procedures, which are the mainstay of quality in manufacturing, can allow the process average to shift within the three sigma limits before declaring that the process is out of control and initiating corrective action.

A conceptual view of the average shift of ±1.5 σ can be viewed when the control charts and the specifications limits are presented together in the same diagram, as in Figure 2.6. The control limits calculated for the manufacturing process are equal to ±3 standard deviations of the process average distribution and are located within the specification limits presented by the nominal ±6 σ. The solid line normal distribution represents the population distribution with average μ and standard deviation σ, and the dashed line normal distribution represents the process distribution of sample averages $\overline{\overline{X}}$, with sample standard deviation ($s$). The two distributions are related by the central limit theorem:

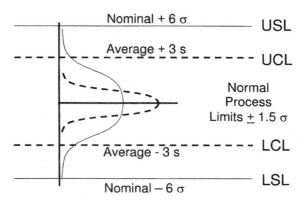

**Figure 2.6**  Specification and control limits.

$$\bar{\bar{X}} = \mu$$

and

$$s = \frac{\sigma}{\sqrt{n}} \tag{2.3}$$

where $n$ is the sample size for each point on the $\bar{X}$ chart.

The $\bar{X}$ control charts work as follows. Each $\bar{X}$ point on the chart represents a sample average of $n$ measurements (as discussed in the next chapter). If the average of a certain sample is calculated with a value just below the 3 $s$ limit in one instance, it is theoretically possible that the control chart will not indicate an out-of-control condition, since the $\bar{X}$ point will be plotted inside the 3 $s$ limit. The factory supplying the parts will not necessarily indicate that an out-of-control condition has occurred in the manufacturing process and will not take corrective action. Assuming a typical sample size of $n = 4$, the 3 $s$ is equal to ±1.5 $\sigma$. Thus, the average of the manufacturing process could theoretically shift by ±1.5 $\sigma$ without triggering the "out-of-control" condition indicated by the SQC process.

## 2.2   The Cpk Approach Versus Six Sigma

Six sigma is focused on the production defect rate or first time yield (FTY) prediction based on the interaction of the process parameters versus the specified tolerance. This ±1.5 $\sigma$ average shift that is allowed under certain definitions of six sigma has led to confusion over defect and FTY calculations. The definition of Cpk attempts to rectify this condition: it is the minimum of the two halves of the distribution interaction of the specifications versus the manufacturing distribution. A capability constant $k$ is provided to calculate Cpk:

$$k = \frac{\text{process shift}}{(\text{USL} - \text{LSL})/2} \quad \text{and} \quad \text{Cpk} = \text{Cp}\,(1 - k) \qquad (2.4)$$

A more direct method for calculating Cpk is to divide the two halves of the distribution as to their interaction with the specification limits:

$$\text{Cpk} = \min \begin{cases} \dfrac{\text{USL} - \text{process average}}{3\sigma} \\[2ex] \dfrac{\text{process average} - \text{LSL}}{3\sigma} \end{cases} \qquad (2.5)$$

When the average shift of the process from specification nominal is equal to zero, then the Cp and Cpk terms are equal.

$$\text{Cpk} = \text{Cp} = \pm\,\frac{\text{SL}}{3\sigma}, \text{ when process average shift from nominal} = 0 \quad (2.6)$$

where
Cp is the process capability index
$k$ is the Cpk constant
USL and LSL are the upper and lower design specifications limits in units of geometry (mm) or output (volts)
SL is the specification limit interval equal to USL or LSL minus the nominal
$\sigma$ is the standard deviation of the manufacturing process

In the design community, Cp = 1 is also called 3 $\sigma$ design, and Cp = 1.33 is called 4 $\sigma$ design.

### 2.2.1  Cpk and process average shift

When there is a manufacturing process average shift, the value of Cpk is not equal to the value of Cp. Using Equation 2.5, Cpk can be calculated for a multitude of conditions, as shown in Figure 2.7. The figure shows specification limits of 27 ± 6, and a varying set of processes, with average $\mu$ and standard $\sigma$ given for each. It can clearly be shown that when the average is equal to the specification nominal, then Cp = Cpk. When the average is shifted, either left or right, then the Cpk value is always less than the Cp.

When this process is reversed—with Cpk given with no information about the process—the amount of average shift with respect to specification nominal cannot be calculated. Table 2.2 is a good illustration of

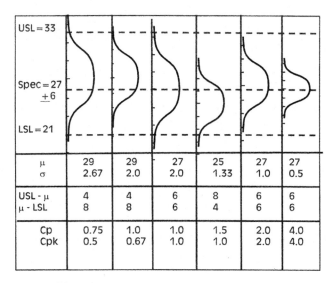

| | | | | | | |
|---|---|---|---|---|---|---|
| μ | 29 | 29 | 27 | 25 | 27 | 27 |
| σ | 2.67 | 2.0 | 2.0 | 1.33 | 1.0 | 0.5 |
| USL - μ | 4 | 4 | 6 | 8 | 6 | 6 |
| μ - LSL | 8 | 8 | 6 | 4 | 6 | 6 |
| Cp | 0.75 | 1.0 | 1.0 | 1.5 | 2.0 | 4.0 |
| Cpk | 0.5 | 0.67 | 1.0 | 1.0 | 2.0 | 4.0 |

(USL = 33, Spec = 27 ± 6, LSL = 21)

**Figure 2.7**  Cp and Cpk sample calculations.

this problem. Several conditions of specification limits are given with varying average shifts. It can be seen that the Cpk = 1.33 could originate from many possible conditions. When the process is centered, the specification limits for Cpk = 1.33 are at ±4 σ. As the process average shifts, the specification limits have to increase to compensate for this shift. For example, if the average shifts by ±1.5 σ, then the specification limits have to increase to ±5.5 σ for the same value of Cpk = 1.33. The easiest condition to achieve Cpk = 1.33 is to design parts specified at 4 σ, and with zero shift of process to the nominal, as shown in Table 2.2.

## 2.2.2  Negative Cpk

Can Cpk be negative? Yes! This is a special condition in which the process average is greater than one of the specification limits. Though

**Table 2.2**  Cpk and process average shift

| Cp | ± SL | No average shift | | ±1 σ Shift | | ±1.5 σ Shift | |
|---|---|---|---|---|---|---|---|
| | | PPM | Cpk | PPM | Cpk | PPM | Cpk |
| 1.33 | ± 4 σ | **64** | **1.33** | 1350.0 | 1.0 | 6210.0 | 0.83 |
| 1.67 | ± 5 σ | 0.6 | 1.67 | **32.0** | **1.33** | 230.0 | 1.17 |
| 1.83 | ± 5.5 σ | 0.02 | 1.83 | 3.4 | 1.5 | **32.0** | **1.33** |
| 2.0 | ± 6 σ | 0.002 | 2.0 | 0.3 | 1.67 | 3.4 | 1.5 |

this is a poor quality design, where more than 50% of the parts made are defective, it is an example of some of the quick indicators that Cpk can provide for prioritizing corrective action for improving products and processes.

### 2.2.3  Choosing six sigma or Cpk

Although both six sigma and Cpk are excellent measurement systems for quality improvements in design and manufacturing, a consensus has not been reached as to which system should be selected based on some of the issues discussed in this section. Currently, major industries and companies have either opted for one or the other, or for their own company brand of six sigma. In the latter case, a combination of rules from both systems is developed to clarify some of the issues, especially when dealing with internal manufacturing and the supply chain. This is important, since the requirements for six sigma or Cpk levels are becoming part of the contractual agreements between companies and their supply chain, as well as performance measures for design and manufacturing centers in modern enterprises.

Some of the issues to be considered when a company plans to launch a quality program based on six sigma or Cpk approaches, and how they can converge, are:

- The classical definition of six sigma corresponds to the last line in Table 2.2. Six sigma is equivalent to Cp = 2 or Cpk = 1.5, while allowing a process average shift to the specification nominal of ±1.5 σ. However, Cpk = 1.5 does not always equate only to six sigma. Many different conditions of specifications tolerance and process average shift can result in Cpk = 1.5, as shown in Table 2.2

- The implication of the six sigma average shift of ±1.5 σ is that the production process variability will not improve beyond the ±1.5 σ shift of the process average. This may be considered as a negative, since it does not encourage those in the supply chain to improve their process variability. By specifying a particular Cpk, a company can encourage its suppliers to minimize their variability, since it is apparent from Table 2.2 that the smaller the average shift, the wider the specification tolerance can be.

- It is widely recognized that older manufacturing processes are more stable than newer processes, and hence are prone to less average shift. This has led to specifying a particular Cpk for new processes, and then a different Cpk when the process matures, in 3 to 6 months after production start-up. In the auto industry, the starting Cpk is set at 1.67 and the mature Cpk at 1.33. This was done to force the supply chain to pay attention to the process in the initial stage of production, a form of learning-curve-based improvements. This issue of time improvements has long been recognized in the supply chain, with commonly used incentives for cost reduction based on time. The six sigma program maintains a constant ±1.5 σ allowable average shift, which is an easier goal to manage irrespective of time. It is the author's opinion that it is better to manage quality with a single number and concept, as opposed to a time-dependant standard. In addition, the reduced life cycle of electronic products, and the emphasis on "doing it right the first time" should encourage the supply chain to set a goal for first production quality and then maintain it. This might prove less costly in the long run.

• The choice of focusing on the process average shift correction to equal the specification nominal or reducing variability or both will be discussed in greater detail together with the quality loss function (QLF), discussed in Chapter 6.

• Cpk and six sigma can have different interpretations when considering attribute processes. These are processes in production, where only the defect rates are determined and there are no applicable specification limits. Examples of attribute processes are assemblies such as printed circuit boards (PCBs) where rejects could be considered to be the result of implied specifications interacting with production variability of materials and processes. In these cases, the quality methodologies are centered around production defect rates and not specifications, thereby clouding the relationships and negotiations between design and manufacturing. Different levels of defect rates based on Cpk levels could be allowed for different processes, resulting in an overall product defect goal setting and test strategy based on these defects. Six sigma quality provides the power of the single 3.4 PPM defect rate as a target for all processes.

• A similar issue arises when using six sigma or Cpk for determining total system or product quality. This is the case when several six sigma designs and parts are assembled together into a system or product. Six sigma practitioners handle this issue by using the concept of rolled yield, that is, the total yield of the product based on the individual yields of the parts. Those using the Cpk terminology can continue to use Cpk throughout the product life cycle, assigning different Cpk targets as the product is going through the design and manufacturing phases. More discussions on this subject are found in Chapter 10.

### 2.2.4    Setting the process capability index

Many companies are beginning to think about the process capability index, be it six sigma or Cpk, as a good method for both design and manufacturing engineers to achieve quality goals jointly, by having both parts work together. Design engineers should open up the specifications to the maximum possible, while permitting the product to operate within customer expectations. Manufacturing engineers should reduce the process variations by maintenance and calibration of processes and materials, training of operators, and by performing design of experiments (DoE) to optimize materials and processing methods.

Another advantage of using the six sigma or Cpk as a quality measure and target is the involvement of the suppliers in the design and development cycle. To achieve the required quality target, the design engineers must know the quality level and specification being delivered by the suppliers and their materials and components. In some cases, the suppliers do not specify certain parameters, such as rise time on integrated circuits, but provide a range. The design engineers must review several samples from different lots from the approved supplier and measure the process variability based on those specifications. A minimum number of 30 samples is recommended.

Many companies use six sigma or a specific Cpk level to set expected design specifications and process variability targets for each part or assembly. Usually, this number has been used to set a particular defect rate such as 64 PPM, which is a Cpk = 1.33 with a centered distribution and specification limit of ±4 σ. The six sigma goal of Cp = 2 results in a defect rate of 3.4 PPM based on a specification limit of ±6 σ and an average shift of ±1.5 σ.

Six sigma or a high Cpk increases the robustness of design and manufacturing. A temporary process average shift does not significantly affect the defect rate. Six sigma (Cp = 2) implies that a shift of the average by as much as ±1.5 σ imparts a defect level of 3.4 PPM to the end product. A comparable shift of the average for a Cp of 1.33 increases the defect rate from 64 PPM to 6210 PPM.

## 2.3   Calculating Defects Using Normal Distribution

Quality defects can be calculated from the defect rate generated by six sigma or Cpk, from the interaction of the production process and the specification limits. The production process characteristics are assumed to be normally distributed. This distribution is also known as the bell curve, and is symmetrical. The area under the curve is equal to 1, and it is much smaller on both ends, as shown in Figure 2.8. Once a process is determined to be normally distributed, it can be characterized by two numbers: a process average μ and a population standard deviation σ. A standard normal curve is one that has

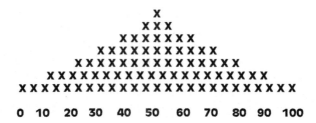

**Figure 2.8**  Graphical presentation of normal distribution.

an average $\mu = 0$ and $\sigma = 1$. For each value $z$ in the x-axis, the area under the curve is given as $f(z)$ in Table 2.3. This area is determined from $x = -\infty$ to $x = z$. Sometimes this normal distribution is called the $z$ distribution, where $z$ is the normalized value of the x-axis intercept.

Since production distributions are not equal to the standard normal distribution, a transformation process is required to convert the specification limits to a form that can be used in the standard normal curve. This is called the $z$-transformation and shown in Figure 2.10. $f(z)$ then determines the defect rate for exceeding the limits of a standard normal curve:

$$z = \frac{\text{SL} - \mu}{\sigma};$$  $f(z)$ is the area under the standard normal    (2.7)
distribution from $-\infty$ to SL

The defect calculations depend on which side of the normal curve is of interest, as shown in Figure 2.11. For the left side of the curve, or the defect rate for product or process values less then the LSL, the defect rate can be calculated directly:

$$z_1 = \frac{\text{LSL} - \mu}{\sigma} \qquad (2.7a)$$

Defects for values of $z < \text{LSL} = f(z_1)$; $z_1$ being negative.

For the right side of the curve, or defects for product values greater then the USL, the defect rate can be derived from the $f(z_2)$ as follows:

$$z_2 = \frac{\text{USL} - \mu}{\sigma} \qquad (2.7b)$$

Defects for value of $z > \text{USL} = 1 - f(z_2)$; $z_2$ being positive.

These $z_2$ defects can be determined quickly, taking advantage of the curve symmetry:

$$\text{defects for value } z > \text{USL} = 1 - f(z_2) = f(-z_2) \qquad (2.8)$$

**Table 2.3**  Standard normal distribution

| z | f(z) | z | f(z) | z | f(z) | z | f(z) | z | f(z) | z | f(z) |
|---|------|---|------|---|------|---|------|---|------|---|------|
| **0** | 0.5 | | | | | | | | | | |
| −0.01 | 0.50399 | 1.01 | 0.84375 | 2.01 | 0.97778 | 3.01 | 0.99869 | 4.01 | 0.99996963 | 5.01 | 0.99999972742 |
| 0.02 | 0.50798 | 1.02 | 0.84614 | 2.02 | 0.97831 | 3.02 | 0.99874 | 4.02 | 0.99997089 | 5.02 | 0.99999974123 |
| 0.03 | 0.51197 | 1.03 | 0.84849 | 2.03 | 0.97882 | 3.03 | 0.99878 | 4.03 | 0.99997210 | 5.03 | 0.99999975436 |
| 0.04 | 0.51595 | 1.04 | 0.85083 | 2.04 | 0.97932 | 3.04 | 0.99882 | 4.04 | 0.99997326 | 5.04 | 0.99999976685 |
| 0.05 | 0.51994 | 1.05 | 0.85314 | 2.05 | 0.97982 | 3.05 | 0.99886 | 4.05 | 0.99997438 | 5.05 | 0.99999977873 |
| 0.06 | 0.52392 | 1.06 | 0.85543 | 2.06 | 0.98030 | 3.06 | 0.99889 | 4.06 | 0.99997545 | 5.06 | 0.99999979002 |
| 0.07 | 0.52790 | 1.07 | 0.85769 | 2.07 | 0.98077 | 3.07 | 0.99893 | 4.07 | 0.99997648 | 5.07 | 0.99999980076 |
| 0.08 | 0.53188 | 1.08 | 0.85993 | 2.08 | 0.98124 | 3.08 | 0.99896 | 4.08 | 0.99997747 | 5.08 | 0.99999981096 |
| 0.09 | 0.53586 | 1.09 | 0.86214 | 2.09 | 0.98169 | 3.09 | 0.99900 | 4.09 | 0.99997842 | 5.09 | 0.99999982066 |
| **0.1** | 0.53983 | **1.1** | 0.86433 | **2.1** | 0.98214 | **3.1** | 0.99903 | **4.1** | 0.99997933 | **5.1** | 0.99999982988 |
| 0.11 | 0.54380 | 1.11 | 0.86650 | 2.11 | 0.98257 | 3.11 | 0.99906 | 4.11 | 0.99998021 | 5.11 | 0.99999983864 |
| 0.12 | 0.54776 | 1.12 | 0.86864 | 2.12 | 0.98300 | 3.12 | 0.99910 | 4.12 | 0.99998105 | 5.12 | 0.99999984696 |
| 0.13 | 0.55172 | 1.13 | 0.87076 | 2.13 | 0.98341 | 3.13 | 0.99913 | 4.13 | 0.99998185 | 5.13 | 0.99999985487 |
| 0.14 | 0.55567 | 1.14 | 0.87286 | 2.14 | 0.98382 | 3.14 | 0.99916 | 4.14 | 0.99998262 | 5.14 | 0.99999986238 |
| 0.15 | 0.55962 | 1.15 | 0.87493 | 2.15 | 0.98422 | 3.15 | 0.99918 | 4.15 | 0.99998337 | 5.15 | 0.99999986952 |
| 0.16 | 0.56356 | 1.16 | 0.87698 | 2.16 | 0.98461 | 3.16 | 0.99921 | 4.16 | 0.99998408 | 5.16 | 0.99999987630 |
| 0.17 | 0.56749 | 1.17 | 0.87900 | 2.17 | 0.98500 | 3.17 | 0.99924 | 4.17 | 0.99998476 | 5.17 | 0.99999988274 |
| 0.18 | 0.57142 | 1.18 | 0.88100 | 2.18 | 0.98537 | 3.18 | 0.99926 | 4.18 | 0.99998542 | 5.18 | 0.99999988885 |
| 0.19 | 0.57535 | 1.19 | 0.88298 | 2.19 | 0.98574 | 3.19 | 0.99929 | 4.19 | 0.99998604 | 5.19 | 0.99999989465 |
| **0.2** | 0.57926 | **1.2** | 0.88493 | **2.2** | 0.98610 | **3.2** | 0.99931 | **4.2** | 0.99998665 | **5.2** | 0.99999990017 |
| 0.21 | 0.58317 | 1.21 | 0.88686 | 2.21 | 0.98645 | 3.21 | 0.99934 | 4.21 | 0.99998722 | 5.21 | 0.99999990540 |
| 0.22 | 0.58706 | 1.22 | 0.88877 | 2.22 | 0.98679 | 3.22 | 0.99936 | 4.22 | 0.99998778 | 5.22 | 0.99999991036 |
| 0.23 | 0.59095 | 1.23 | 0.89065 | 2.23 | 0.98713 | 3.23 | 0.99938 | 4.23 | 0.99998831 | 5.23 | 0.99999991508 |
| 0.24 | 0.59483 | 1.24 | 0.89251 | 2.24 | 0.98745 | 3.24 | 0.99940 | 4.24 | 0.99998882 | 5.24 | 0.99999991955 |
| 0.25 | 0.59871 | 1.25 | 0.89435 | 2.25 | 0.98778 | 3.25 | 0.99942 | 4.25 | 0.99998930 | 5.25 | 0.99999992380 |
| 0.26 | 0.60257 | 1.26 | 0.89617 | 2.26 | 0.98809 | 3.26 | 0.99944 | 4.26 | 0.99998977 | 5.26 | 0.99999992783 |
| 0.27 | 0.60642 | 1.27 | 0.89796 | 2.27 | 0.98840 | 3.27 | 0.99946 | 4.27 | 0.99999022 | 5.27 | 0.99999993165 |
| 0.28 | 0.61026 | 1.28 | 0.89973 | 2.28 | 0.98870 | 3.28 | 0.99948 | 4.28 | 0.99999065 | 5.28 | 0.99999993528 |
| 0.29 | 0.61409 | 1.29 | 0.90147 | 2.29 | 0.98899 | 3.29 | 0.99950 | 4.29 | 0.99999106 | 5.29 | 0.99999993872 |
| **0.3** | 0.61791 | **1.3** | 0.90320 | **2.3** | 0.98928 | **3.3** | 0.99952 | **4.3** | 0.99999145 | **5.3** | 0.99999994198 |
| 0.31 | 0.62172 | 1.31 | 0.90490 | 2.31 | 0.98956 | 3.31 | 0.99953 | 4.31 | 0.99999183 | 5.31 | 0.99999994507 |
| 0.32 | 0.62552 | 1.32 | 0.90658 | 2.32 | 0.98983 | 3.32 | 0.99955 | 4.32 | 0.99999219 | 5.32 | 0.99999994801 |
| 0.33 | 0.62930 | 1.33 | 0.90824 | 2.33 | 0.99010 | 3.33 | 0.99957 | 4.33 | 0.99999254 | 5.33 | 0.99999995079 |
| 0.34 | 0.63307 | 1.34 | 0.90988 | 2.34 | 0.99036 | 3.34 | 0.99958 | 4.34 | 0.99999287 | 5.34 | 0.99999995343 |
| 0.35 | 0.63683 | 1.35 | 0.91149 | 2.35 | 0.99061 | 3.35 | 0.99960 | 4.35 | 0.99999319 | 5.35 | 0.99999995593 |
| 0.36 | 0.64058 | 1.36 | 0.91308 | 2.36 | 0.99086 | 3.36 | 0.99961 | 4.36 | 0.99999349 | 5.36 | 0.99999995830 |
| 0.37 | 0.64431 | 1.37 | 0.91466 | 2.37 | 0.99111 | 3.37 | 0.99962 | 4.37 | 0.99999378 | 5.37 | 0.99999996054 |
| 0.38 | 0.64803 | 1.38 | 0.91621 | 2.38 | 0.99134 | 3.38 | 0.99964 | 4.38 | 0.99999406 | 5.38 | 0.99999996267 |
| 0.39 | 0.65173 | 1.39 | 0.91774 | 2.39 | 0.99158 | 3.39 | 0.99965 | 4.39 | 0.99999433 | 5.39 | 0.99999996469 |
| **0.4** | 0.65542 | **1.4** | 0.91924 | **2.4** | 0.99180 | **3.4** | 0.99966 | **4.4** | 0.99999458 | **5.4** | 0.99999996660 |
| 0.41 | 0.65910 | 1.41 | 0.92073 | 2.41 | 0.99202 | 3.41 | 0.99968 | 4.41 | 0.99999483 | 5.41 | 0.99999996842 |
| 0.42 | 0.66276 | 1.42 | 0.92220 | 2.42 | 0.99224 | 3.42 | 0.99969 | 4.42 | 0.99999506 | 5.42 | 0.99999997013 |
| 0.43 | 0.66640 | 1.43 | 0.92364 | 2.43 | 0.99245 | 3.43 | 0.99970 | 4.43 | 0.99999528 | 5.43 | 0.99999997176 |
| 0.44 | 0.67003 | 1.44 | 0.92507 | 2.44 | 0.99266 | 3.44 | 0.99971 | 4.44 | 0.99999550 | 5.44 | 0.99999997330 |
| 0.45 | 0.67364 | 1.45 | 0.92647 | 2.45 | 0.99286 | 3.45 | 0.99972 | 4.45 | 0.99999570 | 5.45 | 0.99999997476 |
| 0.46 | 0.67724 | 1.46 | 0.92785 | 2.46 | 0.99305 | 3.46 | 0.99973 | 4.46 | 0.99999590 | 5.46 | 0.99999997614 |
| 0.47 | 0.68082 | 1.47 | 0.92922 | 2.47 | 0.99324 | 3.47 | 0.99974 | 4.47 | 0.99999609 | 5.47 | 0.99999997744 |
| 0.48 | 0.68439 | 1.48 | 0.93056 | 2.48 | 0.99343 | 3.48 | 0.99975 | 4.48 | 0.99999626 | 5.48 | 0.99999997868 |
| 0.49 | 0.68793 | 1.49 | 0.93189 | 2.49 | 0.99361 | 3.49 | 0.99976 | 4.49 | 0.99999644 | 5.49 | 0.99999997985 |
| **0.5** | 0.69146 | **1.5** | 0.93319 | **2.5** | 0.99379 | **3.5** | 0.99977 | **4.5** | 0.99999660 | **5.5** | 0.99999998096 |
| 0.51 | 0.69497 | 1.51 | 0.93448 | 2.51 | 0.99396 | 3.51 | 0.999776 | 4.51 | 0.99999676 | 5.51 | 0.99999998201 |
| 0.52 | 0.69847 | 1.52 | 0.93574 | 2.52 | 0.99413 | 3.52 | 0.999784 | 4.52 | 0.99999691 | 5.52 | 0.99999998301 |
| 0.53 | 0.70194 | 1.53 | 0.93699 | 2.53 | 0.99430 | 3.53 | 0.999792 | 4.53 | 0.99999705 | 5.53 | 0.99999998395 |
| 0.54 | 0.70540 | 1.54 | 0.93822 | 2.54 | 0.99446 | 3.54 | 0.999800 | 4.54 | 0.99999718 | 5.54 | 0.99999998484 |
| 0.55 | 0.70884 | 1.55 | 0.93943 | 2.55 | 0.99461 | 3.55 | 0.999807 | 4.55 | 0.99999732 | 5.55 | 0.99999998568 |
| 0.56 | 0.71226 | 1.56 | 0.94062 | 2.56 | 0.99477 | 3.56 | 0.999815 | 4.56 | 0.99999744 | 5.56 | 0.99999998648 |
| 0.57 | 0.71566 | 1.57 | 0.94179 | 2.57 | 0.99492 | 3.57 | 0.999821 | 4.57 | 0.99999756 | 5.57 | 0.99999998723 |

*(continued)*

**Table 2.3**  *Continued*

| z | f(z) | z | f(z) | z | f(z) | z | f(z) | z | f(z) | z | f(z) |
|---|---|---|---|---|---|---|---|---|---|---|---|
| 0.58 | 0.71904 | 1.58 | 0.94295 | 2.58 | 0.99506 | 3.58 | 0.999828 | 4.58 | 0.99999767 | 5.58 | 0.99999998794 |
| 0.59 | 0.72240 | 1.59 | 0.94408 | 2.59 | 0.99520 | 3.59 | 0.999835 | 4.59 | 0.99999778 | 5.59 | 0.99999998862 |
| 0.6 | 0.72575 | 1.6 | 0.94520 | 2.6 | 0.99534 | 3.6 | 0.999841 | 4.6 | 0.99999789 | 5.6 | 0.99999998925 |
| 0.61 | 0.72907 | 1.61 | 0.94630 | 2.61 | 0.99547 | 3.61 | 0.999847 | 4.61 | 0.99999798 | 5.61 | 0.99999998986 |
| 0.62 | 0.73237 | 1.62 | 0.94738 | 2.62 | 0.99560 | 3.62 | 0.999853 | 4.62 | 0.99999808 | 5.62 | 0.99999999043 |
| 0.63 | 0.73565 | 1.63 | 0.94845 | 2.63 | 0.99573 | 3.63 | 0.999858 | 4.63 | 0.99999817 | 5.63 | 0.99999999096 |
| 0.64 | 0.73891 | 1.64 | 0.94950 | 2.64 | 0.99585 | 3.64 | 0.999864 | 4.64 | 0.99999826 | 5.64 | 0.99999999147 |
| 0.65 | 0.74215 | 1.65 | 0.95053 | 2.65 | 0.99598 | 3.65 | 0.999869 | 4.65 | 0.99999834 | 5.65 | 0.99999999196 |
| 0.66 | 0.74537 | 1.66 | 0.95154 | 2.66 | 0.99609 | 3.66 | 0.999874 | 4.66 | 0.99999842 | 5.66 | 0.99999999241 |
| 0.67 | 0.74857 | 1.67 | 0.95254 | 2.67 | 0.99621 | 3.67 | 0.999879 | 4.67 | 0.99999849 | 5.67 | 0.99999999284 |
| 0.68 | 0.75175 | 1.68 | 0.95352 | 2.68 | 0.99632 | 3.68 | 0.999883 | 4.68 | 0.99999856 | 5.68 | 0.99999999325 |
| 0.69 | 0.75490 | 1.69 | 0.95449 | 2.69 | 0.99643 | 3.69 | 0.999888 | 4.69 | 0.99999863 | 5.69 | 0.99999999363 |
| 0.7 | 0.75804 | 1.7 | 0.95543 | 2.7 | 0.99653 | 3.7 | 0.999892 | 4.7 | 0.99999870 | 5.7 | 0.99999999399 |
| 0.71 | 0.76115 | 1.71 | 0.95637 | 2.71 | 0.99664 | 3.71 | 0.999896 | 4.71 | 0.99999876 | 5.71 | 0.99999999433 |
| 0.72 | 0.76424 | 1.72 | 0.95728 | 2.72 | 0.99674 | 3.72 | 0.999900 | 4.72 | 0.99999882 | 5.72 | 0.99999999466 |
| 0.73 | 0.76730 | 1.73 | 0.95818 | 2.73 | 0.99683 | 3.73 | 0.999904 | 4.73 | 0.99999888 | 5.73 | 0.99999999496 |
| 0.74 | 0.77035 | 1.74 | 0.95907 | 2.74 | 0.99693 | 3.74 | 0.999908 | 4.74 | 0.99999893 | 5.74 | 0.99999999525 |
| 0.75 | 0.77337 | 1.75 | 0.95994 | 2.75 | 0.99702 | 3.75 | 0.999912 | 4.75 | 0.99999898 | 5.75 | 0.99999999552 |
| 0.76 | 0.77637 | 1.76 | 0.96080 | 2.76 | 0.99711 | 3.76 | 0.999915 | 4.76 | 0.99999903 | 5.76 | 0.99999999578 |
| 0.77 | 0.77935 | 1.77 | 0.96164 | 2.77 | 0.99720 | 3.77 | 0.999918 | 4.77 | 0.99999908 | 5.77 | 0.99999999602 |
| 0.78 | 0.78230 | 1.78 | 0.96246 | 2.78 | 0.99728 | 3.78 | 0.999922 | 4.78 | 0.99999912 | 5.78 | 0.99999999625 |
| 0.79 | 0.78524 | 1.79 | 0.96327 | 2.79 | 0.99736 | 3.79 | 0.999925 | 4.79 | 0.99999917 | 5.79 | 0.99999999647 |
| 0.8 | 0.78814 | 1.8 | 0.96407 | 2.8 | 0.99744 | 3.8 | 0.999928 | 4.8 | 0.99999921 | 5.8 | 0.99999999667 |
| 0.81 | 0.79103 | 1.81 | 0.96485 | 2.81 | 0.99752 | 3.81 | 0.999930 | 4.81 | 0.99999924 | 5.81 | 0.99999999687 |
| 0.82 | 0.79389 | 1.82 | 0.96562 | 2.82 | 0.99760 | 3.82 | 0.999933 | 4.82 | 0.99999928 | 5.82 | 0.99999999705 |
| 0.83 | 0.79673 | 1.83 | 0.96638 | 2.83 | 0.99767 | 3.83 | 0.999936 | 4.83 | 0.99999932 | 5.83 | 0.99999999722 |
| 0.84 | 0.79955 | 1.84 | 0.96712 | 2.84 | 0.99774 | 3.84 | 0.999938 | 4.84 | 0.99999935 | 5.84 | 0.99999999738 |
| 0.85 | 0.80234 | 1.85 | 0.96784 | 2.85 | 0.99781 | 3.85 | 0.999941 | 4.85 | 0.99999938 | 5.85 | 0.99999999753 |
| 0.86 | 0.80511 | 1.86 | 0.96856 | 2.86 | 0.99788 | 3.86 | 0.999943 | 4.86 | 0.99999941 | 5.86 | 0.99999999768 |
| 0.87 | 0.80785 | 1.87 | 0.96926 | 2.87 | 0.99795 | 3.87 | 0.999946 | 4.87 | 0.99999944 | 5.87 | 0.99999999781 |
| 0.88 | 0.81057 | 1.88 | 0.96995 | 2.88 | 0.99801 | 3.88 | 0.999948 | 4.88 | 0.99999947 | 5.88 | 0.99999999794 |
| 0.89 | 0.81327 | 1.89 | 0.97062 | 2.89 | 0.99807 | 3.89 | 0.999950 | 4.89 | 0.99999950 | 5.89 | 0.99999999806 |
| 0.9 | 0.81594 | 1.9 | 0.97128 | 2.9 | 0.99813 | 3.9 | 0.999952 | 4.9 | 0.99999952 | 5.9 | 0.99999999818 |
| 0.91 | 0.81859 | 1.91 | 0.97193 | 2.91 | 0.99819 | 3.91 | 0.999954 | 4.91 | 0.99999954 | 5.91 | 0.99999999828 |
| 0.92 | 0.82121 | 1.92 | 0.97257 | 2.92 | 0.99825 | 3.92 | 0.999956 | 4.92 | 0.99999957 | 5.92 | 0.99999999838 |
| 0.93 | 0.82381 | 1.93 | 0.97320 | 2.93 | 0.99831 | 3.93 | 0.999958 | 4.93 | 0.99999959 | 5.93 | 0.99999999848 |
| 0.94 | 0.82639 | 1.94 | 0.97381 | 2.94 | 0.99836 | 3.94 | 0.999959 | 4.94 | 0.99999961 | 5.94 | 0.99999999857 |
| 0.95 | 0.82894 | 1.95 | 0.97441 | 2.95 | 0.99841 | 3.95 | 0.999961 | 4.95 | 0.99999963 | 5.95 | 0.99999999865 |
| 0.96 | 0.83147 | 1.96 | 0.97500 | 2.96 | 0.99846 | 3.96 | 0.999963 | 4.96 | 0.99999965 | 5.96 | 0.99999999873 |
| 0.97 | 0.83398 | 1.97 | 0.97558 | 2.97 | 0.99851 | 3.97 | 0.999964 | 4.97 | 0.99999966 | 5.97 | 0.99999999881 |
| 0.98 | 0.83646 | 1.98 | 0.97615 | 2.98 | 0.99856 | 3.98 | 0.999966 | 4.98 | 0.99999968 | 5.98 | 0.99999999888 |
| 0.99 | 0.83891 | 1.99 | 0.97670 | 2.99 | 0.99861 | 3.99 | 0.999967 | 4.99 | 0.99999970 | 5.99 | 0.99999999895 |
| 1 | 0.84134 | 2 | 0.97725 | 3 | 0.99865 | 4 | 0.999968 | 5 | 0.99999971 | 6 | 0.99999999901 |

| z | f(z) | z | f(z) | z | f(z) | z | f(z) | z | f(z) | z | f(z) |
|---|---|---|---|---|---|---|---|---|---|---|---|
| 0 | 0.5 | | | | | | | | | | |
| −0.01 | 0.49601 | −1.01 | 0.15625 | −2.01 | 0.02222 | −3.01 | 0.00131 | −4.01 | 0.00003037 | −5.01 | 0.00000027258 |
| −0.02 | 0.49202 | −1.02 | 0.15386 | −2.02 | 0.02169 | −3.02 | 0.00126 | −4.02 | 0.00002911 | −5.02 | 0.00000025877 |
| −0.03 | 0.48803 | −1.03 | 0.15151 | −2.03 | 0.02118 | −3.03 | 0.00122 | −4.03 | 0.00002790 | −5.03 | 0.00000024564 |
| −0.04 | 0.48405 | −1.04 | 0.14917 | −2.04 | 0.02068 | −3.04 | 0.00118 | −4.04 | 0.00002674 | −5.04 | 0.00000023315 |
| −0.05 | 0.48006 | −1.05 | 0.14686 | −2.05 | 0.02018 | −3.05 | 0.00114 | −4.05 | 0.00002562 | −5.05 | 0.00000022127 |
| −0.06 | 0.47608 | −1.06 | 0.14457 | −2.06 | 0.01970 | −3.06 | 0.00111 | −4.06 | 0.00002455 | −5.06 | 0.00000020998 |
| −0.07 | 0.47210 | −1.07 | 0.14231 | −2.07 | 0.01923 | −3.07 | 0.00107 | −4.07 | 0.00002352 | −5.07 | 0.00000019924 |
| −0.08 | 0.46812 | −1.08 | 0.14007 | −2.08 | 0.01876 | −3.08 | 0.00104 | −4.08 | 0.00002253 | −5.08 | 0.00000018904 |
| −0.09 | 0.46414 | −1.09 | 0.13786 | −2.09 | 0.01831 | −3.09 | 0.00100 | −4.09 | 0.00002158 | −5.09 | 0.00000017934 |
| −0.1 | 0.46017 | −1.1 | 0.13567 | −2.1 | 0.01786 | −3.1 | 0.00097 | −4.1 | 0.00002067 | −5.1 | 0.00000017012 |
| −0.11 | 0.45620 | −1.11 | 0.13350 | −2.11 | 0.01743 | −3.11 | 0.00094 | −4.11 | 0.00001979 | −5.11 | 0.00000016136 |
| −0.12 | 0.45224 | −1.12 | 0.13136 | −2.12 | 0.01700 | −3.12 | 0.00090 | −4.12 | 0.00001895 | −5.12 | 0.00000015304 |

**Table 2.3**  *Continued*

| z | f(z) | z | f(z) | z | f(z) | z | f(z) | z | f(z) | z | f(z) |
|---|------|---|------|---|------|---|------|---|------|---|------|
| −0.13 | 0.44828 | −1.13 | 0.12924 | −2.13 | 0.01659 | −3.13 | 0.00087 | −4.13 | 0.00001815 | −5.13 | 0.00000014513 |
| −0.14 | 0.44433 | −1.14 | 0.12714 | −2.14 | 0.01618 | −3.14 | 0.00084 | −4.14 | 0.00001738 | −5.14 | 0.00000013762 |
| −0.15 | 0.44038 | −1.15 | 0.12507 | −2.15 | 0.01578 | −3.15 | 0.00082 | −4.15 | 0.00001663 | −5.15 | 0.00000013048 |
| −0.16 | 0.43644 | −1.16 | 0.12302 | −2.16 | 0.01539 | −3.16 | 0.00079 | −4.16 | 0.00001592 | −5.16 | 0.00000012370 |
| −0.17 | 0.43251 | −1.17 | 0.12100 | −2.17 | 0.01500 | −3.17 | 0.00076 | −4.17 | 0.00001524 | −5.17 | 0.00000011726 |
| −0.18 | 0.42858 | −1.18 | 0.11900 | −2.18 | 0.01463 | −3.18 | 0.00074 | −4.18 | 0.00001458 | −5.18 | 0.00000011115 |
| −0.19 | 0.42465 | −1.19 | 0.11702 | −2.19 | 0.01426 | −3.19 | 0.00071 | −4.19 | 0.00001396 | −5.19 | 0.00000010535 |
| **−0.2** | 0.42074 | **−1.2** | 0.11507 | **−2.2** | 0.01390 | **−3.2** | 0.00069 | **−4.2** | 0.00001335 | **−5.2** | 0.00000009983 |
| −0.21 | 0.41683 | −1.21 | 0.11314 | −2.21 | 0.01355 | −3.21 | 0.00066 | −4.21 | 0.00001278 | −5.21 | 0.00000009460 |
| −0.22 | 0.41294 | −1.22 | 0.11123 | −2.22 | 0.01321 | −3.22 | 0.00064 | −4.22 | 0.00001222 | −5.22 | 0.00000008964 |
| −0.23 | 0.40905 | −1.23 | 0.10935 | −2.23 | 0.01287 | −3.23 | 0.00062 | −4.23 | 0.00001169 | −5.23 | 0.00000008492 |
| −0.24 | 0.40517 | −1.24 | 0.10749 | −2.24 | 0.01255 | −3.24 | 0.00060 | −4.24 | 0.00001118 | −5.24 | 0.00000008045 |
| −0.25 | 0.40129 | −1.25 | 0.10565 | −2.25 | 0.01222 | −3.25 | 0.00058 | −4.25 | 0.00001070 | −5.25 | 0.00000007620 |
| −0.26 | 0.39743 | −1.26 | 0.10383 | −2.26 | 0.01191 | −3.26 | 0.00056 | −4.26 | 0.00001023 | −5.26 | 0.00000007217 |
| −0.27 | 0.39358 | −1.27 | 0.10204 | −2.27 | 0.01160 | −3.27 | 0.00054 | −4.27 | 0.00000978 | −5.27 | 0.00000006835 |
| −0.28 | 0.38974 | −1.28 | 0.10027 | −2.28 | 0.01130 | −3.28 | 0.00052 | −4.28 | 0.00000935 | −5.28 | 0.00000006472 |
| −0.29 | 0.38591 | −1.29 | 0.09853 | −2.29 | 0.01101 | −3.29 | 0.00050 | −4.29 | 0.00000894 | −5.29 | 0.00000006128 |
| **−0.3** | 0.38209 | **−1.3** | 0.09680 | **−2.3** | 0.01072 | **−3.3** | 0.00048 | **−4.3** | 0.00000855 | **−5.3** | 0.00000005802 |
| −0.31 | 0.37828 | −1.31 | 0.09510 | −2.31 | 0.01044 | −3.31 | 0.00047 | −4.31 | 0.00000817 | −5.31 | 0.00000005493 |
| −0.32 | 0.37448 | −1.32 | 0.09342 | −2.32 | 0.01017 | −3.32 | 0.00045 | −4.32 | 0.00000781 | −5.32 | 0.00000005199 |
| −0.33 | 0.37070 | −1.33 | 0.09176 | −2.33 | 0.00990 | −3.33 | 0.00043 | −4.33 | 0.00000746 | −5.33 | 0.00000004921 |
| −0.34 | 0.36693 | −1.34 | 0.09012 | −2.34 | 0.00964 | −3.34 | 0.00042 | −4.34 | 0.00000713 | −5.34 | 0.00000004657 |
| −0.35 | 0.36317 | −1.35 | 0.08851 | −2.35 | 0.00939 | −3.35 | 0.00040 | −4.35 | 0.00000681 | −5.35 | 0.00000004407 |
| −0.36 | 0.35942 | −1.36 | 0.08692 | −2.36 | 0.00914 | −3.36 | 0.00039 | −4.36 | 0.00000651 | −5.36 | 0.00000004170 |
| −0.37 | 0.35569 | −1.37 | 0.08534 | −2.37 | 0.00889 | −3.37 | 0.00038 | −4.37 | 0.00000622 | −5.37 | 0.00000003946 |
| −0.38 | 0.35197 | −1.38 | 0.08379 | −2.38 | 0.00866 | −3.38 | 0.00036 | −4.38 | 0.00000594 | −5.38 | 0.00000003733 |
| −0.39 | 0.34827 | −1.39 | 0.08226 | −2.39 | 0.00842 | −3.39 | 0.00035 | −4.39 | 0.00000567 | −5.39 | 0.00000003531 |
| **−0.4** | 0.34458 | **−1.4** | 0.08076 | **−2.4** | 0.00820 | **−3.4** | 0.00034 | **−4.4** | 0.00000542 | **−5.4** | 0.00000003340 |
| −0.41 | 0.34090 | −1.41 | 0.07927 | −2.41 | 0.00798 | −3.41 | 0.00032 | −4.41 | 0.00000517 | −5.41 | 0.00000003158 |
| −0.42 | 0.33724 | −1.42 | 0.07780 | −2.42 | 0.00776 | −3.42 | 0.00031 | −4.42 | 0.00000494 | −5.42 | 0.00000002987 |
| −0.43 | 0.33360 | −1.43 | 0.07636 | −2.43 | 0.00755 | −3.43 | 0.00030 | −4.43 | 0.00000472 | −5.43 | 0.00000002824 |
| −0.44 | 0.32997 | −1.44 | 0.07493 | −2.44 | 0.00734 | −3.44 | 0.00029 | −4.44 | 0.00000450 | −5.44 | 0.00000002670 |
| −0.45 | 0.32636 | −1.45 | 0.07353 | −2.45 | 0.00714 | −3.45 | 0.00028 | −4.45 | 0.00000430 | −5.45 | 0.00000002524 |
| −0.46 | 0.32276 | −1.46 | 0.07215 | −2.46 | 0.00695 | −3.46 | 0.00027 | −4.46 | 0.00000410 | −5.46 | 0.00000002386 |
| −0.47 | 0.31918 | −1.47 | 0.07078 | −2.47 | 0.00676 | −3.47 | 0.00026 | −4.47 | 0.00000391 | −5.47 | 0.00000002256 |
| −0.48 | 0.31561 | −1.48 | 0.06944 | −2.48 | 0.00657 | −3.48 | 0.00025 | −4.48 | 0.00000374 | −5.48 | 0.00000002132 |
| −0.49 | 0.31207 | −1.49 | 0.06811 | −2.49 | 0.00639 | −3.49 | 0.00024 | −4.49 | 0.00000356 | −5.49 | 0.00000002015 |
| **−0.5** | 0.30854 | **−1.5** | 0.06681 | **−2.5** | 0.00621 | **−3.5** | 0.00023 | **−4.5** | 0.00000340 | **−5.5** | 0.00000001904 |
| −0.51 | 0.30503 | −1.51 | 0.06552 | −2.51 | 0.00604 | −3.51 | 0.000224 | −4.51 | 0.00000324 | −5.51 | 0.00000001799 |
| −0.52 | 0.30153 | −1.52 | 0.06426 | −2.52 | 0.00587 | −3.52 | 0.000216 | −4.52 | 0.00000309 | −5.52 | 0.00000001699 |
| −0.53 | 0.29806 | −1.53 | 0.06301 | −2.53 | 0.00570 | −3.53 | 0.000208 | −4.53 | 0.00000295 | −5.53 | 0.00000001605 |
| −0.54 | 0.29460 | −1.54 | 0.06178 | −2.54 | 0.00554 | −3.54 | 0.000200 | −4.54 | 0.00000282 | −5.54 | 0.00000001516 |
| −0.55 | 0.29116 | −1.55 | 0.06057 | −2.55 | 0.00539 | −3.55 | 0.000193 | −4.55 | 0.00000268 | −5.55 | 0.00000001432 |
| −0.56 | 0.28774 | −1.56 | 0.05938 | −2.56 | 0.00523 | −3.56 | 0.000185 | −4.56 | 0.00000256 | −5.56 | 0.00000001352 |
| −0.57 | 0.28434 | −1.57 | 0.05821 | −2.57 | 0.00508 | −3.57 | 0.000179 | −4.57 | 0.00000244 | −5.57 | 0.00000001277 |
| −0.58 | 0.28096 | −1.58 | 0.05705 | −2.58 | 0.00494 | −3.58 | 0.000172 | −4.58 | 0.00000233 | −5.58 | 0.00000001206 |
| −0.59 | 0.27760 | −1.59 | 0.05592 | −2.59 | 0.00480 | −3.59 | 0.000165 | −4.59 | 0.00000222 | −5.59 | 0.00000001138 |
| **−0.6** | 0.27425 | **−1.6** | 0.05480 | **−2.6** | 0.00466 | **−3.6** | 0.000159 | **−4.6** | 0.00000211 | **−5.6** | 0.00000001075 |
| −0.61 | 0.27093 | −1.61 | 0.05370 | −2.61 | 0.00453 | −3.61 | 0.000153 | −4.61 | 0.00000202 | −5.61 | 0.00000001014 |
| −0.62 | 0.26763 | −1.62 | 0.05262 | −2.62 | 0.00440 | −3.62 | 0.000147 | −4.62 | 0.00000192 | −5.62 | 0.00000000957 |
| −0.63 | 0.26435 | −1.63 | 0.05155 | −2.63 | 0.00427 | −3.63 | 0.000142 | −4.63 | 0.00000183 | −5.63 | 0.00000000904 |
| −0.64 | 0.26109 | −1.64 | 0.05050 | −2.64 | 0.00415 | −3.64 | 0.000136 | −4.64 | 0.00000174 | −5.64 | 0.00000000853 |
| −0.65 | 0.25785 | −1.65 | 0.04947 | −2.65 | 0.00402 | −3.65 | 0.000131 | −4.65 | 0.00000166 | −5.65 | 0.00000000804 |
| −0.66 | 0.25463 | −1.66 | 0.04846 | −2.66 | 0.00391 | −3.66 | 0.000126 | −4.66 | 0.00000158 | −5.66 | 0.00000000759 |
| −0.67 | 0.25143 | −1.67 | 0.04746 | −2.67 | 0.00379 | −3.67 | 0.000121 | −4.67 | 0.0000015077 | −5.67 | 0.00000000716 |
| −0.68 | 0.24825 | −1.68 | 0.04648 | −2.68 | 0.00368 | −3.68 | 0.000117 | −4.68 | 0.00000144 | −5.68 | 0.00000000675 |
| −0.69 | 0.24510 | −1.69 | 0.04551 | −2.69 | 0.00357 | −3.69 | 0.000112 | −4.69 | 0.00000137 | −5.69 | 0.00000000637 |
| **−0.7** | 0.24196 | **−1.7** | 0.04457 | **−2.7** | 0.00347 | **−3.7** | 0.000108 | **−4.7** | 0.00000130 | **−5.7** | 0.00000000601 |

*(continued)*

**Table 2.3**  *Continued*

| $z$ | $f(z)$ | $z$ | $f(z)$ | $z$ | $f(z)$ | $z$ | $f(z)$ | $z$ | $f(z)$ | $z$ | $f(z)$ |
|---|---|---|---|---|---|---|---|---|---|---|---|
| −0.71 | 0.23885 | −1.71 | 0.04363 | −2.71 | 0.00336 | −3.71 | 0.000104 | −4.71 | 0.00000124 | −5.71 | 0.00000000567 |
| −0.72 | 0.23576 | −1.72 | 0.04272 | −2.72 | 0.00326 | −3.72 | 0.000100 | −4.72 | 0.00000118 | −5.72 | 0.00000000534 |
| −0.73 | 0.23270 | −1.73 | 0.04182 | −2.73 | 0.00317 | −3.73 | 0.000096 | −4.73 | 0.00000112 | −5.73 | 0.00000000504 |
| −0.74 | 0.22965 | −1.74 | 0.04093 | −2.74 | 0.00307 | −3.74 | 0.000092 | −4.74 | 0.00000107 | −5.74 | 0.00000000475 |
| −0.75 | 0.22663 | −1.75 | 0.04006 | −2.75 | 0.00298 | −3.75 | 0.000088 | −4.75 | 0.00000102 | −5.75 | 0.00000000448 |
| −0.76 | 0.22363 | −1.76 | 0.03920 | −2.76 | 0.00289 | −3.76 | 0.000085 | −4.76 | 0.00000097 | −5.76 | 0.00000000422 |
| −0.77 | 0.22065 | −1.77 | 0.03836 | −2.77 | 0.00280 | −3.77 | 0.000082 | −4.77 | 0.00000092 | −5.77 | 0.00000000398 |
| −0.78 | 0.21770 | −1.78 | 0.03754 | −2.78 | 0.00272 | −3.78 | 0.000078 | −4.78 | 0.00000088 | −5.78 | 0.00000000375 |
| −0.79 | 0.21476 | −1.79 | 0.03673 | −2.79 | 0.00264 | −3.79 | 0.000075 | −4.79 | 0.00000083 | −5.79 | 0.00000000353 |
| **−0.8** | 0.21186 | **−1.8** | 0.03593 | **−2.8** | 0.00256 | **−3.8** | 0.000072 | **−4.8** | 0.00000079 | **−5.8** | 0.00000000333 |
| −0.81 | 0.20897 | −1.81 | 0.03515 | −2.81 | 0.00248 | −3.81 | 0.000070 | −4.81 | 0.00000076 | −5.81 | 0.00000000313 |
| −0.82 | 0.20611 | −1.82 | 0.03438 | −2.82 | 0.00240 | −3.82 | 0.000067 | −4.82 | 0.00000072 | −5.82 | 0.00000000295 |
| −0.83 | 0.20327 | −1.83 | 0.03362 | −2.83 | 0.00233 | −3.83 | 0.000064 | −4.83 | 0.00000068 | −5.83 | 0.00000000278 |
| −0.84 | 0.20045 | −1.84 | 0.03288 | −2.84 | 0.00226 | −3.84 | 0.000062 | −4.84 | 0.00000065 | −5.84 | 0.00000000262 |
| −0.85 | 0.19766 | −1.85 | 0.03216 | −2.85 | 0.00219 | −3.85 | 0.000059 | −4.85 | 0.00000062 | −5.85 | 0.00000000247 |
| −0.86 | 0.19489 | −1.86 | 0.03144 | −2.86 | 0.00212 | −3.86 | 0.000057 | −4.86 | 0.00000059 | −5.86 | 0.00000000232 |
| −0.87 | 0.19215 | −1.87 | 0.03074 | −2.87 | 0.00205 | −3.87 | 0.000054 | −4.87 | 0.00000056 | −5.87 | 0.00000000219 |
| −0.88 | 0.18943 | −1.88 | 0.03005 | −2.88 | 0.00199 | −3.88 | 0.000052 | −4.88 | 0.00000053 | −5.88 | 0.00000000206 |
| −0.89 | 0.18673 | −1.89 | 0.02938 | −2.89 | 0.00193 | −3.89 | 0.000050 | −4.89 | 0.00000050 | −5.89 | 0.00000000194 |
| **−0.9** | 0.18406 | **−1.9** | 0.02872 | **−2.9** | 0.00187 | **−3.9** | 0.000048 | **−4.9** | 0.00000048 | **−5.9** | 0.00000000182 |
| −0.91 | 0.18141 | −1.91 | 0.02807 | −2.91 | 0.00181 | −3.91 | 0.000046 | −4.91 | 0.00000046 | −5.91 | 0.00000000172 |
| −0.92 | 0.17879 | −1.92 | 0.02743 | −2.92 | 0.00175 | −3.92 | 0.000044 | −4.92 | 0.00000043 | −5.92 | 0.00000000162 |
| −0.93 | 0.17619 | −1.93 | 0.02680 | −2.93 | 0.00169 | −3.93 | 0.000042 | −4.93 | 0.00000041 | −5.93 | 0.00000000152 |
| −0.94 | 0.17361 | −1.94 | 0.02619 | −2.94 | 0.00164 | −3.94 | 0.000041 | −4.94 | 0.00000039 | −5.94 | 0.00000000143 |
| −0.95 | 0.17106 | −1.95 | 0.02559 | −2.95 | 0.00159 | −3.95 | 0.000039 | −4.95 | 0.00000037 | −5.95 | 0.00000000135 |
| −0.96 | 0.16853 | −1.96 | 0.02500 | −2.96 | 0.00154 | −3.96 | 0.000037 | −4.96 | 0.00000035 | −5.96 | 0.00000000127 |
| −0.97 | 0.16602 | −1.97 | 0.02442 | −2.97 | 0.00149 | −3.97 | 0.000036 | −4.97 | 0.00000034 | −5.97 | 0.00000000119 |
| −0.98 | 0.16354 | −1.98 | 0.02385 | −2.98 | 0.00144 | −3.98 | 0.000034 | −4.98 | 0.00000032 | −5.98 | 0.00000000112 |
| −0.99 | 0.16109 | −1.99 | 0.02330 | −2.99 | 0.00139 | −3.99 | 0.000033 | −4.99 | 0.00000030 | −5.99 | 0.00000000105 |
| **−1** | 0.15866 | **−2** | 0.02275 | **−3** | 0.00135 | **−4** | 0.000032 | **−5** | 0.00000029 | **−6** | 0.00000000099 |

Total defects can thus be calculated for the two sides of the curve. If there is no shift from process average to specification nominal, or the process is centered, then only one side needs to be calculated, then multiplied by two for the total defects:

$$\text{total defects} = f(z_1) + 1 - f(z_2) \qquad (2.9)$$

$$\text{total Defects} = 2 \cdot f(z_1) \text{ when process is centered} \qquad (2.10)$$

The defect rate derived from $f(z)$ in the $z$ table is in terms of a fraction. Since six sigma quality implies very low defect rates, it is shown in parts per million or PPM. PPM can be derived from the defect rate calculations from the standard normal curve as follows:

$$\text{PPM} = \text{defect rate} \cdot 1{,}000{,}000 \qquad (2.11)$$

Figure 2.9 shows part compliance rates outlined for specification limits that are set at multiples of $\sigma$. At specification limits of $\pm 1\ \sigma$, the portion of the curve that is inside the limits (percentage compli-

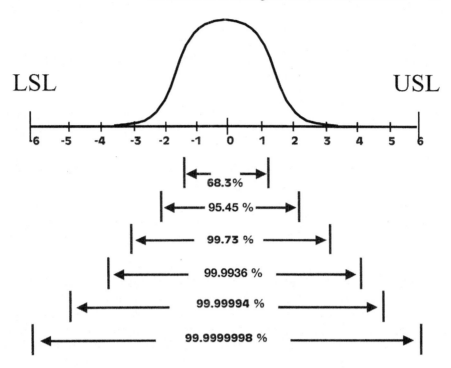

**Figure 2.9**  Graphical presentation of normal distribution with parts compliance percentage and multiple $\sigma$ limits.

ant or OK parts) is 68.3%, for a reject rate of 31.7%. This is equivalent to a $z$ value of 1, whose reject rate $f(-1) = 0.15866$ from the normal distribution tables for one-sided rejects. The total reject rate is $2 \cdot 0.15866 = 0.31732$. For three sigma limits, the area under the curve, or percent compliant parts, is 99.73%, which indicates a reject rate of 0.27% or 2700 PPM, corresponding to a one-sided $z = 3$ and a $2 \cdot f(-3) = 0.00135 \cdot 2 = 0.0027$ reject area under the curve. Sometimes this situation of specification limits at 3 $\sigma$ is also known as 3 $\sigma$ design.

For six sigma limits, it can be seen that the reject rate is equivalent to $2 \cdot (1 - 0.999999999) = 2$ parts per billion. This is the reject rate for six sigma, when there is no shift of the process average with respect to specification nominal. When the ±1.5 $\sigma$ shift is applied, the two-sided $z$ functions become $z_1 = -4.5$ and $z_2 = 7.5$. The reject rate from $z_2$ is too small to be counted, whereas the reject rate for the one-sided $z_1$ is $f(-4.5) = 0.0000034$ or 3.4 PPM, the commonly accepted level of six sigma defects.

### 2.3.1  Relationship between z and Cpk

Since the formulas for $z$ and Cpk are somewhat similar, the two can be derived from each other, especially if the process is centered (no average shift from nominal):

$$\text{Cpk} = [\text{min of } \{z_1, z_2\}]/3 \qquad (2.12)$$

$$\text{Cpk} = \pm \frac{\text{SL}}{3\sigma} = z/3; \qquad \text{when process is centered } (z_1 = z_2) \qquad (2.13)$$

### 2.3.2  Example calculations of defects and Cpk

**Example 2.1**

A check on parts made by a factory indicated that they are made with normal distribution with average = 12.62" and standard deviation of 2.156.

a. What is the probability that parts of the following lengths ($L$) will be made in that factory: $L > 18"$, $L < 8"$, and $10" \le L \le 12"$?
b. If the specifications for the length of the part were $12.62 \pm 3"$, and the factory made parts with a $\mu = 12.62$ and $\sigma = 2.156$, what are Cp and Cpk and the predicted defect or reject rate (RR)? Repeat the above if the process average is shifted with respect to specification nominal by 1" to the left and 0.75" to the right.
c. What should the specifications be if the factory decided on the following: Cp = 1, Cp = 1.5, and Cp = 2 (six sigma), assuming the average is 12.62" and the $\sigma = 2.156$?

**Solutions to Example 2.1**

a. From the standard normal distribution (Table 2.3), the area under the curve is used to determine the answers:

$L > 18"$: $z_2 = (18 - 12.62)/2.156 = 2.5$
$f(z_2) = f(-2.5) = 0.0062$ or 0.62% or 6,200 PPM
$L < 8"$
$z_1 = (8 - 12.62)/2.156 = -2.14$
$f(z_1) = f(-2.14) = 0.0162$ or 1.62% or 16,200 PPM
$10" \le L \le 12"$
$z_2 = (12 - 12.62)/2.156 = -0.29$
$z_1 = (10 - 12.62)/2.156 = -1.22$
$f(z_2) - f(z_1) = f(-0.29) - f(-1.22) = 0.3859 - 0.1112 = 0.2747$

or 27.47% or 274,700 PPM

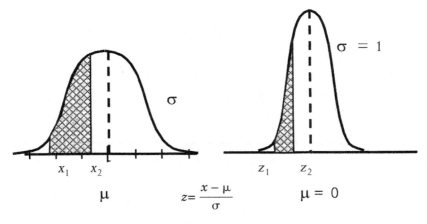

**Figure 2.10**   $z$ transformation.

b.  Nominal = process average

   $Cpk = Cp = \pm \, SL/\pm \, 3 \, \sigma = 3/3 \cdot 2.156 = 0.46$
   $z = 3 \cdot Cp = 1.39$
   $f(z) = f(-1.39) = 0.0823$
   For two-sided defects, $RR = 2 \cdot 0.0823 = 0.1646$ or 16.46% or 164,600 PPM
   Nominal shifted 1″ to the left
   $Cp = 0.46$ (remains the same from above)
   $Cpk = \min(USL - \text{average}/3\sigma)$ or $(\text{average} - LSL/3\sigma)$
   $Cpk = (3 - 1)/3 \cdot 2.156 = 0.31$
   $z_1 = 0.93$ and $z_2 = 1.86$
   $RR = f(-z_1) + f(-z_2) = 0.1762 + 0.0314 = 0.2076$ or 20.76% or 207,600 PPM
   Nominal shifted by 0.75″ to the right
   $Cp = 0.46$
   $Cpk = (3 - 0.75)/3 \cdot 2.156 = 0.35$
   $z_1 = 1.04$ and $z_2 = 1.74$
   $RR = 0.1492 + 0.0409 = 0.1901$ or 19.01% or 190,100 PPM

c.  For $Cp = 1$, specification limits are:

   $12.62 \pm 3 \cdot 2.156 = 12.62 \pm 6.468 = 19.088″$ to $6.152″$
   $Cp = 1.5$, specification limits are:
   $12.62 \pm 4.5 \cdot 2.156 = 12.62 \pm 9.702 = 22.322″$ to $2.918″$

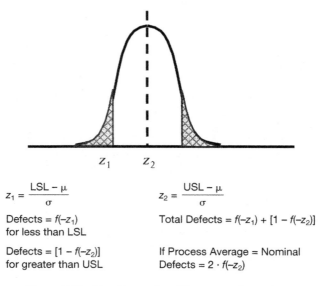

$$z_1 = \frac{LSL - \mu}{\sigma}$$

Defects = $f(-z_1)$
for less than LSL

Defects = $[1 - f(-z_2)]$
for greater than USL

$$z_2 = \frac{USL - \mu}{\sigma}$$

Total Defects = $f(-z_1) + [1 - f(-z_2)]$

If Process Average = Nominal
Defects = $2 \cdot f(-z_2)$

**Figure 2.11**   Negative and positive $z$ transformation.

Cp = 2 or six sigma, specification limits are:

$12.62 \pm 6 \cdot 2.156 = 12.62'' \pm 12.936'' = 25.556''$ to $-0.316''$ or $0''$.

### Example 2.2

Table 2.4 contains a good set of conditions to examine the calculations of Cp, Cpk, and defect rates. It shows that these calculations can be different according to the specification tolerance width or the process distribution as presented by the process average $\mu$ and standard deviation $\sigma$.

### Solutions to Example 2.2

Solutions will be shown for the first two items in Table 2.4 only. The remainder can be solved using the same techniques. For the first

**Table 2.4**   Examples of calculating defect rates, Cp and Cpk

| Specification | | Process | | | | % Above | % Below | | |
| Nominal | Tolerance | $\mu$ | $\sigma$ | % Good | USL | LSL | Cp | Cpk |
|---|---|---|---|---|---|---|---|---|
| 10.00″ | ± .04 | 10.00″ | 0.015 | 99.24 | 0.38 | 0.38 | 0.89 | 0.89 |
| 10.00″ | ± .04 | 9.99″ | 0.015″ | 97.68 | 0.043 | 2.28 | 0.89 | 0.67 |
| 10.00″ | ± .04 | 10.01″ | 0.015″ | 97.68 | 2.28 | 0.043 | 0.89 | 0.67 |
| 10.00″ | ± .05 | 10.00″ | 0.015″ | 99.91 | 0.043 | 0.043 | 1.11 | 1.11 |
| 10.00″ | ± .06 | 10.01″ | 0.015″ | 99.96 | 0.043 | 0.0002 | 1.33 | 1.11 |

item, the specification limits are $10.00'' \pm 0.04''$. The process is characterized by $\mu = 10.00''$ and $\sigma = 0.015$.

Cpk = Cp = $\pm$ SL/$\pm$ 3 $\sigma$ = 0.04/3 · 0.015 = 0.89
$z$ = 3 · Cp = 2.67; $f(-z)$ = .0038; RR = 0.38%; for each side above the
    USL and below the LSL. Percent OK = 1 – total RR = 99.24%.

Note that in this case, the Cp was less than 1, therefore it is expected that the reject rate would be higher than 3 $\sigma$ design defect rates of 2700 PPM or 0.27%.

For the second item, the specification limits remain the same at $10.00'' \pm 0.04''$. The process is shifted from the first item by 0.01" and characterized by $\mu = 9.99''$ and $\sigma = 0.015$.

Cp = 0.89 (remains the same from the first item)
Cpk = min(USL – average/3$\sigma$) or (average – LSL/3$\sigma$)
Cpk = 0.67 = minimum of (10.04 – 9.99/3 · 0.015) = 1.11
or (9.99 – 9.96/3 · 0.015) = 0.67
$z_1$ = 3 · Cpk (low) = 3 · 0.67 = 2.00; $f(-2)$ = 0.0228 or 2.28%
$z_2$ = 3 · Cpk (high) 3 · 1.11 = 3.33; $f(-3.33)$ = 0.00043 or 0.43%
Total RR = $f(-z_1) + f(-z_2)$ = 0.00228 + 0.00043 = 0.02323
Percent OK = 1 – total RR = 1 – 0.2323 = 0.97677 or 97.68%.

It is apparent that if the manufacturing process is not centered with the specification nominal (second case in the table), the total defect rate increases, even if the manufacturing process standard deviation remains the same. Similar increases in the defect rate occur if the manufacturing process standard deviation increases or there is a comparable decrease in the tolerance limits of the design. The table also illustrates the use of Cp or Cpk as indicators of quality, depending on whether the manufacturing process average is equal to the design specification nominal.

### 2.3.3  Attribute processes and reject analysis for six sigma

For attribute processes (those with quality measured in terms of defects in a sample or number defective), an implied Cpk will have to be calculated in the quality assessment of design and manufacturing. It is assumed that defects are occurring because of violation of a particular or a composite specification(s). The composite specification can be one-sided or two-sided, depending on the interpretation of the defects. For example, a wire bond defect could be the result of one-sided specifications, since it is assumed that in specifying the bond, only a minimum value is given. For solder defects, a composite specification can be assumed to be two-sided, since solder defects can be one- or two-sided, as in excessive or insufficient solder. The difference between implied one- or two-sided specifications is that the number of defects representing the $f(z)$ value under the normal curve should be halved for two-sided specifications, or used directly for one-sided specifications, resulting in different implied Cpk interpretations. The decision for one- or two-sided specifications for implied Cpk should be left to the appropriate design and manufacturing engineers.

An example of an attribute process calculation to generate an implied Cpk is for solder defects. They are usually measured in PPM or parts per million of defects obtained in production divided by the total number of solder joints in the product (total number of opportunities for solder defects). Solder defects may result from the combination of several specifications of design parameters such as component pad size, drill hole size, fabrication quality of plated metal surface, and the material and process parameters of the soldering equipment. A 100 PPM solder process (1 solder defect in 10,000 terminations or joints) is calculated to have a Cpk = 1.3 as follows:

1. 100 PPM defects (assuming a two-sided specification), 50 PPM per each tail of the normal curve
2. 50 PPM is $f(z) = 0.00005$ or $z = 3.89$, from standard normal curve tables.
3. Implied Cpk = $z/3$ = 1.3

The assumptions are that the defects can occur on either side of the implied specifications, the process is normally distributed, and the process average is equal to the specification nominal. If this example of Cpk was for a wire bond machine, then it could be assumed that the defects occur due to one side of the specification limits of minimum pull strength. In this case, the Cpk can be calculated as follows:

1. 100 PPM defects (assuming a one-sided specification) is 100 PPM per one tail of the normal curve
2. 100 PPM is $f(z) = 0.0001$ or $z = 3.72$, from standard normal curve tables
3. Implied Cpk = $z/3$ = 1.24, which is lower quality than two-sided defects

It can be seen that the method of implied Cpk could lead to various interpretations of one- versus two-sided specifications when the Cpk methodology is used. If the six sigma interpretation of quality is used, the 100 PPM error rate is significant because it is larger than the target of 3.4 PPM. If a quality team has to report on their progress toward six sigma using 100 PPM current defect rate, then they can present the following arguments:

1. For two-sided specifications, $f(z) = 0.00005$ or $z = 3.89$. If a shift of ±1.5 σ is assumed, then all of the failures result from one side of the distribution, whereas the other side is much lower in defects, and therefore contributes no defects. The design is $3.89 + 1.5 = 5.39$ or 5.39 σ in the classical six sigma definition.
2. For one-sided specifications, $f(z) = 0.0001$ or $z = 3.72$. If we assume a shift of ±1.5 σ, then the design is $3.72 σ + 1.5 σ = 5.22 σ$ or 5.22 σ in the classical six sigma definition.

Attribute processes present more difficulty in calculating and visualizing the reject rates; more on that in upcoming chapters.

## 2.4   Are Manufacturing Processes and Supply Parts Always Normally Distributed?

A very common question regarding the reject rate calculations is whether the normal distribution is always applicable in every part manufacturing or supply case. The answer is a definite no! In some cases, such as high-accuracy resistors, parts are made, then tested and segregated according to the measured accuracy, so that a distribution of supply of low-accuracy parts would look like a disjointed normal curve with the middle of the curve missing. For high-accuracy parts, the distribution is narrow with no trailing edges. Obviously, neither set of parts are normally distributed, since the manufacturing processes have been interfered with.

Several tools are available to design and manufacturing teams to manage this condition. Verifying that the manufacturing process or the supply parts are normally distributed can be accomplished by using simple graphical techniques and, if needed, more complex statistical analysis. If the distribution is not normal, parts can be described in other statistical distributions. Then their data can be transformed into an equivalent normal distribution. All the six sigma calculations can be made, then data can be transformed back to the original distribution.

### 2.4.1   Quick visual check for normality

Using graph paper, spreadsheets, or statistically based software, measurement data from randomly selected samples of parts can be quickly checked for normality as follows:

1. Randomly select a number of parts samples for measurement of the quality characteristic, which is the part attribute of interest to the six sigma effort. Thirty samples are considered statistically significant. However smaller numbers might be used for a quick look at the distribution. (For more on sample sizes, refer to Chapter 5.)

2. Rank the data in ascending order, from 1 to $n$.

3. Generate a normal curve score (NS) corresponding to each data point. Each ranked data point is subtracted by 0.5, then divided by the total number of points $n$ so that it sits in the middle of a box of ranked points. Each data point probability is based on the rank of point $i$, with $i$ ranging from 1 to $n$. The normal score (NS) represents the position of that ranked point versus its equivalent value of the $z$ distribution:

$$P(z) = (i - 0.5)/n \qquad i = 0, 1, \ldots, n \qquad (2.14)$$

$$\text{NS} = z \text{ of } P(z)$$

$N$ = total number of parts to be checked for normality

4. Plot each data point value on the Y axis against its normal score. If the data is normal, it should show as a straight line.

**Example for 5 points: 67, 48, 76, 81, and 93**

| Data | Rank ($i$) | $P(z) = (i - 0.5)/n$ | Normal score (NS) $z$ from $P(z)$ |
|------|------|------|------|
| 67 | 2 | 0.3 | −0.52 |
| 48 | 1 | 0.1 | −1.28 |
| 76 | 3 | 0.5 | 0 |
| 81 | 4 | 0.7 | 0.52 |
| 93 | 5 | 0.9 | 1.28 |

A quick graphical check for normality is given in Figure 2.12. It can be visually determined that the data represents close to a straight line.

An even quicker method to determine normality is to use the same procedure but with seminormal graph paper. This would eliminate the $z$ calculations in step 3 above.

## 2.4.2  Checking for normality using chi-square tests

Chi-square ($\chi^2$) tests can be used to determine whether a set of data can be adequately modeled by a specified distribution. The chi-square test divides the data into nonoverlapping intervals called boundaries. It compares the number of observations in each boundary to the num-

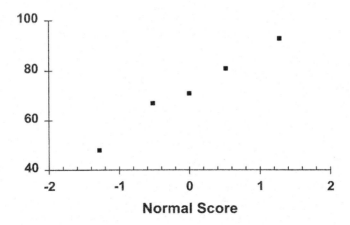

**Figure 2.12**  Quick visual check for normality in Example 2.4.1.

ber expected in the distribution being tested, in this case the normal distribution. Sometimes this test is called "the goodness of fit test."

The boundaries are chosen for convenience, with five being a commonly used number. The boundary limits are used to generate a probability for the expected frequency. This is done in the case of the normal distribution by calculating the $z$ value based on the boundary limit and the average and standard distribution of the data set, in the following manner:

1.  List the data set in ascending order.
2.  Determine the number of boundaries (variable $k$) to be used in this test.
3.  Let $m_i$ be the number of sample values observed in each boundary
4.  Calculate a $z$ value for each boundary. For the two outermost boundaries, there is one single $z$ value. For inside boundaries, there are two $z$ values.
5.  Calculate the expected frequency for each boundary by determining the $P_i = f(z)$ and multiplying that number by the total number in the data set.
6.  Determine the contribution of each boundary to total chi-square value through the formula

$$\chi^2 = \frac{\Sigma(m_i - nP_i)^2}{nP_i}; \qquad \text{with } k-1 \text{ DOF} \qquad (2.16)$$

A hypothesis reject, which indicates that the distribution is not normal is when $\chi^2 \geq \chi^2_\alpha$, which obtained from a $\chi^2$ table for $\alpha = 1 - \text{confi-}$

dence; $k$ is the number of boundaries, and DOF is the degrees of freedom. Selected values of the $\chi^2$ table are given in Table 5.3.

### 2.4.3    Example of $\chi^2$ goodness of fit to normal distribution test

Thirty parts were selected from a production line that was assumed to be normally distributed, and lengths were measured in $\mu$m. The data was sorted in ascending order and five boundaries were created for the sorted data set.

Table 2.5 shows the original as well as the sorted data set of 30 measurements. The average is calculated at 8843.43 and the standard deviation is 743. The boundary limits are a minimum of less then 8000 to a maximum of more than 9500 in 500 increments. The first two boundary calculations are shown for illustration:

*Boundary 1*
$P_1 = P(\chi < 8000) = P\{(\chi - \mu)/\sigma\}$
$z = (8000 - 8843.43)/743 = -1.135$
$P_1 = f(z) = 0.128$ from normal distribution tables
Expected frequency $= NP_i = 30 \cdot 0.128 = 3.84$
$\chi^2$ contribution $= (m_1 - nP_1)^2/nP_1 = (4 - 3.84)^2/3.84 = 0.0067$

*Boundary 2*
$P_2 = P(8000 < \chi < 8500)$
$z_2 = (8500 - \mu)/\sigma = (8500 - 8843.43)/743 = -0.462$
$z_1 = -1.135$ (from previous boundary)
$P_2 = f(z_2) - f(z_1) = 0.3228 - 0.128 = 0.1948$
Expected frequency $= NP_i = 30 \cdot 0.1948 = 6.27$
$\chi^2$ contribution $= (m_i - nP_i)^2/nP_i = (8 - 5.844)^2/5.844 = 0.795$

Other results for the remaining boundaries are shown in Table 2.5. It can be seen that the total number of observed and expected frequencies in all of the boundaries should be equal to the data set total of 30. The total probability $P_i$ should also equal to 1, and the total expected frequency should equal 30.

The total chi-square value for the data set is 2.36, which falls between the limits of $\alpha = 0.10$ (90% confidence) of 1.064 and $\alpha = 0.5$ (50% confidence) of 3.357 for the $\chi^2$ distribution with degrees of freedom DOF = 4 (5 boundaries – 1), from Table 5.3. That implies that the data set is normal since it corresponds with the normal distribution expectations. A plot is shown of the data set values versus their normal

**Table 2.5** $\chi^2$ Goodness of fit test case study

| Original data | Sorted data | Boundaries | Observed frequency $m_i$ | z Terms | $P_i$, $f(z)$ | Expected frequency $30 \cdot P_i$ | Chi-square terms |
|---|---|---|---|---|---|---|---|
| 8146 | 7739 | | | | | | |
| 8956 | 7796 | < 8000 | 4 | −1.135 | 0.128 | 3.84 | 0.0067 |
| 10310 | 7797 | | | | | | |
| 9380 | **7922** | | | | | | |
| 8889 | 8012 | | | | | | |
| 9534 | 8113 | | | | | | |
| 8288 | 8146 | | | | | | |
| 9326 | 8149 | | | | | | |
| 7797 | 8288 | 8000–8500 | 8 | −1.135, −0.46 | 0.1948 | 5.844 | 0.795 |
| 8919 | 8319 | | | | | | |
| 8457 | 8354 | | | | | | |
| 8113 | **8457** | | | | | | |
| 8984 | 8570 | | | | | | |
| 7739 | 8787 | | | | | | |
| 9858 | 8889 | | | | | | |
| 8979 | 8919 | 8500–9000 | 7 | −0.46, 0.21 | 0.2604 | 7.812 | 0.084 |
| 8319 | 8956 | | | | | | |
| 9095 | 8979 | | | | | | |
| 8149 | **8984** | | | | | | |
| 9619 | 9095 | | | | | | |
| 8787 | 9326 | 9000—9500 | 4 | 0.21, 0.88 | 0.2274 | 6.82 | 1.166 |
| 7922 | 9380 | | | | | | |
| 8012 | **9450** | | | | | | |
| 8354 | 9534 | | | | | | |
| 7796 | 9565 | | | | | | |
| 9450 | 9619 | | | | | | |
| 9820 | 9820 | > 9500 | 7 | 0.88 | 0.1894 | 5.682 | 0.305 |
| 8570 | 9858 | | | | | | |
| 10170 | | | | | | | |
| 10170 | 10310 | | | | | | |
| Totals | | | 30 | | 1 | 30 | 2.36 |

Average ($\mu$) = 8843.43.
$\sigma$ = 743.

scores (NS) in Figure 2.13, and it can clearly be seen that the line representing the data versus its normal score equivalent is almost linear. In addition, the expected versus observed frequencies of the data are shown in Figure 2.14. They present a clear adherence to normal curve characteristics.

### 2.4.4   Transformation data into normal distributions

In the cases where the normal distribution cannot be made applicable to the data by using either of the two above methods, then the use of dif-

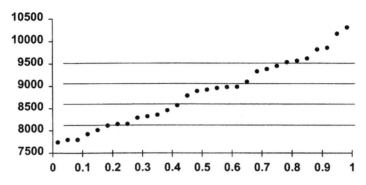

**Figure 2.13**  Normal plot of for data set in Example 2.4.2.

ferent functions to transform data for normality can be attempted. If the distribution is too unsymmetrical or there are data points spread out too far on the ends of the data set, then using functions such as $-1/x$, ln $x$, and $\sqrt{x}$ can be used. If the data points are bunched, then they can be separated using functions such as $x^2$. An example is the following distribution of data that is best described as a lognormal distribution (one that tails off to one side). The data set of 30 values is as follows:

110, 120, 257, 254, 155, 52, 78, 340, 221, 178
55, 450, 185, 222, 138, 89, 398, 156, 69, 385
221, 143, 165, 99, 348, 480, 168, 231, 88, 164

In this case, using a function transform of ln $\sqrt{x}$ for all of the data, it can be seen that the transformed function is much closer to a normal distribution than the original data set, as in Figure 2.15.

In the case of the transformed data, all of the Cp, Cpk, and reject rate calculations are made on the transformed (normal) curve, then

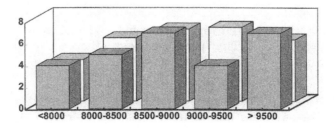

**Figure 2.14**  Plot of observed (dark) versus expected (clear) frequencies.

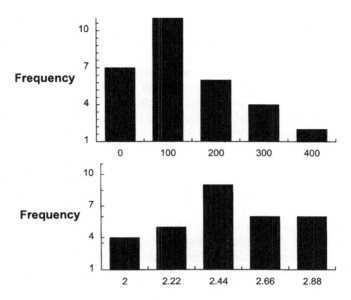

**Figure 2.15**  Plot of Example 2.4 data set original (top) and transformed by log $\sqrt{x}$ on the bottom.

the results are transformed back to the lognormal original. It is recommended that this method only be used when there are critical specifications that have to be set in order to achieve six sigma.

### 2.4.5  The use of statistical software for normality analysis

There are many statistical software packages that can perform the calculation for Section 4 in this chapter. The exercises in this section were provided to show the calculations behind these software packages. When selecting a quality software package, it is important to investigate several and focus on those that offer a broad range of analysis on many of the tools discussed in this book. Many of the leading journals, such as those from professional societies like the American Society of Quality Control (ASQC) offer periodic reviews of quality software in their magazine, *Quality Progress.*

## 2.5  Conclusions

It can be seen from this chapter that the power of the process capability index is the cooperative joining of responsibility for quality improvements between manufacturing and design engineers. Design engineers are responsible for setting the specification limits for new products as broad as possible and still permit the proper functioning of the product. Manufacturing engineers have to narrow the manufacturing process distribution, as measured by the standard deviation of the product characteristics. This can be achieved by more frequent maintenance schedules, improving incoming inspection methods, working with suppliers, increased operator training, and performing design of experiments (DoE) to reduce the variability of the process.

The formal definitions of six sigma and other quality measuring systems such as Cp and Cpk were introduced. In addition, their relationship to determining the defect rate and examples of calculations were also shown, from both variable and attribute manufacturing processes. An important part of these quality systems is the understanding of the assumptions underlying each system. The choice of the proper system should be compatible with the type of business the enterprise is engaged in and its competition.

The assumption that all manufacturing and supply data are normally distributed was examined, and methods to prove normality were shown. In the case of nonnormality, alternate methods for transforming data to normal distribution, performing six sigma calculations, and then converting the data back to the original distribution were also shown.

## 2.6    References and Bibliography

Bowker A. and Lieberman G. *Engineering Statistics.* Engelwood Cliffs, NJ: Prentice-Hall, 1972.

Box, G. and Hunter W. *Statistics for Experimenters.* New York: Wiley, 1978.

Burr, I. *Engineering Statistics and Quality Control.* New York: McGraw Hill, 1953.

Chan, L. et al. "A New Measure for Process Capability: Cpm." *Journal of Quality Technology, 20,* 3, 162–175, July, 1988.

Clausing D. and Simpson H. "Quality by Design." *Quality Progress,* January 1990, 41–44.

Crosby, P. *Quality Is Free.* New York: McGraw Hill, 1979.

Deming, Edwards. *Quality, Productivity and Competitive Position.* Published video lectures and notes. MIT Center for Advanced Engineering Studies. 1982.

Devore, J. *Probability and Statistics for Engineering and the Sciences.* Belmont, CA: Brooks/Cole, 1987.

Dixon W. and Massey, F. *Introduction to Statistical Analysis.* New York: McGraw Hill, 1969.

Ducan, Acheson J. *Quality Control and Industrial Statistics,* 4th ed. Home-wood IL: Irwin. 1995.

Feigenbaum, A. V. *Total Quality Control,* 3rd ed. New York: McGraw Hill, 1983.

Gill, Mark S. "Stalking Six Sigma." *Business Month Journal,* January, 1990.

Ishikawa, K. *Guide to Quality Control* (rev. ed.). Tokyo: Asian Productivity Institute, 1976.

Juran, J. and Gryna, F. *Quality Control Handbook,* 4th ed. New York: Mc-Graw Hill, 1979.

Juran, J. and Gryna, F. *Quality Planning and Analysis.* New York: McGraw Hill. 1970.

Kane, V. "Process Capability Indices." *Quality Technology Journal, 18,* 41–52, 1986.

Kendrick, J. "Hewlett Packard Quest for Quality." *Quality Journal,* November 16–20, 1988.

King, J. *Probability Charts for Decision Making.* New York: Industrial Press. 1971.

Miller, I. and Freund J. E. *Probability and Statistics for Engineers.* Engelwood Cliffs, NJ: Prentice-Hall, 1965.

Moran, J., Talbot, R., and Benson, R. *A Guide to Graphical Problem Solving Processes.* Milwaukee, WI: ASQC Press, 1990.

Nelville, A. and Kennedy J. B. *Basic Statistical Methods for Engineers and Scientists.* Scranton, PA: International Textbook Company, 1964.

Ott, E. *Process Quality Control.* New York. McGraw Hill. 1975.

Ott E. *An Introduction to Statistical Methods and Data Analysis.* North Scituate, MA: Duxbury Press, 1977.

Ryan, T. *Statistical Methods for Quality Improvements.* New York: Wiley, 1989.

Shewhart, W. *Economic Control of Quality of Manufactured Products.* New York: Van Nostrand, 1931.

Smith, B., "Six Sigma Conference." Personal communication, October 1993.

Snedecor G. and Cochran W. *Statistical Methods,* 6th ed. Ames: Iowa State University Press, 1967.

# 3

# Six Sigma and Manufacturing Control Systems

Six sigma originally gained acceptance as a measure of product design for manufacturing (DFM), especially in the process-intensive industries such as integrated circuit (IC) and printed circuit board (PCB) fabrication and assembly. Today, it has become as widely accepted as the traditional measure of quality in manufacturing control systems such as statistical process control (SPC) and total quality management (TQM). Its unique blend of production variability versus design specifications makes it a natural method for setting, communicating, and comparing new product specifications and manufacturing quality levels for competitive manufacturing plants.

By focusing on six sigma, there is a commitment up front to measuring and controlling manufacturing variability through statistical process control (SPC) tools and methods such as control charts. In addition, it is an excellent tool for negotiating and communicating with suppliers to set the appropriate quality level and expectations.

Six sigma focuses on communication between the design, development, and manufacturing parts of an organization. By managing the relationship of design tolerance to manufacturing specifications, it shifts attention away from a possible adversarial relationship between design and manufacturing to a more constructive one, where the common goal of achieving a particular quality level facilitates negotiations and cooperation in new product development.

In this chapter, the relationship between six sigma the early traditions of TQM and SPC will be explored, in the following topics:

1. Manufacturing variability measurement and control (Section 3.1). Statistical process control (SPC) is the key to maintaining and improving the manufacturing process variability. The tools for SPC are presented, with emphasis on control charts and their proper use in the manufacturing environment. These tools can be used collectively for improving quality by collecting and analyzing defects data to determine the most probable causes of defects and counteracting them.

2. Control of variable processes and its relationship to six sigma (Section 3.2). The control of variable processes involves taking periodic or daily actual measurements of the quality characteristic and comparing the measurement to historical values. This section is focused on $\overline{X}$ and $R$ charts. the statistical basis of these charts are examined, as well as their mathematical relationship to six sigma concepts, including various methods of relating the two concepts, with detailed discussions and examples. In addition, the issues of managing the variable control charts and recalculating the chart data are also presented.

3. Control of attribute processes and its relationship to six sigma. In Section 3.3, various types of attribute charts are presented, together with their underlying distributions and relationship to six sigma concepts. Calculations of chart data and their mathematical relationship with six sigma are also presented with formulas and examples. The C chart is shown to be well suited for six sigma applications.

4. Using TQM techniques to maintain six sigma quality in manufacturing (Section 3.4). In factories approaching six sigma quality, the need for sampling techniques such as control charts to maintain and monitor quality are diminished. Individual defects can be analyzed and corrective action taken accordingly on a daily basis. TQM tools can be used in these factories to maintain and even improve quality beyond six sigma. This section presents the TQM tools, their major functions, and how they can be used in the corrective action process.

## 3.1   Manufacturing Variability Measurement and Control

Control charts have been traditionally used as the method of determining the performance of manufacturing processes over time by the statistical characterization of a measured parameter that is dependent on the process. They have been used effectively to determine if the manufacturing process is in statistical control. Control exists when the occurrence of events (failures) follows the statistical laws of the distribution from which the sample was taken.

Control charts are run charts with a centerline drawn at the manufacturing process average and control limit lines drawn at the tail of the distribution at the 3 s points. They are derived from the distribution of sample averages $\overline{X}$, where s is the standard deviation of the production samples taken and is related to the population deviation through the central limit theorem. If the manufacturing process is under statistical control, 99.73% of all observations are within the control limits of the process. Control charts by themselves do not improve quality; they merely indicate that the quality is in statistical "synchronization" with the quality level at the time when the charts were created.

There are two major types of control charts: variable charts, which plot continuous data from the observed parameters, and attribute charts, which are discrete and plot accept/reject data. Variable charts are also known as $\overline{X}$, R charts for high volume and moving range (MR) charts for low volume. Attribute charts tend to show proportion or percent defective. There are four types of attribute charts: P charts, C charts, nP charts, and U charts (see Figure 3.1).

The selection of the parameters to be control charted is an important part of the six sigma quality process. Too many parameters plotted tend to adversely affect the beneficial effect of the control charts, since they will all move in the same direction when the process is out of control. It is very important that the parameters selected for con-

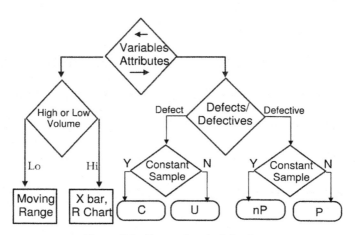

**Figure 3.1**  Types of control charts.

trol charting be independent from each other and directly related to the overall performance of the product.

When introducing control charts to a manufacturing operation, it is beneficial to use elements that are universally recognized, such as temperature and relative humidity, or take readings from a process display monitor. In addition, the production operators have to be directly active in the charting process to increase their awareness and get them involved in the quality output of their jobs. Several shortcomings have been observed when initially introducing control charts. Some of these to avoid are:

- Improper training of production operators. Collecting a daily sample and calculating the average and range of the sample data set might seem to be a simple task. Unfortunately, because of the poor skill set of operators in many manufacturing plants, extensive training has to be provided to make sure the manufacturing operator can perform the required data collection and calculation.

- Using a software program for plotting data removes the focus from the data collection and interpretation of control charting. The issues of training and operating the software tools become the primary factors. Automatic means of plotting control charting should be introduced later in the quality improvement plan for production.

- Selecting variables that are outside of the production group's direct sphere of influence, or are difficult or impossible to control, could result in a negative perception of the quality effort. An example would be to plot the temperature and humidity of the production floor when there are no adequate environmental controls. The change in seasons will always bring an "out-of-control" condition.

In the latter stage of six sigma implementation, the low defect rate impacts the use of these charts. In many cases, successful implementation of six sigma may have rendered control charts obsolete, and the factory might switch over to TQM tools for keeping the quality level at the 3.4 PPM rate. The reason is that the defect rate is so low that only few defects occur in the production day, and the engineers can pay attention to individual defects rather than the sampling plan of the control charts.

## 3.2   Control of Variable Processes and Its Relationship with Six Sigma

Variable processes are those in which direct measurements can be made of the quality characteristic in a periodic or daily sample. The daily samples are then compared with a historical record to see if the manufacturing process for the part is in control. In $\overline{X}$, $R$ charts, the sample measurements taken today are expected to fall within three standard deviations 3 $s$ of the distribution of sample averages taken in the past. In moving range (MR) charts, the sample is compared with the 3 $\sigma$ of the population standard deviation derived from an $\overline{R}$ estimator of $\sigma$. When the sample taken falls outside of the 3 $s$ limits, the process is declared not in control, and a corrective action process is initiated.

Another type of charting for quality in production is the precontrol chart. These charts directly compare the daily measurements to the part specifications. They require operators to make periodic measurements, before the start of each shift, and then at selected time intervals afterward. They require the operator to adjust the production machines if the measurements fall outside a green zone halfway between the nominal and specification limits.

Precontrol charts ignore the natural distribution of process or machine variability. Instead, they require a higher level of operator training and intervention in manufacturing to ensure that production distribution is within halfway of the specification limits, on a daily basis. This is in direct opposition to six sigma concepts of analyzing and matching the process distribution to he specification limits only in the design phase, and thus removing the need to do so every time parts are produced.

Moving range charts (MR) are used in low-volume applications. They take advantage of statistical methodology to reduce the sample size. They will be discussed further in the Chapter 5. In high-volume manufacturing, where several measurements can be taken each day for production samples, $\overline{X}$ and $R$ control charts are used to monitor the average and the standard deviation of production. It is important to note that $\overline{X}$ control charts are derived from the sample average distribution, which is always normal, regardless of the parent distribution of the population $\sigma$, which is used for six sigma calculations of the defect rate, and is not always normal, as discussed in the previous chapter.

The $\overline{X}$ chart shows whether the manufacturing process is centered around or shifted from the historical average. If there is a trend in the plotted data, then the process value, as indicated by the sample average $\overline{X}$, is moving up or down. The causes of $\overline{X}$ chart movements include faulty machine or process settings, improper operator training, and defective materials.

The $R$ chart shows the uniformity or consistency of the manufacturing process. If the $R$ chart is narrow, then the product is uniform. If the $R$ chart is wide or out of control, then there is a nonuniform effect on the process, such as a poor repair or maintenance record, untrained operators, and nonuniform materials.

The variable control charts are generated by taking a historical record of the manufacturing process over a period of time. Shewhart, the father of control charts, recommends that "statistical control cannot be reached until under the same conditions, not less than 25 samples of four each have been taken to satisfy the required criterion." These observations form the historical record of the process. All observations from now on are compared to this baseline.

From these observations, the sample average $\overline{X}$ and the sample range $R$, which is the absolute value of highest value minus the lowest value in the sample, are recorded. At the end of the observation period (25 samples), the average of $\overline{X}$s, designated as $\overline{\overline{X}}$ and the average of $R$'s, designated as $\overline{R}$, are recorded.

### 3.2.1   Variable control chart limits

The control limits for the control charts are calculated using the following formulas and Table 3.1 for control chart factors. The control chart factors were designated with variables such as $A_2$, $D_3$, and $D_4$ to calculate the control limits of $\overline{X}$ and $R$ control charts. The factor $d_2$ is important in linking the average range $\overline{R}$, and hence the standard deviation of the sample ($s$), to the population standard deviation $\sigma$.

The control chart factors shown in Table 3.1 stop at the number 20 of observations of the subgroup. Control charts are based on taking samples to approximate a large production output. If the sample becomes large enough, there is no advantage to using samples and their associated normal distributions to generate variable control charts. Instead, 100% of production could be tested to find out if the parts produced are within specifications.

### 3.2.2   Control chart limits calculations

$\overline{X}$ chart control limits are 3 $s$ of the sample average distribution. This distribution is always normal, with an average equal to the average of sample averages $\overline{\overline{X}}$. The range of each sample is called $R$ and the average of all sample ranges is called $\overline{R}$. The distribution of the ranges is not normal, even if the parent distribution is normal. The control chart factors in Table 3.1 are approximations to convert the $\overline{R}$ to the standard deviation of the sample average distribution $s$ and the population distribution $\sigma$.

### $\overline{X}$ Control limits (3 $s$ limits)

$$\text{Upper control limit (UCL}_X) = \overline{\overline{X}} + 3\,s = \overline{\overline{X}} + A_2 \cdot \overline{R} \qquad (3.1)$$

**Table 3.1**   Control chart factors

| Observations subgroup $n$ | $A_2$ Factor for $\overline{X}$ chart | Lower control $\overline{R}$ limit $D_3$ | Upper control $\overline{R}$ limit $D_4$ | $\overline{R}/\sigma$ $= d_2$ |
|---|---|---|---|---|
| 2  | 1.88 | 0    | 3.27 | 1.128 |
| 3  | 1.02 | 0    | 2.57 | 1.693 |
| 4  | 0.73 | 0    | 2.28 | 2.059 |
| 5  | 0.58 | 0    | 2.11 | 2.326 |
| 6  | 0.48 | 0    | 2.00 | 2.534 |
| 7  | 0.42 | 0.08 | 1.92 | 2.704 |
| 8  | 0.37 | 0.14 | 1.86 | 2.847 |
| 9  | 0.34 | 0.18 | 1.82 | 2.970 |
| 10 | 0.31 | 0.22 | 1.78 | 3.078 |
| 11 | 0.29 | 0.26 | 1.74 | 3.173 |
| 12 | 0.27 | 0.28 | 1.72 | 3.258 |
| 13 | 0.25 | 0.31 | 1.69 | 3.336 |
| 14 | 0.24 | 0.33 | 1.67 | 3.407 |
| 15 | 0.22 | 0.35 | 1.65 | 3.472 |
| 16 | 0.21 | 0.36 | 1.64 | 3.532 |
| 17 | 0.20 | 0.38 | 1.62 | 3.588 |
| 18 | 0.19 | 0.39 | 1.61 | 3.640 |
| 19 | 0.19 | 0.40 | 1.60 | 3.689 |
| 20 | 0.18 | 0.41 | 1.59 | 3.735 |

$$\text{Lower control limit (LCL}_X) = \overline{\overline{X}} - 3\,s = \overline{\overline{X}} - A_2 \cdot \overline{R} \qquad (3.2)$$

## $\overline{R}$ Control limits

$$\text{Upper control limit (UCL}_R) = D_4 \cdot \overline{R} \qquad (3.3)$$

$$\text{Lower control limit (LCL}_R) = D_3 \cdot \overline{R} \qquad (3.4)$$

where
$\overline{X}$ = average of $n$ observation in a subgroup
$\overline{\overline{X}}$ = average of all $\overline{X}$s
$\overline{R}$ = average of all $R$
$R$ = range of $n$ observation in a subgroup (highest to lowest value)
$A_2$ = factor for $X$ chart
$D_3$ = lower control limit factor for $R$ chart
$D_4$ = upper control limit factor for $R$ chart
$d_2$ = estimator for $\sigma$ based on range of samples

### 3.2.3   Control and specification limits

Control chart limits indicate a different set of conditions than the
specification limits. Control limits are based on the distribution of

sample averages, whereas specification limits are related to population distributions of parts. It is desirable to have the specification limits as large as possible compared to the process control limit.

The control limits represent the 3 $s$ points, based on a sample of $n$
observations. To determine the standard deviation of the product population, the central limit theorem can be used:

$$s = \frac{\sigma}{\sqrt{n}} \tag{3.5}$$

where
$s$ = standard deviation the distribution of sample averages
$\sigma$ = population deviation
$n$ = sample size

Multiplying 1/3 the distance from the centerline of the $\overline{X}$ chart to
one of the control limits by $\sqrt{n}$ will determine the total product population deviation. A simpler approximation is the use of the formula
$\sigma = \overline{R}/d_2$ from control chart factors in Table 3.1 to generate the total
product standard deviation directly from the control chart data. $d_2$
can be used as a good estimator for $\sigma$ when using small numbers of
samples and their ranges.

### 3.2.4   $\overline{X}$, R variable control chart calculations example

**Example 3.1**

In this example, a critical dimension for a part is measured as it is being inspected in a machining operation. To set up the control chart,
four measurements were taken every day for 25 successive days, to
approximate the daily production variability. These measurements
were then used to calculate the limits of the control charts. The measurements are shown in Table 3.2.

It should be noted that the value $n$ used in Equation 3.5 is equal to
4, which is the number of observations in each sample. This is not to
be confused with the 25 sets of subgroups or samples for the historical
record of the process. If the 25 samples are taken daily, they represent
approximately a one-month history of production.

During the first day, four samples were taken, measuring 9, 12, 11,
and 14 thousands of an inch. These were recorded in the top of the
four columns of sample #1. The average, or $\overline{X}$, was calculated and entered in column 5, and the $R$ is entered in column 6.

$$\overline{X} \text{ Sample } 1 = (9 + 12 + 11 + 14)/4 = 11.50$$

The range, or $R$, is calculated by taking the highest reading (14 in
this case), minus the lowest reading (9 in this case).

$$R \text{ Sample } 1 = 14 - 9 = 5$$

**Table 3.2**  Control chart limit calculations example

| Sample no. | Parts | | | | Average $\overline{X}$ | Range $R$ |
|---|---|---|---|---|---|---|
| | 1 | 2 | 3 | 4 | | |
| 1 | 9 | 12 | 11 | 14 | 11.50 | 5 |
| 2 | 13 | 16 | 12 | 9 | 12.50 | 7 |
| 3 | 11 | 11 | 10 | 9 | 10.25 | 2 |
| 4 | 14 | 11 | 12 | 12 | 12.25 | 3 |
| 5 | 12 | 14 | 16 | 14 | 14.00 | 4 |
| 6 | 19 | 10 | 13 | 15 | 14.25 | 9 |
| 7 | 13 | 14 | 10 | 13 | 12.50 | 4 |
| 8 | 18 | 11 | 14 | 11 | 13.50 | 7 |
| 9 | 13 | 13 | 11 | 12 | 12.25 | 2 |
| 10 | 12 | 10 | 14 | 12 | 12.00 | 4 |
| 11 | 13 | 10 | 14 | 17 | 13.50 | 7 |
| 12 | 13 | 15 | 10 | 10 | 12.00 | 5 |
| 13 | 16 | 10 | 10 | 11 | 11.75 | 6 |
| 14 | 15 | 15 | 13 | 14 | 14.25 | 2 |
| 15 | 16 | 10 | 14 | 15 | 13.75 | 6 |
| 16 | 12 | 11 | 14 | 9 | 11.50 | 5 |
| 17 | 14 | 10 | 13 | 11 | 12.00 | 4 |
| 18 | 11 | 16 | 13 | 14 | 13.50 | 5 |
| 19 | 12 | 10 | 12 | 13 | 11.75 | 3 |
| 20 | 13 | 10 | 10 | 11 | 11.00 | 3 |
| 21 | 14 | 14 | 10 | 13 | 12.75 | 4 |
| 22 | 13 | 13 | 9 | 10 | 11.25 | 4 |
| 23 | 13 | 13 | 13 | 17 | 14.00 | 4 |
| 24 | 15 | 12 | 15 | 13 | 13.75 | 3 |
| 25 | 15 | 12 | 15 | 13 | 13.75 | 3 |
| Totals | | | | | 315.50 | 111 |

The averages of $\overline{X}$ and $R$ are calculated by dividing the column totals of $\overline{X}$ and $R$ by the number of subgroups.

$$\overline{\overline{X}} = (\text{SUM OF } \overline{X}\text{s})/\text{number of subgroups}$$

$$\overline{\overline{X}} = 315.50/25 = 12.62$$

$$\overline{R} = (\text{SUM OF R's})/\text{number of subgroups}$$

$$\overline{R} = 111/25 = 4.44$$

Using the control chart (Table 3.1), the control limits can be calculated using $n = 4$ as follows:

$\overline{X}$ Control limits

$$\text{UCL}_x = \overline{\overline{X}} + A_2\,\overline{R} = 12.62 + 0.73 \cdot 4.44 = 15.86$$

$$\text{UCL}_x = \bar{\bar{X}} - A_2 \bar{R} = 12.62 - 0.73 \cdot 4.44 = 9.38$$

$\bar{R}$ Control limits

$$\text{Upper control limit (UCL}_R) = D_4 \bar{R} = 2.28 \cdot 4.44 = 10.12$$

$$\text{Lower control limit (LCL}_R) = D_3 \bar{R} = 0$$

Since the measurements were recorded in thousands of an inch, the centerline of the $\bar{X}$ control chart is 0.01262 and the control limits for $\bar{X}$ are 0.01586 and 0.00938. For the $R$ chart, the centerline is set at 0.00444 and the limits are 0.01012 and 0.

These numbers form the control limits of the control chart. After the limits have been calculated, the control chart is ready for use in production. Each production day, four readings of the part dimension are to be taken by the responsible operators, with the average of the four readings plotted on the $\bar{X}$ chart, and the range or difference between the highest and lowest reading to be plotted on the $R$ chart. The daily numbers of $\bar{X}$ and $R$ should plot within the control limits. If they plot outside the limits, the production process is not in control, and immediate corrective action should be initiated.

### 3.2.5   Alternate methods for calculating control limits

The control limits are set to three times standard deviation of the sample distribution ($s$). $s$ can be calculated from $\sigma$ the population standard deviation using the factor $d_2$ according to the central limit theorem:

$$\sigma = \bar{R}/d_2 = 4.44/2.059 = 2.156$$

$$s = \sigma/\sqrt{n} = 2.156/2 = 1.078$$

$\pm 3\ s = 1.078 \cdot 3 = 3.23$, which is close to the $A_2 \cdot \bar{R}$ value of 3.24, which corresponds to the distance from the centerline to one of the control limits in the variable control charts.

It is interesting to note that of the total population of 100 numbers (Table 3.2), then the standard deviation is $\sigma = 2.156$, which is exactly the one predicted by the $\bar{R}$ estimator. If the specifications limits are given, then the Cp, Cpk, and reject rates can be calculated as in the example in the previous chapter.

### 3.2.6   Control chart guidelines, out of control conditions, and corrective action procedures and examples

Figures 3.2 and 3.3 are examples of $\bar{X}$ and $R$ charts showing the solder paste height deposition process for a surface mount technology (SMT)

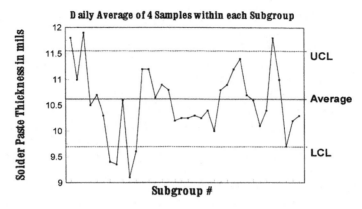

**Figure 3.2** $\overline{X}$ control chart example.

process. Several observations can be made from examining these charts:

1. In Figures 3.2 and 3.3, the two charts, $\overline{X}$ and $R$, are measuring process average and process variability, respectively. Although one might be out of control, the other one is not, or vice versa. This is due to the independence of the two attribute of the process.

2. The two charts are related mathematically, since the distance from the $\overline{\overline{X}}$ to one of the control limits is equal to $3\ s$ or $A_2 \cdot \overline{R}$. The $\overline{R}$ number in the chart (1.25 in Figure 3.3) can be multiplied by 0.73

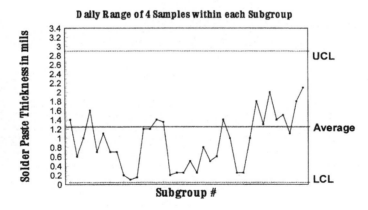

**Figure 3.3**   R control chart example

(the $A_2$ factor for $n = 4$ from Table 3.1), resulting in 0.9125. This is the approximate distance from the $\overline{\overline{X}}$ (sometimes called the centerline or CL) to one of the control limits in the $\overline{X}$ chart.

3. The frequency of taking samples for control charts is left up to the manufacturing process quality status controller. For high-quality processes, a daily sample for each shift is adequate to ensure conformance. For production lines with frequent quality problems, more sampling might be required, depending on the number of parts being produced or the number of hours since the last sample. This is necessary if reworking out-of-control parts is required. In this case, material or parts produced since the last good sample plot on the chart has to be reworked. In addition, The problem has to be investigated by production engineers and possible causes recorded on the chart. The production engineer may require that more frequent samples be taken until the process is more stable. Figure 3.4 is an example of such a condition for a bonding process for plastic parts.

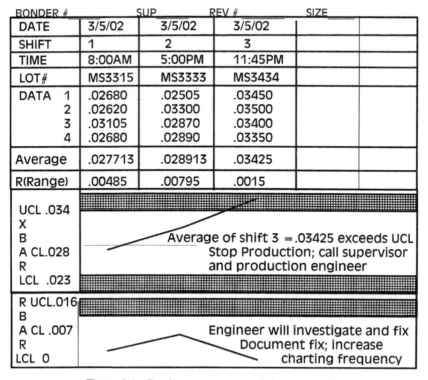

| BONDER # | | SUP | REV # | SIZE | |
|---|---|---|---|---|---|
| DATE | 3/5/02 | 3/5/02 | 3/5/02 | | |
| SHIFT | 1 | 2 | 3 | | |
| TIME | 8:00AM | 5:00PM | 11:45PM | | |
| LOT# | MS3315 | MS3333 | MS3434 | | |
| DATA   1 | .02680 | .02505 | .03450 | | |
| 2 | .02620 | .03300 | .03500 | | |
| 3 | .03105 | .02870 | .03400 | | |
| 4 | .02680 | .02890 | .03350 | | |
| Average | .027713 | .028913 | .03425 | | |
| R(Range) | .00485 | .00795 | .0015 | | |

UCL .034
X
B
A CL.028      Average of shift 3 = .03425 exceeds UCL
R               Stop Production; call supervisor
LCL .023          and production engineer

R UCL.016
B
A CL .007        Engineer will investigate and fix
R                  Document fix; increase
LCL  0             charting frequency

**Figure 3.4**  Bonding process control chart example.

4. The control limits should not be recalculated unless there is a change in the manufacturing process. Examples could be new materials, machinery, operators, or process improvement projects. When a chart shows an out-of-control condition, the process should be investigated and the reason for the problem identified on the chart. Figure 3.5 shows a typical scenario of plotting a parameter (in this case the surface cleanliness measurements on PCBs), which was necessitated by a defective laminate lot. Note that the new lot has significantly increased the resistance value, which would necessitate recalculating the control limits.

5. In the $\overline{X}$ chart, the upper and lower control limits are usually symmetrical around the $\overline{\overline{X}}$ or the centerline, as shown in Figure 3.3. In the case of a maximum specification, only one control limit is sufficient. In the $\overline{R}$ chart, symmetry is not necessary when the sample size is less than 7, since $D_3$ (the control factor for the lower limit) is equal to zero.

6. In many six sigma manufacturing plants, manufacturing has added additional information such as the specification limits, and then calculated the Cp or Cpk on the control charts. This can easily be done, as shown in examples earlier in this chapter, by deriving $\sigma$ either from the $\overline{R}$ or s calculation in step 2, using the formulas $\sigma = s \cdot \sqrt{n}$ or $\sigma = \overline{R}/d_2$.

7. The most common indicator of out-of-control condition is that one sample average is plotted outside the $\overline{X}$ chart control limits, or one sample range is outside the $R$ chart control limits. If these observations are confined to one portion of the chart, then many other indi-

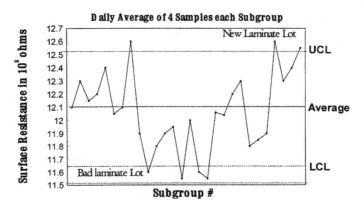

**Figure 3.5**   Surface cleanliness control chart example.

cators of out-of-control conditions can be used as well. These indicators have a probability approaching that of the one $\overline{X}$ point outside the control limits, whose probability is equal to 0.00135. Each half of the $\overline{X}$ chart can be divided into three segments, being one standard deviation ($s$) wide. The probability of an $\overline{X}$ point occurring outside the 2 $s$ limit or beyond is $f(-2) = 0.0228$, and the probability of $\overline{X}$ point occurring outside the 1 $s$ limit is $f(-1) = 0.1587$ from the standard normal distribution or $z$ table (Table 2.3).

The probability of multiple $\overline{X}$ points occurring in succession might equal that of the one point outside the 3 $s$ limits. For example, two successive points in the zone beyond 2 $s$ (the outer one-third zone in the upper half of the chart) is $0.0228 \cdot 0.0228$ or 0.00052. A combination of points inside and outside the zones can be used. For this zone, two out three $\overline{X}$ points can be used. The probability of the third point is $1 - 0.0228 = 0.9772$. Since this point can occur anywhere within the sequence, the total probability has to be multiplied by 3 or $0.0228 \cdot 0.0228 \cdot 0.9772 \cdot 3 = 0.00152$, which is comparable with the 0.00135 probability of a single point outside the control limit. Table 3.3 shows the out-of-control conditions for several successions of points in one-half of the $\overline{X}$, $R$ control charts.

### 3.2.7  Examples of variable control chart calculations and their relationship to six sigma

These examples were developed to show the relationship of variable control charts and six sigma. They can be used as guidelines for communications between an enterprise and its suppliers.

### Example 3.2a

A variable control chart for PCB surface resistance was created. There is only one minimum specification for resistance. The $\overline{X}$ bar was

**Table 3.3**  Probabilities for out-of-control conditions

| | | Probability of out of control | |
| --- | --- | --- | --- |
| Upper half of the control chart | Zone | $\overline{X}$ | Ranges of samples of 5 |
| One point beyond upper control limit | $> 3\,s$ | 0.00135 | 0.0046 |
| 2 out of 3 | $2\,s - 3\,s$ | 0.0015 | 0.0033 |
| 4 out of 5 | $1\,s - 2\,s$ | 0.0027 | 0.0026 |
| 8 in a row | $CL - 1\,s^*$ | 0.0039 | 0.0023 |

*CL = Centerline or $\overline{\overline{X}}$.

20 megaohms (MH) and the $UCL_x$ was 23 MH, with a sample size of 9. A new specification was adopted to keep resistance at a minimum of 16 MH. Assuming that the resistance measurement or process average = specification nominal ($N$), describe the Cp and Cpk reject rates and show the $R$ chart limits.

### Example 3.2a solution

Since the process is centered, Cp = Cpk. The distance from the $\bar{X}$ to $UCL_x = 3\ s = 3$, therefore:

$s = 1$

$\sigma = s \cdot \sqrt{n} = 3$

$LSL = 16$ MH

Process average = 20 MH

$Cp = Cpk = (LSL - \text{process average})/3\sigma = (20 - 16)/3 \cdot 3 = 4/9 = 0.444$

$z = (SL - \text{average})/\sigma = (16 - 20)/3 = 1.33$ or $z = 3 \cdot Cpk = 1.33$

Reject rate $= f(-z) = 0.0976 = 91,760$ PPM (one-sided rejects only, below LSL)

$\bar{R} = \sigma \cdot d_2\ (n = 9) = 3 \cdot 2.97 = 8.91$

$UCL_R = 1.82 \cdot 8.91 = 16.22$ MH

$LCL_R = 0.18 \cdot 8.91 = 1.60$ MH

### Example 3.2b

A four sigma program was introduced at the company in Example 3.2a. For the surface resistance process, the lower specification limit (LSL) remained at 16 MH and the process $\sigma$ remained the same. Describe the Cp and Cpk reject rates and show the $\bar{X}$ and $R$ chart limits, using the same sample size of 9. Repeat for a six sigma program, with 1.5 $\sigma$ shift, with the process average and sigma remaining the same.

### Example 3.2b solution

The four sigma program implies a specification limit of $N \pm 4\ \sigma = N \pm 4 \cdot 3 = N \pm 12$. The process average ($\bar{\bar{X}}$), which is equal to the nominal $N$, is 4 $\sigma$ away from the LSL, and is $16 + 12 = 28$ MH, given LSL = 16 MH. $Cp = Cpk = \pm 4\ \sigma / \pm 3\ \sigma = 1.33$ and two-sided reject rate from the $z$ table (Table 2.3) = 64 PPM.

The $\bar{R}$ chart remains the same as Example 3.4a, since the process variability $\sigma$ did not change. The $X$ chart is centered on $X = 28$ MH; $LCL_x = 28 - 3s = 25$ MH; $UCL_x = 31$ MH.

For six sigma, the same methodology applies, except that there is a $\pm 1.5\ \sigma$ shift. The specification limits are $N \pm 6\ \sigma = N \pm 6 \cdot 3 = N \pm 18$.

Given the LSL = 16 MH, the specification nominal $N$ is $16 + 18 = 34$ MH. Therefore, Cp = 2; Cpk = 1.5; reject rate from previous tables ($\pm 1.5\ \sigma$ shift) = 3.4 PPM.

Assuming that the shift is toward the lower specification, then the process average could be $+4.5\ \sigma$ from the LSL or $-1.5\ \sigma$ from the nominal: $34 - 1.5 \cdot 3 = 29.5$ MH; or $16 + 4.5 \cdot 3 = 29.5$ MH.

The $\overline{R}$ chart remains the same as Example 3.4a, since the process variability $\sigma$ did not change. If the $\overline{X}$ chart is centered on $\overline{\overline{X}} = 29.5$, then $\text{LCL}_x = 29.5 - 3\ s = 26.5$ MH and $\text{UCL}_x = 32.5$ MH.

### Example 3.2c
A new contract was written with a supplier to deliver parts with the following stipulation: Cpk = 0.85 and part specifications = 10 ± 2 mils. The supplier found that part process average was 11 mils. The supplier kept control of their process by using a variable control chart with sample size $n = 4$. Calculate the Cp and the variable chart limits.

### Example 3.2c solution

Cpk is the minimum of $[(\text{USL} - \mu)/3\sigma]$ or $[(\mu - \text{LSL})/3\sigma]$
Cpk = Min $(12 - 11)/3\sigma$ vs. $(11 - 8)/3\sigma$

The first alternative is chosen because it is the minimum of the two:

Cpk = $0.85 = (\text{USL} - \mu)/3\sigma = 1/3\sigma$
$\sigma = 0.392$ and $s = \sigma/\sqrt{n} = 0.196$
Cp = $\pm\text{SL}/\pm3\ \sigma = \pm2/(3 \cdot 0.392) = 1.7$
$z = (\text{SL} - \mu)/\sigma$
$z_2 = 1/0.392 = 2.55$; $f(-z_2) = 0.0054$
$z_1 = 3/0.392 = 7.65$; $f(-z_1) = 0$
Total reject rate = $0.0054 = 5400$ PPM
$\text{UCL}_x$ = process average + 3 $s$ = $11 + 3 \cdot 0.196 = 11.588$ and $\text{LCL}_x =$
    $11 - 3 \cdot 0.196 = 10.412$
$\overline{R} = d_2 \cdot \sigma = 2.059 \cdot 0.392 = 0.807$
$\text{UCL}_R = 2.28 \cdot 0.807 = 0.84$
$\text{LCL}_R = 0$

## 3.3   Attribute Charts and Their Relationship with Six Sigma

Attribute charts directly measure the rejects in the production operation, as opposed to measuring a particular value of the quality characteristic as in variable processes. They are more common in manufacturing because of the following:

1. Attribute or pass–fail test data are easier to measure than actual variable measurement. They can be obtained by devices or tools such as go/no-go gauges, calibrated for only the specification measurements, as opposed to measuring the full operating spectrum of parts.
2. Attribute data require much less operator training, since they only have to observe a reject indicator or light, as opposed to making several measurements on gauges or test equipment.
3. Attribute data can be directly collected from the manufacturing equipment, especially if there is a high degree of automation.
4. Storage and dissemination of attribute data is also much easier, since there is only the reject rate to store versus the actual measurements for variable data.

Attribute charts use different probability distributions than the normal distribution used in variable charts, depending on whether the sample size is constant or changing, as shown in Figure 3.1. For C and U charts, the Poisson distribution is used, whereas the P and nP charts use the binomial distribution.

### 3.3.1 The binomial distribution

The binomial distribution is characterized by the outcome of each manufacturing event: each operation can result in a pass or fail. The probability of a pass is equal to 1 minus probability of a failure. The failure can occur for many reasons, but the outcome is counted as one "defective" unit, possibly containing more than one "defect." The binomial distribution has "memory," that is, successive failures are connected in the distribution formula. Therefore, when a failure occurs, the probability of the next failure is related to this failure. The binomial distribution formulas are as follows:

$$\beta(x; n, p) = Cnx \cdot p^x(1 - p)^{n-x} \tag{3.6}$$

where
$x$ = number of failures (or successes)
$n$ = number of trials
$p$ = probability of one failure (or success)

$$\text{Average} = \text{Expected value} = \mu = E(x) = n \cdot p$$

$$\text{Standard deviation} = \sqrt{\text{variance}} = \sqrt{n \cdot p \cdot (1 - p)}$$

### 3.3.2   Examples of using the binomial distribution

## Example 3.3
If the probability of failure of one part is 25%, what is the probability that the next two parts out of four are also failures?

$$B(2, 4, 0.25) = (4!/2!2!)(0.25)^2(0.75)^2 = 21\%$$

## Example 3.4
The probability of a failed part is 5%. If 20 parts are made from the same machine, what is the average (expected value) and standard deviation of a failure? What is the probability that the first four parts will fail?

$$E(x) = 20 \cdot 0.05 = 1, \text{Standard deviation} = \sqrt{20 \cdot 0.05 \cdot 0.95} = 0.975$$

Probability of the first four parts failing = $\Sigma$(probability of part 1 fail + probability of part 2 fail + part 3 + part 4)

$$P(x = 1, 2, 3, 4) = \Sigma\beta(1, 2, 3, 4; n, p) = 0.64 \text{ or } 64\%$$

### 3.3.3   The Poisson distribution

The Poisson distribution approximates the binomial distribution when the number of trials ($n$) is large and the probability of each trial ($p$) is small. In this case the variable $\lambda$, sometimes called the outcome parameter of the distribution is equal to $np$. The formula for the Poisson distribution is as follows:

$$p(x, \lambda) = e^{-\lambda}(\lambda^x/x!) \tag{3.7}$$

where $x$ is the outcome during a specific time or region and $\lambda$ is the average number of outcomes in the time interval or region and

$$\text{Average} = \text{Variance} = np = \lambda$$

Use of the Poisson distribution is more appropriate when each event has an equal probability of failure, producing a "defect." It is especially useful in complex production operations, where the possibilities or opportunities of defects increase very rapidly, and the probability of getting a single defect at a specific place or time is small. The Poisson-distribution-based charts (C or U charts) should be used when the area of opportunity or boundary of finding defects is kept constant. Examples are:

- Defects in a one-shift operation
- Solder defects in one electronic product
- Defect in one PCB

- Defects in 20,000 units of production
- Total number of defects in a computer system per month

The Poisson distribution implies occurrences of events or defects within a boundary of time, space, or region. It has no "memory"; that is, the outcome or defect during one interval is in proportion to its length, and independent of other intervals. In addition, the probability of two or more outcomes or failures in a single time interval is zero.

### 3.3.4 Examples of using the Poisson distribution

**Example 3.5**

Assuming the number of defects in a part is $\lambda = 5$, What is the expected number of defects in a part? What is the probability of two defects? Up to two defects?

Expected number of defects = 5
Probability of two defects = $P(x = 2, \lambda = 5) = e^{-5}5^2/2! = 0.0842$
Probability of up to two defects = $P(0, 1, 2) = e^{-5}(1 + 5 + 25/2) = 0.12$

**Example 3.6**

Assuming that the number of defects in a production line during a single hour is $\lambda = 4$. What is the probability that six defects will occur in that hour?

$$P(x = 6, \lambda = 4) = e^{-4}4^6/6! = 0.1042$$

**Example 3.7**

Assuming the probability of obtaining a defective product is 0.01, what is the probability of obtaining at least three defective products out of a lot of 100, using binomial and Poisson distributions?

For binomial distribution:

$$P(0, 1, 2, 3) = C_{100,0}(0.01)^0(0.99)^{100} + C_{100,1}(0.01)^1(0.99)^{99}$$
$$+ C_{100,2}(0.01)^2(0.99)^{98} + C_{100,3}(0.01)^3(0.99)^{97} = 0.9816$$

For Poisson distribution:

$$\lambda = np = 1$$

$$P(0, 1, 2, 3) = e^{-1}(1^0/0!) + e^{-1}(1^1/1!) + e^{-1}(1^2/2!) + e^{-1}(1^3/3!)$$

$$P(0, 1, 2, 3) = e^{-1}(1 + 1 + \tfrac{1}{2} + 1/6) = 0.9810$$

The result of the Poisson distribution is in good agreement with the value of the binomial distribution for small $p$ and large $n$, but much easier to compute.

### 3.3.5   Attribute control charts limit calculations

All attribute control charts follow the same three sigma control limit away from the centerline methodology of the variable control charts:

$$\text{Control limits for attribute charts} = \text{centerline} \pm 3\,s \qquad (3.8)$$

## For constant samples ($C$ or $nP$ charts)

For Poisson distribution:

$$\text{Centerline} = \text{Poisson average } \lambda \text{ or } \bar{c}$$

$$s = \sqrt{\lambda} = \sqrt{\bar{c}} \text{ for Poisson standard deviation}$$

and

$$\text{CL}_c = \bar{c} \pm 3 \cdot \sqrt{\bar{c}}$$

$$\overline{np} = \frac{\Sigma np}{k} \qquad (3.9)$$

For binomial distribution:

$$\text{Centerline} = \text{Binomial average } \overline{np}$$

$$s = \sqrt{\overline{np} \cdot (1 - \bar{p})}$$

$$\text{CL}_{np} = \overline{np} \pm 3 \cdot \sqrt{\overline{np} \cdot (1 - \bar{p})} \qquad (3.10)$$

where
$\bar{c}$ = number of defects in a unit
$\overline{np}$ = number of defectives found in each constant sample $n$
$n$ is the number of units in sample
$k$ is the number of samples

## For changing sample sizes (U or P charts)

For Poisson distribution:

$$\text{Centerline} = \text{Poisson average number of defects in a sample } \bar{u}$$

$$\bar{u} = \frac{\Sigma c}{\Sigma n}$$

$$s = \sqrt{\frac{\bar{u}}{n}} = \text{ for Poisson standard deviation}$$

and

$$\text{CL}_u = \bar{u} \pm 3 \cdot \sqrt{\frac{\bar{u}}{n}}; \qquad (3.11)$$

For binomial distribution:

$$\overline{p} = \frac{\Sigma np}{\Sigma n} = \frac{np_1 + np_2 + \ldots + np_k}{n_1 + n_2 + \ldots + n_k} = s$$

$$= \sqrt{\frac{\overline{p}(1-\overline{p})}{n}} \text{ for binomial standard deviation}$$

and

$$\text{CL}_p = \overline{p} \pm 3 \cdot \sqrt{\frac{\overline{p}(1-\overline{p})}{n}} \qquad (3.12)$$

where
$\overline{u}$ = average number of defects in a sample
$\overline{p}$ = fraction or percent defective = number of defective units in sample $n$

C and U charts are considered as a special form of control charts in which the possibility of defects is much larger, and the probability of getting a defect at any specific point, place, or time is much smaller.

The relationship of attribute charts to the six sigma concept is through the defects implied in the charts. The centerline represents the defect rate. These defect rates can be translated into an implied Cpk, as shown in the previous chapter.

Several assumptions have to be made in the case of the attribute chart connections to six sigma:

1. There is one or a complex set of specifications that are not readily discernible that govern the manufacturing process for the parts.
2. These specifications are either one- or two-sided, resulting in one- or two-sided defects (defects < LSL and defects > USL).
3. The manufacturing process is assumed to be normally distributed.
4. There is a relationship between the process average and the specification nominal. In some definitions of six sigma, an assumption is made that there is a 1.5 σ shift from process average to specification nominal.

The control limits of the attribute charts are not related to the population distribution. Therefore, the method of finding the population standard deviation σ is quite different from that used in variable control charts, as shown in the examples below.

### 3.3.6  Examples of attribute control chart calculations and their relationship to six sigma

### Example 3.8
Fuses are tested in sample lots of 100 and defectives are found to be 1%. To control the quality, the company takes hourly samples of 100

fuses and counts the number of defectives. Calculate the Cp, Cpk, population $\sigma$, and the control limits of the nP chart. Assume that the fuses were made in a normally distributed manufacturing process and that the process is centered (process average = specification nominal) and has a two-sided distribution of defects.

**Example 3.8 solution**
One-sided RR = 0.01/2 = 0.005 = $f(-z)$; therefore $z$ = 2.575 from the $z$ table (Table 2.3). Cp = Cpk = $z$/3 = 0.86 = ± SL/3$\sigma$ for no shift.

Using the formula for the nP charts:

$$s = (\sqrt{1 \cdot 0.99}) = 0.995$$

$$\text{UCL}_{np} = \overline{np} + 3\,s = 1 + 3 \cdot 0.995 = 3.985$$

$$\text{LCL}_{np} = \overline{np} - 3\,s = 1 - 3 \cdot 0.995 = 0$$

Note that in this example, the specification limits were not given, yet the implied Cp and Cpk could be calculated. If a process average shift to the specification limits is given (such as ±1.5 $\sigma$), it is still possible to calculate Cpk if we assume that the rejects are mostly generated by one side of the distribution.

**Example 3.9**
A company's quality team was sent to audit a supplier plant making their parts given specifications of 8 ± 3. They read a variable control chart with $n$ = 4 and $\overline{X}$ = 8.1, $\text{UCL}_x$ = 11.1 and $\text{LCL}_x$ = 5.1. What are the quality data for the population of parts delivered to the company?

**Example 3.9 solution**
From the $\overline{X}$ control chart:

$3\,s = 3$

$s = 1$

$\sigma = s \cdot \sqrt{n} = 2$

average shift = 0.1

Cp = 3/(3 · 2) = 0.5

Cpk = min of [(3 − 0.1)/6 = 0.48 or (3 + 0.1)/6 = 0.517] = 0.48

$z_1 = 1.55$

$f(-z_1) = 0.0606$

$z_2 = 1.45$

$f(-z_2) = 0.0735$

Total rejects = 0.0606 + 0.0735 = 0.1341 or 13.41% or 134,100 PPM

### 3.3.7    Use of control charts in factories that are approaching six sigma

The C chart is the most widely used chart in factories that are approaching six sigma. Since the defect rates are very low, binomial-based control charts would require a very large sample, and hence are impractical to use. For six sigma quality, a defect rate of 3.4 PPM would result in a $nP$ chart with a centerline probability 0.0000034. Such a chart would require a very large sample to determine if the process indeed has gone out of control.

Using C charts with well-defined areas of opportunity, such as defects per shift or defects per 10,000 units, can be effective for monitoring quality control in production. In some factories, the discussion has shifted to the number of possibilities of defects, or the number of opportunities. The electronics industry has defined a new C chart metric, the DPMO (defects per million opportunities) chart. A discussion of DPMO concepts and calculations is given in Chapter 4.

A more realistic way to achieve quality control in factories that approach six sigma is to closely couple the total defect reporting to the continuous quality improvement team. The low defect rate of six sigma manufacturing operation would produce a small number of total defects per day, even in a large factory. For example if we assume that a factory produces 5000 PCBs per day, and each PCB requires 2000 operations, that is a total defects opportunity of 10 million operations per day. For the six sigma defect rate of 3.4 defects per million, the total expected defects is 34. The management of the factory can review these defects individually each day, then decide what corrective action is needed, whether immediate, short, or long term. They can use the tools of TQM to monitor, organize, and rank defects and initiate a corrective action plan to reduce them further.

### 3.4    Using TQM Techniques to Maintain Six Sigma Quality in Manufacturing

When factories approach six sigma quality, the need for control charts with their sampling-based methods is reduced. The quality team can review all of the defects that occurred each day in production, using the TQM tools to effectively manage the corrective action process.

Table 3.4 shows a list of TQM tools grouped into three major areas according to their use: including tools for data analysis and display of problems, then tools for generating ideas and information about a likely solution, and then tools for decision making and consensus for the TQM team to resolve the problems. In the example of the factory in the last section that generates 34 defects per day the procedure for corrective action could be as follows:

**Table 3.4**   TQM tool usage

---

Tools for data analysis and display
1. Cause and effect
2. Histograms
3. Pareto analysis
4. Pie/time charts
5. Scatter diagrams
6. Spider diagrams
7. Flowcharting
8. Cost–benefit analysis

*Tools for generating ideas and information*
1. Brainstorming
2. Checksheets
3. Interviewing/surveying

*Tools for decision making and consensus*
1. Balance sheet
2. Weighted voting
3. Criteria rating
4. Paired comparison

---

1. The six sigma or corrective action teams use these tools to list and rank the defects, displaying or plotting them in an ordered rank, and then identify the defects that need priority resolutions.
2. The teams generate ideas about the most likely cause for the top defects, from the previous paragraph, using information from within and outside of the team.
3. The team makes a decision as to the most probable cause for the defects, using some of the decision making and consensus tools mentioned in Table 3.4.
4. The team recommends the most appropriate method for removing the causes of defects. An adverse consequence analysis has to be made to insure that the proposed solution does not generate new or additional problems.

### 3.4.1   TQM tools definitions and examples

The following set of tools, which were developed for continuous process improvements, have been used successfully by different companies and organizations. They include techniques for collecting defect data, manipulating and plotting them, prioritizing and identifying defect causes, and removing the most probable ones.

**3.4.1.1  Brainstorming.** A technique used to get a group to generate the maximum number of ideas on a topic or a problem, brainstorming is useful in opening discussions by involving all group members to generate as many ideas as possible without bias to any single idea.

Brainstorming is a good tool to use for group discussion trying to solve a problem or initiate an action. It has been used extensively in developing and focusing teams of engineers to solve problems or generate ideas for initiating and completing tasks.

The group members should be knowledgeable on the topic to be discussed. Every member should participate in brainstorming. The ideas should be promptly recorded without any arguments and no one person should dominate the discussion.

There are three phases of brainstorming:

1. *Idea generation*
   - Create as many ideas as possible. List these ideas on a flip chart or sticky paper.
   - All ideas are permitted; the team should be as freewheeling as possible. One good idea can trigger another.
   - The team members should not interrupt each other or analyze ideas presented; there should be no jumping to conclusions. They should only ask questions to clarify issues when ideas are recorded.
   - The team should adapt or build on ideas already listed.
2. *Clarification*
   - Team facilitator should repeat all items on the list and have every team member agree and understand each idea.
   - Remove duplications and add any new ideas.
   - Record the list as necessary.
3. *Evaluation*
   - Narrow down the list by allowing discussions.
   - Agree on a final list of ideas acceptable to the group.

It is advisable to use simulated training sessions for brainstorming. A group could attempt to tackle a problem, such as the design of a paper airplane or improving a golf swing or a tennis game, before embarking on brainstorming the problem at hand.

**3.4.1.2  The cause and effect diagrams.** This tool shows the relationship between the effect (reject) and its possible causes. It is used to logically group and identify all possible problems. It is also referred to as the "fishbone" or "Ishakawa" diagram.

To construct a cause and effect diagram:

- Use brainstorming to identify all possible causes for the effect. Ask outside experts to add to the list produced by brainstorming.
- Review the list and look for any interrelationships between the possible causes. Define three to six major categories that can be grouped together and categorize them. Common categories are sometimes referred to as the four M's: Materials, Machines, Methods and Manpower.
- Within each category, further subdivision might be required based on relationship or cause. They can ultimately be divided into subgroups.
- Draw the diagram, using arrows and names of each group, subgroup, and individual cause.
- Evaluate and select the most probable cause(s), based on the problem solving group decision tools.

An example of a cause and effect diagram is given in Figure 3.6, the shipment integrity cause and effect diagram. Another chart for PCB assembly is shown in Figure 8.2. Once the most probable cause has been identified, problem solving techniques such as design of experiments (DoE) can be used to verify the problem cause and institute corrective action.

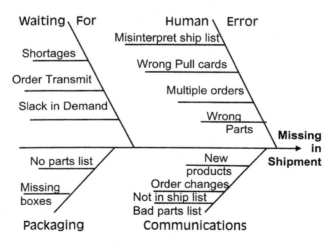

**Figure 3.6**   Shipment integrity cause and effect diagram.

**3.4.1.3   Checksheets.** A checksheet is a form used to identify, gather, organize, and evaluate data. A well-designed checksheet can eliminate confusion, enhance accuracy, and reduce time needed to take data.

There are two types of checksheets:

1. Recording check sheet. This is used for recording data on types of defects. The types should be listed in categories, and a mark made each time a defect is found in the sample. It is important not to collect too many types of defects.

   It is difficult to properly train production operators to distinguish between very similarly worded types of defects, even if photographs and other methods of graphically presenting them are used. Count the total number of checks for each defect.
2. Location check sheet. This is used to collect the location of the defects, and list how often they occur. This technique is useful to identify concentration of defects on a printed circuit board (PCB).

Other information should be included when available, such as date, part number, lot number, supplier name, supplier date code, area location, etc. Using automatic means of collecting and categorizing data, such as bar code readers and scanners, can speed up the recording of data. The defects data categories could be arranged in bar code format so that an operator with a bar code wand could enter all the data without writing down any information by hand.

**3.4.1.4   Flowcharts.** A flowchart is a picture of a process. It represents a step-by-step sequence. It can help in reaching a common understanding of how the manufacturing process is run and can act as a base for enhancing or changing the process. It can also be used as a documentation and training tool for pointing out areas for data collection and control, and as the basis of brainstorming for enhancing and troubleshooting the manufacturing process. Recently, it has been mostly replaced with process mapping. Figure 3.6 is flowchart representation of control charts.

The flowcharting process consists of these steps:

- Identify the first and last steps of the process.
- Fill in each process step. Include any time the product is handled, transferred, joined, or changed in form.
- Show feedback loops such as rework paths; they indicate inefficiency and possible low quality.

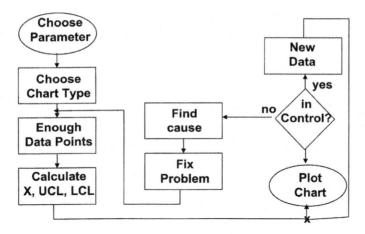

**Figure 3.7**   Control chart flow diagram.

- Choose symbols that are well understood or previously used: oblongs for start/end of process, diamonds for steps, and squares for decision points.
- Use structured analysis (SA) to simplify charts. Break down each major step into a box in the upper-level chart. Make sure all lines in the charts connect to at least one process step.
- Keep charts up to date as process evolves.

**3.4.1.5   Pareto charts.** Pareto charts have data plotted in bar graph form and display the number of times each defect has occurred, in ascending order. They plot the relative contribution of each defect cause, and tell at a glance the largest causes.

The Pareto charting process consists of these steps:

- Decide how many categories to plot. This will be equal to the total number of bars.
- Draw an axis, which could be in either direction—horizontal (x) or vertical (y) axis. Label each category. Draw the vertical axis with percentage and total number of occurrences for each category shown for each bar.
- Use same-width bars, arranged from tallest to shortest.
- Add information: title, preparer, date, and so on.

The Pareto principle is similar to the "80–20" rule: 20% of the problems cause 80% of the defects. It could be used to focus on the probable causes of defects, as well as prioritize them. Ideally, a Pareto chart should have all small bars.

Figure 3.8 is a Pareto chart presentation of the percent reasons for production downtime, showing the relative distribution of defect sources in terms of their occurrences.

**3.4.1.6  Scatter diagrams.** Scatter diagrams are simple graphical methods used to study relationships between two variables. They can quickly determine if a relationship exists (positive or negative) and the strength of that relationship (correlation).

Scatter diagram procedures are:

- Decide how many points to plot. A minimum of 30 points is needed to make conclusions significant.
- Arrange the pairs of measurements in ascending value of $x$. Divide data into subgroups of $x$.
- Draw and label horizontal and vertical axes. Choose the proper scale to fit all points.
- If the diagram shows an upward trend, there is a positive correlation. A downtrend is negative, and a level trend implies no correlation between variables.
- It might be necessary to plot logarithmic scales or many y points to a single x point to show data.

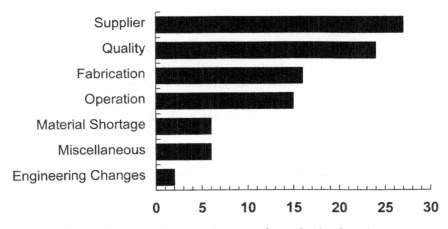

**Figure 3.8**  Pareto diagram—% reasons for production downtime.

Use regression analysis to accurately determine degree of correlation and the "best fit line." Significance can be determined using techniques similar to those shown on the next chapter.

**3.4.1.7  Histograms.** Histograms are pictures of the frequency distribution of a process. They are bar graphs with the height of each column representing the number of occurrences in each step for the process or measurement.

3.4.1.7.1  Information from histograms. By drawing the process specification on the axes, histograms clearly show the position of the process relative to desired performance. It becomes clear whether the process is performing as desired, or that the process average needs to be shifted, or the distribution needs to be narrowed.

One of the problems in plotting histograms is determining the best fit for a probability distribution. The $\chi^2$ Goodness of fit test, discussed in Chapter 2, is a good method to test the histogram to a specific distribution.

Figure 8.4 is an example of a histogram presentation of data for the improvement of a PCB soldering process, before and after a DoE was performed to improve the process.

**3.4.1.8  Time series graphs.** Time series graphs are sometimes called run charts. They are line charts used to monitor process quality measures over time. Run charts identify how process parameters change with time and indicate trends, shifts, and process cycling. They should be used to set quality process measures and goals.

To obtain information from run charts:

• Decide on quality units; a universal one such as defects per unit (DPU), expressed in parts per million (PPM) can be used. DPU (PPM) goals are universal, they can be benchmarked with similar processes in other companies or locations.

• Show goal line if appropriate. These goals should be set aggressively. However, they should not be set if they are impossible to meet, and must be met in too short a time. Realistic goals should be reached first, then they can be set for higher quality when current ones are met. A run chart is shown in Figure 8.1, representing a run chart of the use of quality tools for improving the PCB soldering process. The run chart shows the performance of process quality over a period of two years.

## 3.5    Conclusions

This chapter reviewed the different methodologies for controlling production. Historically, they originated from statistical sampling techniques and three standard deviation limits. The relationship of these classical techniques to the concepts of six sigma was determined directly, using the central limit theorem for variable charts. For attribute charts, an implied Cpk concept was introduced to translate the defect rate into six sigma terminology. As factories approach six sigma quality, the use of control charts can be reduced, since the number of total defects are few and sampling techniques to represent these defects are not required. In this case, a corrective action process based on TQM can be implemented to review and manage the six sigma quality on the factory floor.

## 3.6    References and Bibliography

AT&T. *Statistical Quality Control Handbook,* 9th ed. Easton PA: Mack Printing Company, 1984.

Afifi, A. and Azen, S. *Statistical Analysis, A Computer Oriented Approach,* 2nd ed. New York: Academic Press, 1979.

American National Standards Institute (ANSI). "Control Charts Methods of Analyzing Data." *ASQC Standard B2/ANSI 21.2.*

American National Standards Institute (ANSI). "Control Charts Method of Controlling Quality During Production." *ASQC Standard B3/ANSI 21.3.*

American National Standards Institute (ANSI). "Guide for Quality Control Charts." *ASQC Standard B1/ANSI 21.1.*

American National Standards Institute (ANSI). *ANSI/IPC-PC-90 Standard.* Developed by the IPC, Lincolnwood, IL.

American Society for Quality Control (ASQC). "Definitions, Symbols, Formulas and Tables for Control Charts." *ANSI/ASQC A1.*

Burr, I. *Engineering Statistics and Quality Control.* New York: McGraw-Hill, 1953.

Ducan, A. J. *Quality Control and Industrial Statistics,* 4th ed. Homewood, IL: Irwin, 1995.

Feigenbaum, A. V. *Total Quality Control,* 3rd ed. New York: McGraw-Hill, 1983.

Grant E. and Leavenworth R. *Statistical Quality Control,* 5th ed. New York: McGraw-Hill, 1980.

Johnson, R. *Miller and Freund's Probability and Statistics for Engineers.* Englewood Cliffs, NJ: Prentice Hall, 1994.

Moran, J., Talbot, R., and Benson, R. *A Guide to graphical Problem Solving Processes.* Milwaukee, WI: ASQC Press, 1990.

Smith, G. *Statistical Process Control and Quality Improvements.* Upper Saddle River, NJ: Prentice Hall, 1995.

Walpole R. and Myers, R. *Probability and Statistics for Engineers and Scientists.* New York: Macmillan, 1993.

Western Electric Company. *Statistical Quality Control Handbook.* Easton, PA: Mack Printing Company, 1956.

# The Use of Six Sigma in Determining the Manufacturing Yield and Test Strategy

Manufacturing is a multistep process, with each step generating its own variability, and therefore contributing to the overall defect rate. In a large multistep operation, individual process quality has to be very high in order for the overall yield to be reasonably acceptable. Otherwise, the probability of producing one good part is very low. In the case of PCB or IC fabrication, with 30–50 steps each, there are usually several in-process inspections or tests to cull out the intermediate defects, so that good parts can be produced when all production steps are completed. This chapter will examine methods to allocate for and plan these tests based on the expected quality of production.

It is important to measure quality in terms of the total number of defects found anywhere in the manufacturing process, and prior to any test or inspection. This will reduce confusion when setting quality targets or benchmarking similar operations in different plants. In addition, it will result in a true measure of quality that is not masked by the test or inspection costs.

Units of these quality measures are expressed in terms of first time yield (FTY) and defects per unit (DPU), expressed in parts per million (PPM). Recently the term defects per million opportunities (DPMO) has been used to reduce confusion on how to calculate defect rates in a complex multistep process such as PCB fabrication and assembly. Repairs are not considered as part of the definition of first time yield (FTY).

The issues of calculating FTY have become important in light of the increase in subcontracting the manufacturing of high-technology electronic products. Project teams and their leaders need accurate estimates of new product yields to plan and budget for test and troubleshooting equipment and personnel. In addition, management needs to benchmark potential suppliers in terms of their manufacturing quality. The results have been beneficial in several categories, and will be further highlighted in this chapter:

- By rolling up the yields of its various product components and manufacturing operations, the total product yield can be estimated. Project teams are thus able to manage carefully where additional resources are needed in terms of improving particular designs or manufacturing capabilities. By using these yield estimates, the new product team can also increase the accuracy of the new product cost estimates.
- Design for manufacture (DFM) principles, as championed by manufacturing engineers, can be emphasized to the design team in order to increase the FTY of new products, since a direct relationship can be made between the two concepts.
- FTY yield calculations can influence the focus of quality improvement teams.
- Yield calculation can clarify the best test strategy for reducing the overall test and troubleshooting costs.

In this chapter, the issues of yield and test strategy will be examined in a hierarchy of steps:

1. Determining units of defects
2. Determining manufacturing yield on a single operation or a part with multiple similar operations
3. Determining design or manufacturing yield of multiple parts with multiple manufacturing operations or design specifications
4. Determining overall product testing strategy

## 4.1    Determining Units of Defects

The basic definition of a defect is one that is based on the Poisson distribution. The defect rate, or defects per unit (DPU), is calculated based on defects, opportunities, and units. Defects are any deviation of the product functions that causes customer dissatisfaction or nonconformance to specifications. Units are the number of parts, subassemblies, assemblies, or systems that are inspected or tested. Opportunities are the characteristics that are inspected or tested. DPU is traditionally based on the opportunities of defects provided in one unit.

Defects can be attributes of units, as defined by a time or region. Units can be incoming materials, individual designs, transistors in an IC, repetitive manufacturing processes such as welds in a joint, etc. They can be individual units in a product, such as printed circuit boards (PCBs), or a single product. Defects represent the total defects found on that unit, expressed as a number called defects per unit (DPU). Since six sigma quality implies a very low DPU of 3.4 parts in a million operations, this definition has been converted to units of parts per million (PPM) in order to make it easier to communicate six sigma quality requirements. The following are the equations used to describe these units and their relationships:

$$\text{DPU} = \frac{\text{number of defects found anywhere}}{\text{number of units processed}} \qquad (4.1)$$

$$\text{DPU (PPM)} = \text{DPU (fractional)} \cdot 1{,}000{,}000 \qquad (4.2)$$

DPU (PPM) is the normalization of the DPU by a factor of 1,000,000 in order to facilitate equating a lower number with lower defects and driving it down to zero. Sometimes it is shortened to just PPM.

The definition of units is sometimes confusing. A unit could be a single transistor on an IC chip containing a million transistors. A unit could also be the IC itself, or it could the PCB containing many ICs, or the product containing many PCBs. In addition, the manufacturing steps needed to produce the transistors up to the final product have their own defect rate. Clearly, a uniform approach to these situations needs to be taken.

A historical approach to this dilemma has been to declare that six sigma or Cpk targets have to be achieved in "everything that we do." That means every material part or manufacturing operation has a six sigma goal. The collective aggregation of six sigma parts or operations will also have to be equal to six sigma. This approach would logically lead to the following strategy:

- Divide the manufacturing process into the smallest defined operations, each with its own DPU.
- Each manufacturing operation or material part represents a distinct transformation of product or material.
- In order for the next level of part aggregation (assembly or fabrication) to achieve six sigma quality without test, the individual DPUs have to be much greater in quality than the aggregation output.

- There is a need to translate the DPU of each operation into a DPU for the next level.
- For product design and manufacturing process engineers, it is much more useful to communicate and plan manufacturing tests using process and product yields as opposed to DPUs for the higher levels of product.
- There is a need to manage the conversion of DPUs into yields.

To address these issues, particular industries have developed the concept of defects per million opportunities (DPMO). These are standards that define the total defect opportunities per particular product or assembly. They use specific methods to combine the DPUs of parts and manufacturing operations, to arrive at the total number of opportunities. Opportunities can be defined in terms such as:

- Opportunities are characteristics or features of the product or the manufacturing process.
- Opportunities must be measurable and have a standard or specification with which they can be compared.
- Opportunities must be appraised. If a product has features that are not appraised, they should not be counted as opportunities.
- Opportunities are assumed to be independent.
- There cannot be more defects in a unit than opportunities.
- The opportunity count for a product is constant until the design or the manufacturing process changes.

An example of a DPMO methodology is the Institute of Printed Circuits (IPC) Standard 7912 for calculations of DPMO for PCB assemblies, which will be discussed later in Section 4.3.

## 4.2    Determining Manufacturing Yield on a Single Operation or a Part with Multiple Similar Operations

The manufacturing yield determination is based on the definition of the probability of obtaining a defect. The FTY is the percentage number of units produced without defects, prior to test or inspection. It is different than the traditional yield, which includes rework and repair.

The Poisson distribution, as discussed in the previous chapter, is a good basis for calculations of defects, especially when the number of possibilities or outcomes of defects is large and the probability of getting a defect at any time or region is small. In this case, the Poisson distribution can be simplified from Equation 3.7 as follows:

$$p(x, \lambda) = e^{-\lambda}(\lambda^x/x!)$$

$$P \text{ (at least 1 defect)} = 1 - P \text{ (no defects or } X = 0) = 1 - e^{-\lambda} (\lambda^0/0!) \quad (4.3)$$

$$\text{First time yield} = \text{FTY} = 1 - P \text{ (at least 1 defect)} = 1 - (1 - e^{-\lambda}) = e^{-\lambda}$$
$$(4.4)$$

Since $\lambda = np = \text{DPU}$

$$\text{FTY} = e^{-\text{DPU}} \qquad (4.5)$$

When an assembly is made from similar parts or operations, such as the transistors in an IC or soldering in a PCB, then the FTY for the assembly can be derived from the total DPUs of the individual operations. Sometimes, this yield is referred to as total yield ($Y_T$) or assembly estimated yield ($Y_A$) to distinguish it from FTY. It can be derived a follows:

$$Y_T = Y_A = e^{-\Sigma \text{DPU}} \qquad (4.6)$$

In six sigma quality, the DPUs are very small, and approximations can be performed without sacrificing the accuracy of the yield estimates. In this case, the general equation for yield can be further simplified by the power series expansion of exponential functions:

$$\text{FTY} = e^{-\text{DPU}} = 1 - \text{DPU}/1! + \text{DPU}^2/2! - \text{DPU}^3/3! + \text{DPU}^4/4! + \ldots$$
$$+ (-1)^{n+1} \text{DPU}^n/n! \qquad (4.7)$$

Since the DPU is small in six sigma quality (0.000034), we can ignore all the terms beyond the first two:

$$\text{FTY} = 1 - \text{DPU} = 1 - (\text{\# of defects/\# of opportunities}) \qquad (4.8)$$

and

$$Y_T = Y_A = (1 - \text{DPU})^n$$

where $n$ is the number of operations to be analyzed for defects.

### 4.2.1  Example of calculating yield in a part with multiple operations

In Figure 4.1, the wire bonding of an IC is shown. The chip is centered in the middle of the IC package frame, and wires are bonded from the chip to the frame. There are two bonds per IC termination. If there are 256 connections in the IC frame, and the bonding operation DPU is 100 PPM, what is the FTY for the bonding of an IC?

There are three methods of calculating the FTY, either by using the Poisson distribution [Equation (4.6)], the first two terms of the expo-

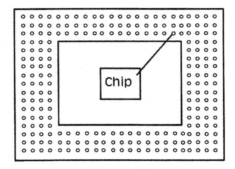

Figure 4.1    First-time yield (FTY) IC wire bonding example.

nential power expansion [Equation (4.8)], or by independently calculating the total defects in 100 ICs.

1. Process defects per wire bond = 100 PPM/1,000,000 = 0.0001
   Total wire bond opportunities/IC = 512 bonds
   $\Sigma_{DPU}$/IC = 512 · 0.0001 = 0.0512
   FTY = $e^{-\Sigma DPU}$ = $e^{-0.0512}$ = 0.95 or 95% FTY using the Poisson distribution
2. $Y_T = Y_A = (1 - \Sigma DPU)^n = (1 - 0.0512)^1 = 0.9488$ or 94.88% using power expansion
3. FTY actual for 100 ICs = 51,200 bonds @ 0.0001 = 5.12 defects per 100 ICs or 94.88% FTY

It can be seen that the FTY actual, which is the most accurate, is closely approximated by the Poisson distribution, and is exactly equal to the power expansion. These differences are small at the 100 PPM level, which is approximately four sigma quality. In the case of poor quality, such as those below two sigma, the differences in the calculated yield among the three methods become large. In that case, using the actual calculations is the most prudent way to obtain the yield. Note that the resultant five defects do not necessarily imply that five ICs are defective; one IC could have more than one defect.

### 4.2.2   Determining assembly yield and PCB and product test levels in electronic products

In typical electronic manufacturing lines, printed circuit boards (PCBs) are assembled and tested individually. Multiple PCBs are then assembled into finished products, which are tested. The test engineers will set goals for each type of test in order to plan for test and troubleshooting equipment and train operators for the production phase. Usually, PCBs are tested on automatic in-circuit testers (ICTs), which remove some of the assembly or part defects. The PCB test programs and the effort to develop them depend on these goals. A high yield in PCB test will reflect a higher turn-on ratio at the product level, saving the company valuable product test time and resources. Final assembly of the electronic product is accomplished from these tested PCBs and other components, such as power supplies and display devices, and turned on for final test. Yield calculations for PCBs and final product are similar to the ones discussed in this section.

### 4.2.3   PCB yield example

A product contains 10 PCBs, and a goal of 95% turn on yield was set for each PCB at in-circuit test (ICT). The product final test turn-on yield will be as follows:

$$DPU\ (PCB) = 0.05$$

$$\text{Product turn-on yield} = Y_T = e^{-\Sigma DPU} = e^{-10 \cdot 0.05} = 0.606 = 61\%$$

A turn-on yield of 61% is disappointing, especially when 95% in-circuit PCB yield could be difficult to achieve. To achieve a 95% final product turn-on, what should the PCB ICT test goal be?

$$\text{Expected product turn-on yield} = Y_T = e^{-\Sigma DPU} = e^{-10 \cdot (DPU)}$$
$$= 95\% \text{ or } 0.95$$

$$10 \cdot DPU = -\ln\ (0.95) = 0.05$$

$$DPU\ (\text{of each PCB test}) = 0.005$$

$$\text{PCB individual test yield} = 1 - DPU = 0.995 = 99.5\%$$

When a final test DPU of 95% is required for a product of 10 PCBs, the individual PCB ICT yield goals should be set at 99.5%.

It can be seen that the test yield for each component making the final product has to increase substantially in order to increase the turn-on yield of a large electronic product. The manufacturing process has to increase its quality level in order to match increased product complexity. Several methodologies and tools can be used for each part of the PCB assembly process. These steps do not necessarily require the use of more sophisticated inspection methods and equipment, but simple problem solving techniques such as:

- Incoming electronic component quality can be improved with better supplier certification and supplier process control methods.

- PCB assembly process quality can be enhanced with better employee training, the use of more automation such as autoinsertion of through hole (TH) and auto placement of SMT components, and improving the design guidelines of PCBs. Design guidelines might include standards for component polarity indicators, component placement and orientation in one axis, pad, hole and line geometry, and graphic placement aids.
- Soldering quality can be improved by continuously upgrading soldering materials and processes with the latest technology available, and performing design of experiment (DoE) techniques to optimally meet the soldering process parameters.

### 4.3   Determining Design or Manufacturing Yield on Multiple Parts with Multiple Manufacturing Operations or Design Specifications

A typical production line consists of multiple sources of materials and multiple distinct operations for fabrication and assembly of parts into the next-higher level of product assembly. Figure 4.2 is an example of a multistep manufacturing process line. Some of the issues pertaining to six sigma quality for this line are as follows:

- If the line is to be upgraded to six sigma quality, it is logical to assume that, at a minimum, all of the incoming parts and the individual operations of the line are to be upgraded to six sigma.
- The goal of six sigma quality for each incoming part and operation is a good management tool, since the individual part or operation can be analyzed or upgraded, independently of other parts.
- The output quality of the line, even if all of the incoming component parts and operations are of six sigma quality, is not at six sigma.

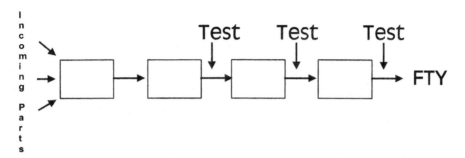

**Figure 4.2**   An example of a multistep manufacturing process line.

The defects from all operations add up to reduce line output quality from the six sigma target.

- The yield of the line is dependent on the complexity of the parts and manufacturing operations. The more parts and operations, the lower the yield. In addition, more operations require a much higher level of quality for each operation in order to obtain a reasonable overall line yield.

- Although each operation or an incoming part could be evaluated for six sigma or a targeted Cpk quality, the evaluation of the total line quality is not readily apparent, and there can be many different options to do so. This section will explore different approaches to this condition.

- The yield of the line can be calculated using different methodologies, as shown in the previous section. This yield can result in different test strategies, depending on the economics of the alternative test methods to be used to bring up the final line quality to the specified level.

Treating the line yield as a Poisson distribution can result in quickly estimating the line FTY by adding the DPUs of each of the different processes. For example, in a line with three steps process—A, B, and C—the FTY calculations would be as shown in Table 4.1. Total line yield can be calculated from either the multiplication of the individual yields of each step or the addition of the individual DPUs of each step, then converting the total DPUs to the total yield using the Poisson distribution. The results should be the same, since the probability of the defects in each process step is assumed to be independent.

An alternate method for calculating the yield is to use the approximation $FTY_a = 1 - a$ instead of the $e^{-a}$ calculations shown in Table 4.1. When several parts are made in each operation, then the total yield can be calculated using either of the above two methods, as shown in Table 4.2, using $n$ parts through the three-step process line.

**Table 4.1**   Yield calculation in a three-step production line

| Process steps | A | B | C |
|---|---|---|---|
| Yield for each step | $Y(A)$ | $Y(B)$ | $Y(C)$ |
| DPU at each process step | $a$ | $b$ | $c$ |
| Process yield (FTY) in each step | $e^{-a}$ | $e^{-b}$ | $e^{-c}$ |
| Total process yield $Y_T$ | $Y\{A\} \cdot Y\{B\} \cdot Y\{C\}$ | | |
| Or use FTY {total} | $e^{-a+b+c}$ | | |

**Table 4.2**   Yield calculation in a line with $n$ parts in a three-step production line

| Process steps | $A$ | $B$ | $C$ |
|---|---|---|---|
| Yield for each step | $Y(A)$ | $Y(B)$ | $Y(C)$ |
| DPU at each process step | $a$ | $b$ | $c$ |
| Process yield (FTY) in each step | $e^{-na}$ | $e^{-nb}$ | $e^{-nc}$ |
| Or process yield (FTY) in each step | $(1-a)^n$ | $(1-b)^n$ | $(1-a)^n$ |
| Total process yield $Y_T$ | $Y\{A\} \cdot Y\{B\} \cdot Y\{C\}$ | | |
| Or use FTY {total} | $e^{-n(a+b+c)}$ | | |

### 4.3.1   Determining first-time yield at the electronic product turn-on level

The electronic products being developed today are more complex than previous products. The number of components on each printed circuit board (PCB) is increasing, as well as the total number of PCBs in the product. In the following example, the effects of these complexities on the final product turn-on will be demonstrated. The historical quality level that sustained the production process for older products is not adequate for new complex products. The in-process manufacturing quality of components and PCBs will have to be improved significantly to counteract the increased number of assemblies and components.

### 4.3.2   Example of yield calculations at the PCB assembly level

The defect rate for new PCBs is usually calculated based on process observations for existing PCBs. Assuming a PCB with through-hole technology, defects are usually obtained from three sources: incoming materials and components; assembly defects of missing, wrong, or reversed components; and soldering or termination defects. If it is assumed that each component has 2.5 solder connections per PCB, the quality level for multiple component PCBs can be calculated as follows, assuming reasonable PCB assembly process quality:

   Solder defect rate DPU = 100 PPM

   Component assembly defect rate DPU = 500 PPM

   Incoming component defect rate DPU = 300 PPM

Assuming 2.5 solder connections per component, what is the total process yield at the PCB test level for 100, 500, and 1000 component PCBs?

## Solution method 1. Calculating total yield using nDPU

$$\text{FTY \{total\}} = e^{-\{(\text{solder DPU} \cdot n \cdot 2.5) + \text{assembly DPU} \cdot n + \text{component DPU} \cdot n\}}$$

| # Parts $n$ | Solder defects | Assembly defects | Component defects | Total nDPU defects | FTY $= e^{-n\text{DPU}}$ yield |
|---|---|---|---|---|---|
| 100 | 0.025 | 0.05 | 0.03 | 0.105 | 90% |
| 500 | 0.125 | 0.25 | 0.15 | 0.525 | 59% |
| 1000 | 0.25 | 0.5 | 0.3 | 1.05 | 35% |

## Solution method 2. Calculating the total yield by multiplying individual process yields

| # Parts $n$ | Solder yield $e^{-n\text{dpu}}$ | Assembly yield $e^{-n\text{dpu}}$ | Component yield $e^{-n\text{dpu}}$ | Total yield $Y(\text{solder}) \cdot Y(\text{assembly}) \cdot Y(\text{component})$ |
|---|---|---|---|---|
| 100 | 0.975 | 0.951 | 0.97 | 90% |
| 500 | 0.882 | 0.779 | 0.861 | 59% |
| 1000 | 0.779 | 0.606 | 0.741 | 35% |

## Solution method 3. Calculating the total yield using power series expansion.

In this method, the solution is derived by calculating the total yield by multiplying individual process yields based on $1 - \text{DPU}_{\text{component}}$ expansion, where DPU is the process defect rate for one component. Note that the defect rate for the PCB operations should not be used, because some of the values are too high (i.e., the DPU for total assembly defects is 0.5) to ignore the higher-order terms in the power expansion.

| # Parts $n$ | Solder yield $(1 - \text{DPU})^n$ | Assembly yield $(1 - \text{DPU})^n$ | Component yield $(1 - \text{DPU})^n$ | Total yield $Y(\text{solder}) \cdot Y(\text{assembly}) \cdot Y(\text{component})$ |
|---|---|---|---|---|
| 100 | 0.975 | 0.951 | 0.97 | 90% |
| 500 | 0.882 | 0.779 | 0.861 | 59% |
| 1000 | 0.779 | 0.606 | 0.741 | 35% |

The total yield results using all three methods of calculations mentioned above were approximately equal in values.

It can be shown that as the number of components increases in the PCBs, first-time yields decrease significantly, assuming that the quality level of the assembly process remains the same. In order to achieve higher first-time yields for complex PCBs of more than 500 parts, the quality level of the assembly process steps has to be improved from hundreds of PPM defects to tens of PPM defects.

### 4.3.3    DPMO methods for standardizing
### defect measurements

In the previous example, the yield calculations involved two types of opportunities—components and solder joints or terminations. This is similar to the problem presented in Figure 4.1, where the IC was the component and the bonding was used for the terminations. A common problem in electronics manufacturing quality has been to decide which of the two choices, components or terminations, should be the basis for defect opportunities when calculating the yield of assemblies.

An additional problem is defining the cause for termination failures. If the IC in Figure 4.1 was not placed properly in the frame, some of the terminations could become defective, even if the bonding process was completed successfully. If one IC chip was misplaced in the assembly step of the process, it could lead to 256 defects in the bonding process. This would falsely penalize the bonding process, even if it was functioning properly. Obviously, a set of rules need to be applied in order to clarify the quality of the assembly operation and to benchmark it with similar operations in the supply chain.

The defects per million opportunities (DPMO) concept was developed for the PCB assembly operation to tackle the problems outlined above. Developed as IPC Standards 7912 and 9261, they set the rules for counting opportunities and defects. They define a mix of defects and opportunities for components and assembly operations consisting or placements and terminations. Table 4.3 shows a basic grouping of defects and opportunities for PCB assemblies. A number of defects and a number of opportunities are defined for each operation. The defects for each operation could be influenced by prior operations. For example, a misaligned component in the placement operation might

**Table 4.3**  DPMO grouping of defects and opportunities for PCB assemblies

| Source | Opportunities | Causes |
|---|---|---|
| Components | Number of components | Bent leads<br>Wrong value<br>Cracked<br>Wrong label |
| Placements | Number of Components | Missing, loose<br>skewed, reversed |
| Terminations | Number of leads soldered | Solder deposition,<br>improper reflow |
| Total defect opportunities | Total of three items above | |

cause many termination defects, as discussed earlier. In this method-ology, it would be counted as one placement defect and zero termina-tion defects for the PCB. The number of opportunities for components include all of the components plus the fabricated (raw) PCB. The number of termination is the actual number of solder joints on the PCB. Some the definitions are as follows:

$$\text{DPMO}_{\text{operation}} = \frac{\text{number of defects}}{\text{number of opportunities}} \cdot 10^6 \qquad (4.9)$$

$$\text{DPMO index} = \frac{\Sigma \text{ operation defects}}{\Sigma \text{ opportunities defects}} \cdot 10^6 \qquad (4.10)$$

$$\text{OMI} = \left[ 1 - \left\{ \left( 1 - \frac{\text{defects}_1}{\text{opportunities}_1} \right) \cdot \left( 1 - \frac{\text{defects}_2}{\text{opportunities}_2} \right) \cdots \right\} \right] \cdot 10^6 \qquad (4.11)$$

The DPMO for each operation is equivalent to DPU (PPM) defined earlier in this chapter. The DPMO index is a useful tool for calculat-ing the actual yield of the PCB, since it is based on the total number of defects divided by the total number of opportunities. It is usually dominated by the termination count. The DPMO index is the basis for DPMO charts, discussed in the next section.

The overall manufacturing index (OMI) is an attempt to equalize the weight of all three basic operations in PCB assembly. The yield of each operation is calculated using the power expansion formula 4.8, then the yields are multiplied together to form a multiplier yield for the assembly line. A multiplier defect rate for the assembly line is de-rived from the one-multiplier yield, and then multiplied by 1 million to obtain the OMI index.

The OMI index represents an overall theoretical defect rate in which each operation is given equal weight, based on the its own cal-culated yield. The OMI index is independent of the number of oppor-tunities of each operation, and therefore can be used to compare the quality of alternate PCB assembly lines.

### 4.3.4 DPMO charts

DPMO charts are attribute charts used to monitor the quality of PCB assembly lines. They are best used instead of attribute defect charts such as U or C charts. Each type of PCB can be charted every time it is run through the assembly line. A multiplication factor (MF) is pro-vided in the calculations to make the conversion to million opportuni-ties. DPMO charts can be used with defects codes for quality tracking and continuous improvements.

The following definitions apply to DPMO charts:

DPU = Defects found in PCB lot sample/total number of PCBs in lot sample

MF = 1,000,000/total defect opportunities

DPMO = DPU × MF

$\overline{\text{DPMO}}$ = Average DPMO over time (20 samples minimum)

Control limits = $\overline{\text{DPMO}} \pm 3 \cdot \sqrt{\overline{\text{DPMO}}}$/number in lot sample (U charts)

Control limits = $\overline{\text{DPMO}} \pm 3 \cdot \sqrt{\overline{\text{DPMO}}}$ (C charts)

Table 4.4 is an example of U chart DPMO-based calculations. The DPMO chart is displayed in Figure 4.3. It is plotted by the daily activity of the assembly line for a particular PCB. The PCB was assembled on different shifts and on different days by different operators. The process seems to be out of control if two or more defects are found in any point plots of the assembly line operation.

The control limits appear too narrow for the fluctuation of the pattern, and the fluctuations are erratic. This called a pattern of instability. Either more data is required for each DPMO point or the pattern must be simplified before the data can be analyzed. Simplification might involve some of the following steps:

- Complex patterns might mean that the variable used as the basis for plotting the point on the chart in sequence is not the most significant variable. For example, the defects might vary according to the shift or the operator manning the assembly line. The chart can be replotted with the x axis data arranged according to these possi-

**Table 4.4**   DPMO chart data

PCB with 84 components, 298 leads, with varying sample or lot sizes to be plotted on DPMO U chart

| PCB's inspected | = 10 | 7 | 12 | 11 | 12 | 4 | 10 | 7 | 7 |
|---|---|---|---|---|---|---|---|---|---|
| Defects | = 3 | 1 | 3 | 0 | 2 | 0 | 2 | 0 | 1 |
| Defects per PCB | = 0.3 | 0.14 | 0.25 | 0 | 0.17 | 0 | 0.2 | 0 | 0.14 |

Average DPU = 12/80 = 0.15

Total defect opportunities = 85 Components (including raw PCB) + 84 Placements + 297 Solder = 466

MF = 1,000,000/466 = 2146

DPMO = DPU × MF = 0.15 · 2146 = 322

UCL = 322 + 3 · $\sqrt{322/8}$ = 341

LCL = 322 − 3 · $\sqrt{322/8}$ = 303

| DATE | | SUP | | | PART # | | | | REV # | | |
|---|---|---|---|---|---|---|---|---|---|---|---|
| # Defects | 3 | 1 | 3 | 0 | 2 | 0 | 2 | 0 | 1 | | |
| PWB Lot | 10 | 7 | 12 | 11 | 12 | 4 | 10 | 7 | 7 | | |
| DPU | 0.3 | 0.14 | 0.25 | 0 | 0.17 | 0 | 0.2 | 0 | 0.14 | | |
| MF | 2146 | 2146 | 2146 | 2146 | 2146 | 2146 | 2146 | 2146 | 2146 | | |
| DPMO | 644 | 300 | 536 | 0 | 365 | 0 | 429 | 0 | 300 | | |
| Operator | JS | MB | FA | JS | MB | FA | JS | FA | MB | | |
| Date | 2/21 | 2/22 | 2/22 | 2/22 | 2/23 | 2/23 | 3/10 | 3/10 | 5/2 | | |

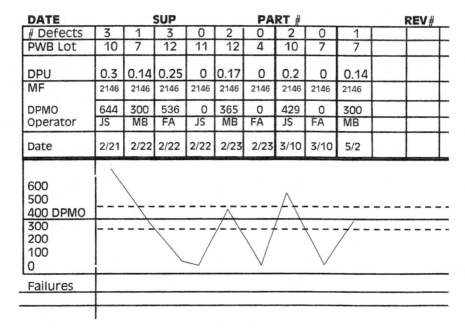

**Figure 4.3**  DPMO chart example.

bilities. It can be seen that operator MB is close to control limits, whereas the other two operators, FA and JS, are always not in control.

- If the pattern does not become simpler, then some other variable might be selected such as length of service for the operators. If the pattern becomes simpler but more simplification is desired, then the $x$ axis can still be further divided according to other significant variables.
- The procedure outlined above should be repeated until the pattern is a simple shift in level or a simple trend.

DPMO as well as C and U charts can be good tools for monitoring the assembly process, but care must be taken to achieve good charting and interpretation of data.

### 4.3.5   Critique of DMPO methods

DPMO and OMI are good tools to calculate PCB assembly line yield and to compare and benchmark electronic PCB assembly in the supply chain. Issues that arise with the implementation of the DPMO and OMI indices might be as follows:

- Confusion over the utility of both functions. DPMO is easier to calculate than OMI and therefore will become the more commonly used function.

- DPMO/OMI deployment will require extensive training of assembly labor as well as management and support staff such as process and quality engineers to interpret the rules for calculating defects.

- Guidelines will have to be defined for certain defect conditions to assure the independence of component, placement, and termination defects

- Some components might have different defect rates than others. For example, mechanical, through-hole (TH), and surface mount technology (SMT) components can all be part of the assembly line process. Each will have a different defect rate, and they should not be lumped together in one defect number.

- DPMO concepts require knowledge of the actual number of termination opportunities, which are readily available in manufacturing but do not get finalized until late in the design and development process for electronic products (after PCB layout). Intermediate metrics such as the ratio of components versus termination opportunities might be more useful in the design stage, especially for design for manufacturing (DFM) input, before the design in "hardened" after PCB layout. This intermediate metric was shown in Example 4.3.2.

- DPMO is an example of the attribute quality problem in six sigma. The notion of striving for "six sigma in everything that we do" is not directly shown with the use of one or two indices such as DPMO and OMI. Individual process quality as well as total assembly line quality should be examined. In DPMO, the emphasis is on a modified defect rate. In the next section, an alternate method for calculating and comparing quality of assembly lines using back-calculated or "implied" Cpk is discussed with examples.

### 4.3.6  The use of implied Cpk in product and assembly line manufacturing and planning activities

As discussed earlier, some industries have adopted a form of six sigma that is based on target values of Cpk. Examples are the auto industries with the QS 9000 (Cpk 1.67 for new and 1.33 for old products), and the defense industry with various Cpk values for weapon systems (Cpk = 1.33 for the F22 jet fighter). In these cases, an "implied Cpk" value is used to characterize the quality of the process or the product being evaluated.

When using implied Cpk, it is assumed that defects are occurring because of violation of a particular or a composite set of specifications. The composite specification can be one-sided or two-sided, depending on the interpretation of the defects. For example, a wire bond could be treated as one-sided, since it is assumed that in testing the bond, only a minimum specification value is given. For solder defects, a composite specification can be assumed to be two-sided, since solder defects can be caused by too much solder (solder shorts), or too little solder (insufficient solder defects). The difference between implied one- or two-sided specifications is that the number of defects representing the $f(z)$ value under the normal curve should be halved for two-sided specifications or used directly for one-sided specifications, resulting in different implied Cpk interpretations. The decision for one- or two-sided specifications for implied Cpk should be left to the appropriate design and manufacturing engineers. A description of the use of such an implied Cpk process is given in Chapter 2.

The use of an implied Cpk process in assembly line activities is similar to the DPMO process. Individual manufacturing processes are analyzed for quality, with a DPU (PPM) and an implied Cpk calculated for each. For each assembly, such as a PCB running through the line, the parts counts and process steps are calculated, then multiplied by the DPU rate to obtain the defects for each step in the assembly line. Finally the defects are added and then reflected in a total yield using Equation 4.8 and an implied Cpk. Alternately, the yields could be calculated for each step, then multiplied together to form a total yield. A decision has to be made for each process as to the type of quality data to be collected. In the PCB assembly line case, the choices of the quality data for each process can be as follows:

- The use of a particular defect parameter for each process step. For component types, defect data can be collected on the following: axial insertion for through-hole components, pick and place operations for SMT components, odd-shape components for automatic as well as manual placement, and mechanical parts assembly such as with screws and special connectors. For terminations, defect data can be collected on manual as well as automatic soldering.

- This division of defect data according to the process used can help in identifying lower-quality process steps and in targeting these processes for quality improvements.

- Data collection can be based on a selected attribute. For example, placement quality data can be based on components, leads, or a combination of both.

- Guidelines have to be established in order to handle defects from prior operations that might influence defects in subsequent operations; for example, a placement defect that can cause multiple terminations defects. This can skew the termination data. Decisions have to be made and training programs offered to operators in order to follow guidelines on apportioning and analyzing defects according to source.

### 4.3.7 Example and discussion of implied Cpk in IC assembly line defect projections

Figure 4.4 is an example of a portion of an IC fabrication line. Only a few operations are shown in order to demonstrate the utility of using Cpk-based analysis for the line. This analysis can be used to determine defect projections for all different IC types that are made by the line, based on the number of manufacturing steps required by the IC for each operation. Note that by using the Cpk approach, the 1.5 $\sigma$ shift of the average to the specification nominal is not considered in the defect calculations.

For each operation, several attributes are shown by rows in Figure 4.4 to classify their quality:

• The process specification. Each operation is characterized by one- or two-sided implied specifications that cause defects to occur when they interact with the variability of the process. This information is

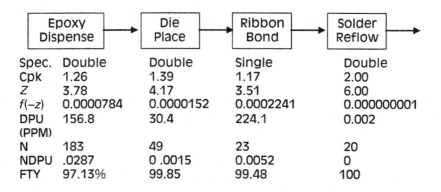

| | Epoxy Dispense | Die Place | Ribbon Bond | Solder Reflow |
|---|---|---|---|---|
| Spec. | Double | Double | Single | Double |
| Cpk | 1.26 | 1.39 | 1.17 | 2.00 |
| Z | 3.78 | 4.17 | 3.51 | 6.00 |
| $f(-z)$ | 0.0000784 | 0.0000152 | 0.0002241 | 0.000000001 |
| DPU (PPM) | 156.8 | 30.4 | 224.1 | 0.002 |
| N | 183 | 49 | 23 | 20 |
| NDPU | .0287 | 0 .0015 | 0.0052 | 0 |
| FTY | 97.13% | 99.85 | 99.48 | 100 |

Total NDPU = 0.0354    One sided NDPU = 0.0177
Total FTY = 96.48%      Z = 2.1 = Cpk/3
Line Cpk (2 sided) = 0.70

**Figure 4.4**    IC assembly line Cpk example.

required to make the decision when back-calculating the Cpk from the defect data. The specifications are assumed to be either single (one-sided) or double (two-sided)

- The Cpk for each operation. This Cpk is calculated from previous historical data when the process capability of each operation was determined. They are recorded as the current quality level of that operation. Note that in the last operation, solder reflow has achieved six sigma quality of Cpk = 2.

- The next two attributes convert this Cpk number to the more familiar DPU number for defect measurement in PPM. The DPU number could alternately be used to record the quality instead of the Cpk number.

- $z$ is the variable from the standard normal distribution, derived from Cpk by Equation 2.13 ($z = 3 \cdot$ Cpk). The next line is the $f(-z)$ to determine the one-sided probability of defects that can be found directly.

- DPU (PPM) is the defect rate of the operation. It is derived from the $f(-z)$ and then multiplied by 1,000,000. If the implied specifications of the operation in this section are two-sided, then the defect rate is multiplied by two. The DPU can be used as a substitute for defining the quality of the operation, instead of the Cpk if so desired.

- $N$ is the number of operations required for the IC being assessed for quality. In this case, the IC has to undergo 183 epoxy dispense operations. NDPU, or total defects for producing the IC in this operation, is calculated by multiplying $N$ by the DPU to produce NPDU for that IC.

- The operation FTY is calculated by subtracting the NDPU from 1 for each operation.

- When all of the data for each operation have been determined, then the total line information can be calculated. Depending on the goals set for the IC manufacturing line, three indicators can be determined for each IC type that is produced on that line:

  1. Total line NDPU—the total manufacturing defects for the line resulting from making a particular IC. This is calculated by adding the defects (NDPU) from each operation.

  2. Total line FTY—the total yield for a particular IC made in the line. It can be calculated either by multiplying the yield of each operation or from subtracting the total NDPU from 1.

  3. Total line Cpk—the quality index for the IC being made in the line. This is back-calculated from the defect rate, assuming two-sided specifications and no process average shift.

The information gathered from this example can be used by different parts of the organization, helping them achieve their individual goals. Management can use this information to document the production lines progress toward six sigma. Test engineers can use this information to plan for test and troubleshooting stations. Production and process engineers can use this information to focus on which manufacturing operations most need quality improvements. In this example, the ribbon bonding operation has the lowest Cpk and highest DPU, and therefore should be the first operation to be targeted for quality improvements.

## 4.4   Determining Overall Product Testing Strategy

Ultimately, all defects have to be removed by testing the individual assemblies that make up the product, and then finally testing the product. Test engineers are concerned about the yield of the product, in order to budget and plan for test and troubleshooting equipment and operators. The six sigma quality defect rate and yield calculations are excellent tools to help in the planning of electronic product test strategy.

It is common knowledge in the test industry that the cost of inspecting for and removing defects can be as high as 30% of the overall manufacturing cost. In addition, the earlier a defect is caught and removed in the manufacturing cycle, the cheaper it is in terms of equipment cost and operator skills. The best alternative to expensive test equipment and skilled operators is achieving six sigma quality and the resultant assembly yield goals.

As shown in the examples in this chapter, the quality of the individual elements of an assembly can be linked to its total quality performance. In Example 4.2.3, 10 PCB assemblies, each with 95% test yield, can result in the next level of assembly (final product made up of the 10 PCBs) having a yield of only 61%. If a higher yield for the next step in the assembly is desired, then the yield of the individual components have to be improved further.

In Example 4.3.2, it was shown that increasing the number of components or steps in the assembly have a similar effect on reducing the yield. The yield for an assembly of 90% based on 100 components or steps quickly drops to 59% yield with 500 components, and then to 35% yield with 1000 components.

This combined effect of setting the yield goal and the number of the underlying steps in the assembly operations have led test engineers to examine the test strategy based on the ability of various test equipment to remove certain level of defects.

### 4.4.1    PCB test strategy

Figure 4.5 is an example of the test methodologies available for PCB assemblies. These methodologies can be summarized as follows:

1. Visual test and inspection. These tests use trained operators to inspect PCBs for defects using the naked eye or visual magnification such as microscopes or enlarging lenses. They concentrate on geometrical defects that are easily observed by the human eye, but may be difficult for machines to detect, such as solder shapes and shorts.

2. In-circuit test (ICT). This type of test is used to eliminate defects that result from individual components not meeting their specifications. The defects either due to the components being defective as supplied, or becoming defective through the PCB assembly operations. They could either be missing, wrong, placed or inserted into the PCB incorrectly, or become defective because of PCB assembly operations exceeding manufacturing specifications.

   The ICT test consists of a machine with electronic means of comparing the components to a preprogrammed value. The components, already soldered in place on the PCBs, are reached through a bed of nails fixture that provides contact of the component pads on the PCB to pins in the fixture. Many sources of electronic noise may be present, such as stray capacitance and resistance in the fixture and its wires. In addition, some components in the circuit are used in parallel with other components, so that it is difficult for the tester to isolate the individual component to be tested.

   The ICT is not always capable of detecting all component defects because of the tester connections to the circuit. This inability to detect all of the component defects is called defect or test coverage. A

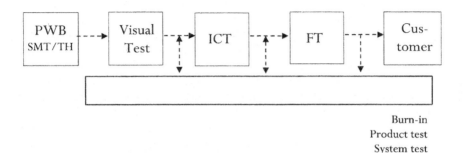

Burn-in
Product test
System test

**Figure 4.5**    PCB test alternatives.

low test coverage, under 90% yield of good PCB's into the next test cycle, will result in the need for a functional test of the PCB.

3. Functional test (FT). This type of test is used to eliminate design-based as well as assembly defects. The latter occur when the set of components being assembled meet their individual specifications and are assembled correctly, but the assembled PCB does not meet its systems specifications. Functional testers use the PCBs in a similar fashion to their intended use in the product. In its simplest form, FT is called a box test, which consists of testing the PCBs in a fully functional product.

4. The customer of PCB test is the next level of production. It usually is the product assembly process, where the PCBs are combined with mechanical parts to form the product. The product might undergo additional testing such as burn-in, product test, and system test.

   Product test occurs after the product has been assembled with PCBs and other mechanical and input/output modules. Burn-in occurs when the product is subjected to environmental stress conditions, and then tested to see if there were any "infant mortality" failures. System test occurs when the product is combined with other products with maximum-length cabling to form a system configuration similar to those in customer sites.

The different types of PCB tests have different expectation of defects removal. Test engineers usually communicate quality through using the PCB yields from each of the test methods mentioned above, whereas the assembly community communicates through DPU or DPMO. The management sets enterprise goals at certain Cpk levels or six sigma. Using the examples in this chapter, it was shown that quality communications could be just as effective using any of these common methods outlined above.

The PCB test strategy is formulated based on the lowest-cost alternative for removing defects, given the current quality level of the design and manufacturing process. To achieve a successful strategy, the costs of each type of test as well as visual inspection should be known, including nonrecurring costs such as test development and programming, fixture design, and troubleshooting. A test strategy could be developed to provide the proper balance between investing in improving the PCB assembly process capability versus performing test and troubleshooting to remove defects generated in design and assembly. This can be accomplished given the availability of alternative test method costs, and having the defects and their sources quantified using methods outlined in this chapter.

## 4.4.2    PCB test strategy example

A typical PCB test method comparison is given in Table 4.5. It can be seen that the test time increases with the complexity of the test performed. More complicated tests allow for a higher removal of defects, resulting in greater yields from these test methods. The cost of the repair cycle for each method is also shown; it increases geometrically as the test complexity increases. The defects that are not culled out at test could escape to the customer, be it the next-higher level in the manufacturing operation, or the actual paying customer.

Table 4.5 shows three scenarios of PCB test strategy. Scenario 1 is that of typical three sigma company that is performing a good job of manufacturing control through control charting, but has not yet implemented the goals of six sigma quality improvement programs. The assembly yield of 60% prior to test is typical of in-control but not capable assembly operations, as shown in Example 4.3.2 for PCBs with 500 components. In many of these operations, visual inspection is used in order not to overwhelm the in-circuit (IC) test operations. Visual tests bring up the assembly yield to 80% by removing 50% of the defects in the PCBs. The in-circuit test design in three sigma operations is targeted at 95% yield into the functional test (FT). The FT test produces PCBs with 99.8% yield, resulting in a defect rate of 0.2% that will escape to the customer. This defect rate is close to the three sigma assembly process output of 2700 PPM or 0.27%. Table 4.6 shows two different strategies using scenario 1. One test strategy

**Table 4.5**  PCB test methods comparison

|  | Visual test | In-circuit test | Functional test | At-customer failures |
|---|---|---|---|---|
| Test time (minutes) | 2 | 3 | 10 | |
| Test cost/PCB ($) | 1 | 3 | 10 | |
| Repair ratio | 1 × | 10 × | 100 × | 1000 × |
| Repair cost ($) | 1 | 6 | 50 | 500 |
| Scenario 1 (typical three sigma company) | | | | |
| Expected yield before test | 60% | 80% | 95% | |
| Expected yield after test | 80% | 95% | 99.8% | 0.2% |
| Scenario 2 (four sigma company) | | | | |
| Expected yield before test | | 80% | 95% | |
| Expected yield after test | | 95% | 99.99% | 0.01% |
| Scenario 3 (six sigma company) | | | | |
| Expected yield before test | | 95% | 99.8% | |
| Expected yield after test | | 99.8% | 99.9999% | 0.00034% |

Table 4.6   PCB test methods scenario 1 (two strategies)

| | Visual test | In-circuit test | Functional test | At-customer failures | Totals |
|---|---|---|---|---|---|
| Test cost/PCB ($) | 1 | 3 | 10 | | |
| Repair cost ($) | 1 | 6 | 50 | 500 | |
| **Scenario 1 (Strategy 1)** | | | | | |
| Expected yield before test | 60% | 80% | 95% | | |
| Expected yield after test | 80% | 95% | 99.8% | 0.2% | |
| **100,000 PCBs @ 500 components** | | | | | |
| Test costs ($) | 100,000 | 300,000 | 1,000,000 | | 1,400,000 |
| Defective PCBs before test | 40,000 | 20,000 | 5,000 | 200 | |
| Defective PCBs after test | 20,000 | 5,000 | 200 | | |
| PCBs repaired | 20,000 | 15,000 | 4,800 | 200 | |
| Repair cost ($) | 20,000 | 90,000 | 240,000 | 100,000 | 450,000 |
| Total test and repair cost ($) | | | | | 1,850,000 |
| Cost/PCB ($) | | | | | 18.50/PCB |
| **Scenario 1 (Strategy 2)** Omit visual test | | | | | |
| Expected yield before test | | 60% | 95% | | |
| Expected yield after test | | 95% | 99.8% | 0.2% | |
| **100,000 PCBs @ 500 components** | | | | | |
| Test costs ($) | | 300,000 | 1,000,000 | | 1,300,000 |
| Defective PCBs before test | | 40,000 | 5,000 | 200 | |
| Defective PCBs after test | | 5,000 | 200 | | |
| PCBs repaired | | 35,000 | 4,800 | 500 | |
| Repair cost ($) | | 210,000 | 240,000 | 100,000 | 550,000 |
| Total test and repair cost ($) | | | | | 1,850,000 |
| Cost/PCB ($) | | | | | 18.50/PCB |

uses visual inspection, the other does not, taking the PCB's directly into in-circuit testing. The second strategy removes the high labor cost and low job satisfaction of visual test, and shifts the burden of removing defects to in-circuit testing. It can be seen from the two strategies that the operational costs are the same, resulting in a cost of $18.50 per PCB when the production rate is assumed to be at 100,000

PCBs. Therefore, most companies would opt for the non-visual test strategy, because automatic testing is usually more predictable than manual inspection. In addition, in-circuit testing can be improved with better equipment, whereas visual testing would not greatly increase in efficiency with increased operator experience.

In order to properly devise the best strategy for scenario 1, more information will have to be collected. This would include the capital and depreciation costs of the in-circuit equipment and fixtures, as well as the resources needed to maintain and repair them. More discussion is given on that in Chapter 6.

Scenario 2 is that of four sigma company. The test method summary is given in Table 4.7. In this case, the PCB assembly area yield increases to 80%. This is based on a PCB with 500 components, having 2250 opportunities for defects at the four sigma level, at $f(z) = 0.9999$ for a 1.3 Cpk process capability. These opportunities result from 500 components, 500 placements, and 1250 terminations, or $0.9999^{2250} = 80\%$. The defects escape rate to the customer from a four sigma assembly operation is equivalent to 1 minus 0.9999 or 0.01%. This number is equivalent to 100 PPM, which is close to the four sigma error rate of 64 PPM. It can also be described as Cpk = 1.3. If the same level of in-circuit test design is used, the test cost per PCB drops to $16.45.

Scenario 3 is that of a six sigma company. The test method summary is given in Table 4.8. In this case, the PCB assembly area yield increases to 95%, and the defects from the assembly line escaping to the

Table 4.7     PCB test methods scenario 2 (four sigma company)

| | Visual test | In-circuit test | Functional test | At-customer failures | Totals |
|---|---|---|---|---|---|
| Test cost/PCB ($) | 1 | 3 | 10 | | |
| Repair cost ($) | 1 | 6 | 50 | 500 | |
| Scenario 2 (four sigma company) | | | | | |
| Expected yield before test | | 80% | 95% | | |
| Expected yield after test | | 95% | 99.99% | 0.01% | |
| 100,000 PCBs @ 500 components | | | | | |
| Test cost ($) | | 300,000 | 1,000,000 | | 1,300,000 |
| Defective PCBs before test | | 20,000 | 5,000 | 10 | |
| Defective PCBs after test | | 5,000 | 10 | | |
| PCBs repaired | | 15,000 | 4,990 | 10 | |
| Repair cost ($) | | 90,000 | 249,500 | 5,000 | 344,500 |
| Total test and repair cost ($) | | | | | 1,644,500 |
| Cost/PCB ($) | | | | | 16.45/PCB |

**Table 4.8**   PCB test methods scenario 3 (six sigma company), 3 strategies

| | Visual test | In-circuit test | Functional test | At-customer failures | Totals |
|---|---|---|---|---|---|
| Test cost/PCB ($) | 1 | 3 | 10 | | |
| Repair cost ($) | 1 | 6 | 50 | 500 | |
| Scenario 3 (strategy 1) | | | | | |
| Expected yield before test | | 95% | 99.8% | | |
| Expected yield after test | | 99.8% | 99.9999% | 0.00034% | |
| 100,000 PCBs @ 500 components | | | | | |
| Test cost ($) | | 300,000 | 1,000,000 | | 1,300,000 |
| Defective PCBs before test | | 5,000 | 200 | | |
| Defective PCBs after test | | 200 | 0 | | |
| PCBs repaired | | 4,800 | 200 | 0 | |
| Repair cost ($) | | 28,800 | 10,000 | 0 | 38,800 |
| Total test and repair cost ($) | | | | | 1,338,800 |
| Cost/PCB ($) | | | | | 13.39/PCB |
| Scenario 3 (strategy 2) | | | | | |
| Expected yield before test | | | 95% | | |
| Expected yield after test | | | 99.9999% | 0.00034% | |
| Omit in-circuit test | | | | | |
| Test cost ($) | | | 1,000,000 | | 1,000,000 |
| Defective PCBs before test | | | 5000 | | |
| Defective PCBs after test | | | | 0 | |
| PCBs repaired | | | 5000 | 0 | |
| Repair cost | | | 250,000 | 0 | 250,000 |
| Total test and repair cost | | | | | 1,250,000 |
| Cost/PCB ($) | | | | | 12.50/PCB |
| Scenario 3 (strategy 3) | | | | | |
| Expected yield before test | | 95% | | | |
| Expected yield after test | | 99.8% | | 0.2% | |
| 100,000 PCBs @ 500 components | | | | | |
| Omit functional test, four sigma to customer | 300,000 | | | | 300,000 |
| Defective PCBs before test | | 5,000 | | 200 | |
| Defective PCBs after test | | 200 | | | |
| PCBs repaired | | 4,800 | | 200 | |
| Repair cost ($) | | 28,800 | | 100,000 | 128,800 |
| Total test and repair cost | | | | | 428,800 |
| Cost/PCB ($) | | | | | 4.29/PCB |

customer are reduced to six sigma levels. The in-circuit test has a current limit of four sigma or 99.8%. For this scenario, three strategies are given:

- Strategy 1 is a follow-on from scenario 2, in which the cost is reduced with less in-circuit and functional testing because of the higher quality coming from PCB assembly.
- Strategy 2 eliminates the in-circuit test, allowing only for a functional test, and the cost is further reduced.
- Strategy 3 is the obvious lowest-cost one, in which functional testing is omitted and only in-circuit testing is used. However, strategy 3 violates the six sigma and TQM tenets of not passing on defects to the customer, and will await further improvements in in-circuit test technology.

It can be readily seen that achieving six sigma in assembly can have a great impact on reducing the cost of test and repair.

### 4.4.3    In-circuit test effectiveness

In the previous section, the in-circuit test was deemed the most important for achieving the six sigma level. As shown in Table 4.8, significant savings in cost could be achieved if this test method could deliver PCBs directly to the customer, without having to undergo functional testing. A brief review of some of the terms and strategies of in-circuit testing are given to help in outlining a six sigma quality plan for testing. The plan is based on investigating the defect removal functions and rating their efficiency. This plan can also be used for any type of testing after assembly in manufacturing.

The functions that can be performed by in-circuit testing can comprise some or all of the following:

- Shorts and opens
- Polarity check
- Analog and digital component testing
- Analog, digital, and mixed signal in-circuit testing
- Analog, digital, and functional (powered-on) testing
- Digital pattern rate
- Interconnect and in-circuit boundary scan

The measures of a tester's ability to correctly distinguish between bad and good PCBs are the test operation parameters: test coverage, bad test effectiveness, and good test effectiveness. They are measured as percentage values:

1. Test coverage (%): the test coverage for a given fault. Coverage of 0 for a defect category means that this defect is not tested.
2. Bad test effectiveness (%): the percentage of bad components that fail a test. Thus, a tester with 100% bad test effectiveness will fail all bad items, whereas one with 0% bad test effectiveness will pass all bad items.
3. Good test effectiveness (%): the percentage of good parts that pass a test. Thus, a tester with 100% good test effectiveness will pass all good items, whereas one with 0% good test effectiveness will fail all good items.

### 4.4.4   Factors affecting test operation parameters

Factors that affect test effectiveness can be divided into three broad categories: technology, management decisions, and design for test (DFT) efforts. They are listed in Table 4.9 and further explained in the next section. A factor-based model could be created in order to make PCB design decisions during the development stage. The model could help the design team investigate the effect that different design choices would have on the test effectiveness.

### 4.4.5   Test coverage

Test coverage (also called defect coverage) is a measure of the ability of a tester to detect defects. It is the percentage of those defects that

**Table 4.9**   Factors that affect test effectiveness

| Category | Examples |
| --- | --- |
| Circuits tested | Microwave circuits (require shielding) |
| | Digital versus analog versus mixed |
| Manufacturing | Through-hole versus SMT |
| | Test pad size |
| | Pitch size |
| | Nodal access |
| | Fixture design and fit |
| Management decisions | Time and resources for test and fixture development |
| | Time planned for in-line test |
| Design for test (DFT) | Design review for DFT |
| | Use of built-in self-test (BIST) |
| | Unit under test memory space dedicated to test |

are "covered" by a test, with 100% representing test coverage of all possible defects within a particular PCB defect category. There are many factors affecting test coverage:

- "Nodal access" refers to physical access to the nodes of a circuit by a test probe. When there is less than 100% nodal access, the coverage of circuit functionality is lessened. It is dependent upon the technology of the circuitry of the PCB. Coverage of analog circuitry increases approximately linearly with nodal access, whereas coverage of digital circuitry increases in step increments, depending on whether the node controls important digital pins such as reset pins. Thus, digital circuits have a higher number of critical nodes, i.e., nodes that control or affect a large amount of functionality.

- The manufacturing technology of a PCB includes features that can affect nodal access, such as the use of surface mount technology (SMT) and the component population density of the PCB. Through-hole (TH) circuits have about 100% nodal access; SMT PCBs can have significantly less. Double-sided PCBs impede nodal access, since using the underside of the PCB for circuitry imposes a competition for PCB "real estate" between that circuitry and the test routes needed for accessing the top of the PCB. High-density PCBs result in less access, due to difficulty in probing the test pads.

- Strategic business decisions concerning the amount of time and financial resources budgeted for test and fixture development. A model for this effect would involve two stages: the first would assume a minimal test development time of approximately two weeks to develop 60–70% of test programs; the second stage would allow additional time of two to four weeks to complete the remaining tests.

- Design for test efforts (DFT). Test coverage can be increased by DFT efforts and built-in self-test (BIST) features. These are tests embedded inside the PCBs. The amount of memory allotted for BIST is a good indicator of good test coverage.

### 4.4.6   Bad and good test effectiveness

Bad and good test effectiveness values are the percentage of PCBs that are properly distinguished as bad or good. This measure differs from test coverage, which is determined by the percentage of defects covered. Since both are measures of defect detection, factors that increase test coverage will also increase bad test effectiveness.

Good test effectiveness is a measure of properly passing good PCBs. Factors that affect good test effectiveness include proper fixturing and appropriate test-target size and spacing. A bad fixture may result in incorrectly failing a good PCB, due to improper fit or contact. In addition, very small target size and inadequate spacing between targets may result in false failures, due to improper contact between the tester and the PCB. A small target size of less than about 35 mil, or with less than 50 mil between targets, is not considered adequate.

### 4.4.7    Future trends in testing

The increased use of higher-density PCB component technology might change some of the analysis performed in the test strategy shown in this chapter. Nodal access to some components, such as the ball grid array (BGA), is limited, since the higher number of leads has resulted in the leads being placed underneath the body of the component. These leads could be placed on the top side of the PCBs with no access to test pins, hence in-circuit testing could not be performed for the BGA connections to test whether the terminations were successfully completed. Newer testing technologies that are currently available, such as x-ray machines that can detect solder defects through the PCBs, might have to be added to the test methods and strategies. The most efficient method to reduce test costs is to increase the quality of the assemblies being tested, as shown earlier in the examples in Tables 4-5 to 4-8.

## 4.5    Conclusions

This chapter discussed the various methods of measuring defects through the product design and manufacturing cycle. Different terms were examined, such as DPU (PPM), DPMO, and FTY. The formulation for each term, how it can be derived, and where it is used to measure quality were also shown with examples. In addition, some emerging standards were discussed, and different strengths and weaknesses shown for these standards. These terms were then referred back to six sigma and Cpk as discussed in earlier chapters.

The development of a good test strategy for product assembly was shown for PCBs and ICs. Different methods for testing and removing defects were analyzed, and the potential savings from higher-quality assemblies quantified in terms of various scenarios. In addition, a discussion on the effectiveness of in-circuit testing was presented, because it can be the most financially rewarding system for reducing testing cost.

Six sigma offers an excellent system of designing for and controlling quality in product assemblies. It provides a target for each design and manufacturing operation in terms of very low defect rates. Subsequently, it allows for a system to manage the removal of these defects through good testing strategies in a large product or system with the tools mentioned in this chapter.

## 4.6   References and Bibliography

Byle, F. "Using Industry DPMO Standards—An In-depth Look at IPC 9261 and IPC 7912." In *Proceedings of SMTI International,* Chicago, IL, September 2001, pp. 507–510.

Fink, D. and Beaty, H. *Standard Handbook for Electrical Engineers,* 12th ed. New York: McGraw-Hill, 1997.

Hewlett-Packard Company. "HP 3070 Series II Board Test Family, Test Methods and Specifications," 1995.

Higaki, W. "Minesweeper Project Proposal." *Measurement Systems Newsletter, 15,* 1997.

IPC-7912. "Calculation of DPMO and Manufacturing Indices for Printed Board Assemblies." Northbrook, IL: IPC, June 2000.

IPC-9261. "Calculation of Defects per Million Opportunities (DPMO) in Electronic Assembly Operations." Northbrook, IL: IPC, January 2002.

Phung, N. "Control Charts for DPMO." *Circuits Assembly,* September 1995, p. 40.

Ungar, L., "Board Level Built-in Test: The Natural Next Step." *Nepcon West Proceedings,* 1997.

Rowland, R., "DPMO and IPC 7912." *Surface Mount Technology Magazine,* March 2001.

Texas Instruments. "Six-Sigma—Reaching Our Goal, Guiding Principles for Counting Opportunities." Corporate statement, September 1995.

Woody, T. "DPMO, A Key Metric for Process Improvement in PCB Assembly." *Proceedings of SMTI International,* Chicago, IL, September 2001, pp. 503–506.

# 5

# The Use of Six Sigma with High- and Low-Volume Products and Processes

One of the concerns about using six sigma is the volume of production. There are two parts to this concern. The immediate reaction is that the 3.4 PPM defect rate associated with six sigma might imply that the volume of production has to be very large in order to properly assess this high level of quality. The other concern is that the tools of six sigma used for quality control and defect rate prediction might not apply because of the difficulty of properly obtaining statistical information such as the standard deviation of the manufacturing variability of the production process. Low-volume industries including defense, aerospace, and medical, as well as their suppliers, share these concerns.

Several statistical tools will be discussed in this chapter in order to allow for the use of six sigma in low-production environments, with minimum uncertainties. They are based on sampling theory and distribution, and the relationships between samples and populations. These tools are:

1. Process average and standard deviation calculations for samples and populations. Section 5.1 will discuss the sample probability distribution and its relationship to the parent population distribution. It gives examples of determining population standard deviation and error based on sample sizes.

2. Determining process capability. Section 5.2 will discuss the amount of data required to properly determine process capability.

The data volume is important in increasing the accuracy and the amount of effort necessary to correct the design or the manufacturing process to meet the process capability goals. This section will also examine moving range control charts as a means of controlling quality in low-volume production.

3. Determining gauge capability. The use of gauge repeatability and reproducibility (GR&R) to quantify measurement variability will be presented in Section 5.3. In addition, The relationship of GR&R to six sigma concepts and calculations will be examined.

4. Determining short- and long-term process capability. Section 5.4 will discuss the issues of determining process capability during the different stages of the product lifecycle, beginning with multiple specifications of the product and prototype quantities manufactured, and continuing with production volume. The strategies of setting different quality expectations during prototype versus volume production will also be examined.

## 5.1  Process Average and Standard Deviation Calculations for Samples and Populations

The knowledge of certain properties of a subset (sample), can be used to draw conclusions about the properties of the whole set (populations). Properties can be of two types, as discussed in earlier chapters:

1. Quantitative (variable). These properties can be observed and recorded in units of measure such as the diameter of shafts. The units are all produced under replicating conditions in production.

2. Qualitative (attribute). These properties can be observed when units are being tested with the same set of gauges or test equipment; for example, the set of all shafts produced under the same conditions, either fitting or not fitting into a tester consisting of a dual set of collars. A shaft with a diameter within specifications should fit into one of the collars whose diameter is equal to the shaft upper specification limit, and the shaft should not fit into the other collars whose diameter is equal to the lower specification limit.

The sample size $(n)$ is the random choice of $n$ objects from a population, each independent of each other. As $n$ approaches $\infty$, the sample distribution values of average and standard deviation become equal to that of the population.

It has been shown in Chapter 3 that variable control charts constitute a distribution of sample averages, with constant sample size $n$.

This distribution is always normal, even if the parent population distribution is not normal. It has also been shown the standard deviation $s$ of the distribution of sample averages is related to the parent distribution standard deviation $\sigma$ by the central limit theorem, which states that $s = \sigma/\sqrt{n}$ (Equation 3.5). The number of samples needed to construct the variable chart control limits was also set at a high level of 20 successive samples to ensure that the population $\sigma$ will be known.

When the total number in the samples ($n$) is small, very little can be determined by the sampling distribution for small values of $n$, unless an assumption is made that the sample comes from a normal distribution. The normal distribution assumes an infinite number of occurrences that are represented by the process average $\mu$ and standard deviation $\sigma$. The Student's $t$ distribution is used when $n$ is small. The data needed to construct this distribution are the sample average $\overline{X}$ and sample standard deviation $s$, as well as the parent normal distribution average $\mu$:

$$t = \frac{\overline{X} - \mu}{S/\sqrt{n}} \qquad (5.1)$$

where $t$ is a random variable having the $t$ distribution with $\nu = n - 1$.

$$\nu = \text{degrees of freedom (DOF)} = n - 1 \qquad (5.2)$$

It can be seen from Figure 5.1 that the shape of the $t$ distribution is similar to the normal distribution. Both are bell-shaped and distributed symmetrically around the average. The $t$ distribution average is equal to zero and the number of degrees of freedom governs each $t$ dis-

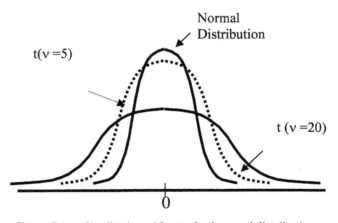

Figure 5.1   $t$ distribution with standard normal distribution.

tribution. The spread of the distribution decreases as the number of degrees of freedom increases. The variance of the $t$ distribution always exceeds 1, but it approaches 1 when the number $n$ approaches infinity. At that time, the $t$ distribution becomes equal to the normal distribution.

The $t$ distribution can be used to determine the area under the curve, called significance or $\alpha$ given a $t$ value. However, the $t$ distribution is different from the normal distribution in that the number in the sample or degrees of freedom $v$ have to be considered. The table output value of variable $t$, called $t_\alpha$, is given, corresponding to each area under the $t$ distribution curve to the right of $\alpha$ and with $v$ degrees of freedom. Figure 5.2 shows an example of how the $t_\alpha$ is related to the significance. The term "significance" is not commonly used, but its complement is called confidence, which is set to 1 minus significance and expressed as a percent value:

$$\text{confidence } (\%) = 1 - \text{significance} = 1 - \alpha \qquad (5.3)$$

Table 5.1 shows a selected set of the values of $t_\alpha$. The $t$ distribution is used in statistics to confirm or refute a particular claim about a sample versus the population average. It is always assumed that the parent distribution of the $t$ distribution is normal. This is not easily verified using the formal methods discussed in Chapter 2, since the sample size is small. In most cases, the graphical plot method of the sample data discussed in Chapter 2 is the only tool available.

Historically, the confidence percentage used depended on the particular products being made. For commercial products, a 95% confidence level is sufficient, whereas for medical and defense products, which require higher reliability, 99% confidence has been used. The

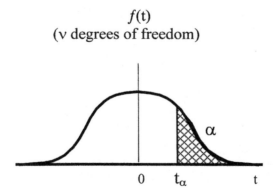

$f(t)$
($v$ degrees of freedom)

**Figure 5.2**   $t$ distribution with significance $\alpha$.

**Table 5.1**  Selected values of $t_{\alpha,\nu}$ of student's $t$ distribution

| $\nu$ | $\alpha = 0.10$ | $\alpha = 0.05$ | $\alpha = 0.025$ | $\alpha = 0.01$ | $\alpha = 0.005$ | $\alpha = 0.001$ | $\alpha = 0.0005$ |
|---|---|---|---|---|---|---|---|
| 1 | 3.078 | 6.314 | 12.706 | 31.821 | 63.657 | 318.3 | 636.6 |
| 2 | 1.886 | 2.920 | 4.303 | 6.965 | 9.925 | 22.327 | 31.600 |
| 3 | 1.638 | 2.353 | 3.182 | 4.541 | 5.841 | 10.214 | 12.922 |
| 4 | 1.533 | 2.132 | 2.776 | 3.747 | 4.604 | 7.173 | 8.610 |
| 5 | 1.476 | 2.015 | 2.571 | 3.365 | 4.032 | 5.893 | 6.869 |
| 6 | 1.440 | 1.943 | 2.447 | 3.143 | 3.707 | 5.208 | 5.959 |
| 7 | 1.415 | 1.895 | 2.365 | 2.998 | 3.499 | 4.785 | 5.408 |
| 8 | 1.397 | 1.860 | 2.306 | 2.896 | 3.355 | 4.501 | 5.041 |
| 9 | 1.383 | 1.833 | 2.262 | 2.821 | 3.250 | 4.297 | 4.781 |
| 10 | 1.372 | 1.812 | 2.228 | 2.764 | 3.169 | 4.144 | 4.587 |
| 20 | 1.325 | 1.725 | 2.086 | 2.528 | 2.845 | 3.552 | 3.849 |
| 30 | 1.310 | 1.697 | 2.042 | 2.457 | 2.750 | 3.386 | 3.646 |
| $\infty$ | 1.282 | 1.645 | 1.960 | 2.326 | 2.576 | 3.090 | 3.290 |
| Confidence or $(1 - \alpha)$ | 90% | 95% | 97.5% | 99% | 99.5% | 99.9% | 99.95% |

higher the confidence percentage, the larger the span of the confidence interval and its endpoints, the confidence limits. For low-volume production data, the confidence limits for the population average $\mu$ and standard deviation $\sigma$ estimates are used to give an estimate of the span of these two variables. The 95% confidence limits can be used for calculating six sigma data (Cpk, defect rates, FTY), whereas higher confidence numbers (99% and 99.9%) can be used as worst-case conditions checks on the base calculations.

### 5.1.1  Examples of the use of the $t$-distribution for sample and population averages

### Example 5.1

A manufacturing line produces resistors in a normal process with an average value of 500 ohms. A Sample of nine resistors were taken from yesterday's production, with sample average = 540 ohms and sample standard deviation = 60. Does the sample indicate that the production process was out of control yesterday?

### Solution to Example 5.1

$$t = \frac{540 - 500}{60/\sqrt{9}} = 2.0 \text{ and } \nu = 8$$

In the $t$-distribution table (Table 5.1), the number 2 falls between $t_{\alpha,8}$ values of 95% and 97.5% confidence (1.860 and 2.306, respectively). Hence, the yesterday's production process can be assumed to be in control within 95% confidence but not within 97.5% confidence. The sample process average taken yesterday results in $t = 2$, and this number can be used to compare the variability in production to a normally occurring variability. The probability that $t$ will exceed 1.860 is 0.05 (1 in 20 times will occur in this manner naturally), whereas the probability that $t$ will be greater than 2.306 is 0.025 (1 in 40 times will occur in this manner naturally).

## Example 5.2

A manufacturing process for batteries has an average battery voltage output of 12 volts, with production assumed to be normally distributed. It has been decided that if a sample of 21 batteries taken from production has a sample average of 11 and sample standard deviation of 1.23, then production is declared out of control and the line is stopped. What is the confidence that this decision is a proper one to take?

$$t = \frac{11 - 12}{1.23/\sqrt{21}} = -3.726 \text{ and } v = 20$$

Since the $t$ distribution is symmetrical, the absolute value of $t$ can be used. The calculated value of 3.726 falls between the $t_{\alpha,20}$ for $\alpha = 0.001$ and $\alpha = 0.0005$. The probability that $t$ will exceed $-3.552$ is 0.001, and the probability that $t$ will be greater than $-3.849$ is 0.0005. Thus, the decision is proper, since the significance of the sample occurring from the normal distribution is less 0.001 or 99.9% confidence.

### 5.1.2  Other statistical tools: Point and interval estimation

The previous section has introduced some statistical terms that are not widely used by engineers but are very familiar to statisticians. This section is a review of some of the statistical terms and procedures dealing with error estimation for the average and standard deviation as well as their confidence limits.

A good number to use for statistically significant data is 30. It is a good threshold when using some of the six sigma processes such as calculating defect rates. This is based on the fact that a $t$ distribution with $v$ degrees of freedom = 29 approaches the normal distribution. It can be from Table 5.2 that the data for the value of $t_{\alpha,30}$ is close to the value of the standard normal distribution. The error $E$ is calculated as the difference between the $t_{\alpha,30}$ value and the $z$ value from the normal distribution. For a significance of 0.025, or confidence of 97.5%, the error is less than 5%. Note that this point of $z = 1.96$ is close to the $z = 2$

**Table 5.2**   Error of the $t_{\alpha,\nu}$ of student's $t$ distribution

| | $\alpha$ or $f(z)$ = 0.05 | $\alpha$ or $f(z)$ = 0.025 | $\alpha$ or $f(z)$ 0.01 | $\alpha$ or $f(z)$ = 0.005 | $\alpha$ or $f(z)$ = 0.001 | $\alpha$ or $f(z)$ = 0.0005 |
|---|---|---|---|---|---|---|
| $\nu = 30$ | 1.697 | 2.042 | 2.457 | 2.750 | 3.386 | 3.646 |
| $\nu = \infty$ or $z$ | 1.645 | 1.960 | 2.326 | 2.576 | 3.090 | 3.290 |
| Error = $t_{\alpha,30} - z$ | 3.2% | 4.2% | 5.6% | 6.8% | 9.6% | 10.8% |
| Confidence $(1 - \alpha)$ or probability for $z$ | 95% | 97.5% | 99% | 99.5% | 99.9% | 99.95% |

or the 2 $\sigma$ point. For the 3 $\sigma$ point, or 99.9%, the error approaches 10%. The defect rate can thus be calculated using the $t$-distribution with small samples and known errors.

The relationship between the error and the sample size can be expanded to include the general conditions in which the standard deviation $\sigma$ is known from the sample and the number of the sample taken is large (>30). The maximum error $E$ produced when sample average $\overline{X}$ is used to estimate $\mu$, the population average, can be calculated in the following equation. In addition, the random sample size needed to estimate the average of a population, with a confidence of $(1 - \alpha)\%$ can also be shown as:

$$E = z_{\alpha/2} \cdot \sigma/\sqrt{n} \tag{5.4}$$

and

$$n = \left[ \frac{z_{\alpha/2} \cdot \sigma}{E} \right]^2$$

Where $E$ is the error, $\sigma$ is the standard deviations of the population, and $n$ is the sample size used in calculating the error.

If the sample size $n$ is small (<30), and the sample is drawn from a normal distribution of the population, the standard deviation of population $\sigma$ is not known, but the sample standard deviation $s$ can be calculated from the sample. In this case, the error made when the sample average $\overline{X}$ is used to estimate population average $\mu$ is as follows:

$$E = t_{\alpha/2} \cdot \frac{s}{\sqrt{n}} \tag{5.5}$$

### 5.1.3   Examples of point estimation of the average

**Example 5.3**

An engineer uses 100 samples to check the average noise output of amplifiers (in dB) produced in the production line. If it is known that the line is normally distributed with standard deviation of the noise measurements equal to 10, what is the maximum error (in dB) of the noise measurement population average given that the engineer wants to express it with a probability of 99%?

Probability of 0.99% implies a significance $(\alpha) = 0.01$

$$z_{\alpha/2} = z \text{ corresponding to } \{f(z) = 0.005\} = 2.575$$

$$E = 2.575 \cdot 10/\sqrt{100} = 2.575 \text{ dB}$$

The engineer can state with 99% probability that error between the sample average and the population average is less then 2.575 dB.

## Example 5.4
A factory makes PCBs and the gold plating thickness on the PCB fingers is expected to meet a minimum value of 20 mils prior to shipping. The gold thickness population is normal, with an average equal to 10 mils and standard deviation $\sigma$ equal to 3.0. Process improvements were made to reduce variability, and hence less gold can be plated on average to ensure conformance to specifications. How many units must be made with the new process to ensure with 95% probability $(\alpha = 0.025)$ that new population average is within ±1 mil?

$$n = (z_{\alpha/2} \cdot \sigma/E) = (1.96 \cdot 3/1)^2 = 34.6 \text{ or } 35 \text{ sample size}$$

## Example 5.5
A sample of nine measurements was taken for turn-on rise time of an IC. The average of the sample was 51 units and the sample standard deviation was 6. Given that this sample is derived from a population with normal distribution, calculate the maximum error of the population average with 95% confidence.

$$E = t_{\alpha/2, \nu=n-1} \cdot s/\sqrt{n} = t_{0.025,8} \cdot 6/3 = 2.306 \cdot 2 = 4.612, \quad \text{or } 4.612/51 = 9\%$$

$E$ is the maximum error between the sample average and the population average, with 95% confidence.

### 5.1.4    Confidence interval estimation for the average

Engineers have found the use of the confidence percentage discussed in the last section for estimating the average or average rather unfamiliar. They are more comfortable with the concept of the confidence interval. This term shows the range of the average having the degree of confidence $(1 - \alpha)\%$. The endpoints are referred to as the confidence limits. The formulas for the interval of the average estimation are for high- and low-volume samples, respectively:

$$\overline{X} - z_{\alpha/2} \cdot \frac{\sigma}{\sqrt{n}} < \mu < \overline{X} + z_{\alpha/2} \cdot \frac{\sigma}{\sqrt{n}} \qquad (5.6)$$

and

$$\overline{X} - t_{\alpha/2} \cdot \frac{s}{\sqrt{n}} < \mu < \overline{X} + t_{\alpha/2} \cdot \frac{s}{\sqrt{n}} \qquad (5.7)$$

Figure 5.3 shows an interpretation of the confidence interval for 13 samples from the same population with a known $\sigma$. The different samples produce different values for $\overline{X}$ and, consequently, the interval spans are centered at different points. When the population $\sigma$ is known, the confidence interval is the same for all samples, because all their confidence limits are derived from $\sigma$. If the population $\sigma$ is unknown, then the sample standard deviations ($s$) are used to calculate the confidence interval for each sample from Equation 5.7, and the span is different for different samples.

If the confidence limit was at 95% (or $z = 2$ $\sigma$ away from the average) then it is expected that the probability of at least one interval span falling outside the population average is 5%, or one out of 20 samples. Therefore, a sample whose average is outside the population average is considered unlikely to happen. In Figure 5.3, the unlikely sample is shown highlighted third from the top.

**Example 5.6**
A sample has the following characteristics: $n = 81$, sample average = 20, and standard deviation = 5. Find 95% and 99.9% confidence intervals, assuming that the population is normally distributed.

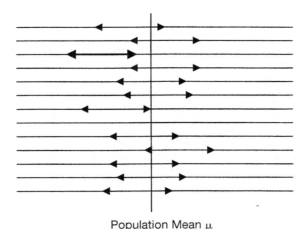

Population Mean $\mu$

**Figure 5.3**  Confidence interval around the mean $\mu$ and $\sigma$ is known.

The sample is large enough to use $z$ tables. From equation 5.6:

95% confidence ($\alpha = 0.25$) = 20 ± 1.960 · 5/9 = 20 ± 1.09

99.9% confidence ($\alpha = 0.0005$)= 20 ± 3.290 · 5/9 = 20 ± 1.83

Note that the confidence interval for 99.9% is almost double the one for 95%.

## Example 5.7

For a sample of the following values, 2.6, 2.1, 2.4, 2.5, 2.7, 2.2, 2.3, 2.4, and 1.9, find the confidence interval of the population average, assuming that it is normal, for 90%, 95%, and 99.9% confidence.

For the sample data: $n = 9$; sample average $\overline{X} = 2.34$, and sample standard deviation $s = 0.25$. Using the $t$ distribution with $t_{\alpha/2,8}$ and Equation (5.7):

$$90\% \text{ confidence } (\alpha = 0.05) = 2.34 \pm 1.860 \cdot 0.25/3$$
$$= 2.34 \pm 0.16 \ (2.18 - 2.5)$$

$$95\% \text{ confidence } (\alpha = 0.025) = 2.34 \pm 2.306 \cdot 0.25/3$$
$$= 2.34 \pm 0.19 \ (2.15 - 2.53)$$

$$99.9\% \text{ confidence } (\alpha = 0.0005) = 2.34 \pm 5.041 \cdot 0.25/3$$
$$= 2.34 \pm 0.42 \ (2.76 - 1.92)$$

In every case, the sample point 1.9 falls outside the lower confidence limit, making it an unusual event. At 99.9% confidence, the point has a probability of less than 0.005.

### 5.1.5   Standard deviation for samples and populations

The statistical relationships of the sample and population averages have been discussed in previous sections. There is a similar distribution for the sample variability $s^2$, which can be used to learn about its parametric counterpart, the population variance or $\sigma^2$. This distribution is called the chi square or $\chi^2$. Since the distribution cannot be negative, it is not symmetrical, but is in fact related to the gamma distribution. The $\chi^2$ distribution is shown in Figure 5.4. The probability that that a random sample produces a $\chi^2$ greater than some specified value is equal to the area of the curve to the right of the value. The variable $\chi\alpha^2$ represents the value of $\chi^2$ above which there is the area $\alpha$. The equation for the distribution variable is as follows:

$$\chi^2 = \frac{(n-1)^2 s^2}{\sigma^2} \tag{5.8}$$

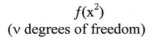

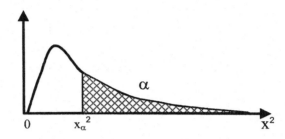

**Figure 5.4**  $\chi^2$ distribution with significance $\alpha$.

where $s^2$ is the variance of a random sample of size $n$ taken from a normal population having the variance $\sigma^2$, and $\chi^2$ is a random variable having the distribution with degrees of freedom $v = n - 1$.

Table 5.3 contains selected values of the $\chi^2$ distribution. Since it is not symmetrical, two $\chi^2$ values will have to be returned when confidence percentages are needed for two-sided limits, as can be seen in Figure 5.5. As in the $t$ distribution, the $\chi^2$ distribution can be used in two cases:

1. When the population variance $\sigma^2$ is known, and therefore the probability that the sample variance $s^2$ can be tested to see if it is related to the population variance $\sigma^2$

**Table 5.3**    Selected values of $\chi^2$ distribution

| $v$ | $\alpha = 0.995$ | $\alpha = 0.975$ | $\alpha = 0.95$ | $\alpha = 0.90$ | $\alpha = 0.50$ | $\alpha = 0.10$ | $\alpha = 0.05$ | $\alpha = 0.025$ | $\alpha = 0.005$ |
|---|---|---|---|---|---|---|---|---|---|
| 1 | 0.0000393 | 0.000982 | 0.00393 | 0.0158 | 0.455 | 2.706 | 3.841 | 5.024 | 7.879 |
| 2 | 0.0100 | 0.0506 | 0.103 | 0.211 | 1.386 | 4.605 | 5.991 | 7.378 | 10.597 |
| 3 | 0.0717 | 0.216 | 0.352 | 0.584 | 2.366 | 6.251 | 7.815 | 9.348 | 12.838 |
| 4 | 0.207 | 0.484 | 0.711 | 1.064 | 3.357 | 7.779 | 9.488 | 11.143 | 14.860 |
| 5 | 0.412 | 0.831 | 1.145 | 1.610 | 4.351 | 9.236 | 11.070 | 12.832 | 16.750 |
| 6 | 0.676 | 1.237 | 1.635 | 2.204 | 5.348 | 10.645 | 12.592 | 14.449 | 18.548 |
| 7 | 0.989 | 1.690 | 2.167 | 2.833 | 6.346 | 12.017 | 14.067 | 16.013 | 20.278 |
| 8 | 1.344 | 2.180 | 2.733 | 3.490 | 7.344 | 13.362 | 15.507 | 17.535 | 21.955 |
| 9 | 1.735 | 2.700 | 3.325 | 4.168 | 8.343 | 14.684 | 16.919 | 19.023 | 23.589 |
| 10 | 2.156 | 3.247 | 3.940 | 4.865 | 9.342 | 15.987 | 18.307 | 20.483 | 25.188 |
| 15 | 4.601 | 6.262 | 7.261 | 8.547 | 14.339 | 22.307 | 24.996 | 27.488 | 32.801 |
| 20 | 7.434 | 9.591 | 10.851 | 12.443 | 19.337 | 28.412 | 31.410 | 34.170 | 39.997 |
| 25 | 10.520 | 13.120 | 14.611 | 16.473 | 24.337 | 34.382 | 37.652 | 40.646 | 46.928 |
| 30 | 13.787 | 16.791 | 18.493 | 20.559 | 29.336 | 40.256 | 43.773 | 46.979 | 53.672 |

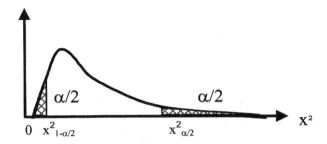

**Figure 5.5**   Obtaining confidence limits from $\chi^2$ distribution with confidence $(1 - \alpha)\%$.

2. When the population variance $\sigma^2$ is not known, and the sample variance $s^2$ is used to determine $\sigma^2$, with confidence limits and confidence intervals. The equation for this case is as follows:

$$\frac{(n - 1)s^2}{\chi^2_{\alpha/2}} < \sigma^2 < \frac{(n - 1)s^2}{\chi^2_{1-\alpha/2}} \tag{5.9}$$

where $s^2$ is the variance of a random sample of size $n$ from a normal population, confidence interval for $\sigma^2$ is $(1 - \alpha)\%$, and $\chi^2_{1-\alpha/2}$ and $\chi^2_{-\alpha/2}$ are values having areas of $\alpha/2$ and $-\alpha/2$ to the right and left of the distribution average.

### 5.1.6   Examples of population variance determination

**Example 5.8**
Five samples are taken from a normal population of parts from a factory with average = 3 and $\sigma$ = 1. The samples are 2.0, 2.5, 3.0, 3.5, and 4.0. Does this sample of parts support the belief that the sample came from the factory with $\sigma$ equal to 1?
$\overline{X}$ of sample = 3 and $s$ of the sample = 0.79. From Equation (5.8)

$$\chi^2 = 4 \cdot 0.79^2/1 = 2.50$$

The calculated value of $\chi^2$ (2.50) with $\nu$ = 4 is close to 50% confidence (3.357) and is in between the 90% and 10% (1.064–7.779) confidences. Therefore, based on variance, it is highly likely that the sample was made at that factory.

**Example 5.9**
Nine samples (from Example 5.7) were taken from an assumed normal population with the following values from example: 5.7: 2.6, 2.1, 2.4, 2.5, 2.7, 2.2, 2.3, 2.4, and 1.9. What are the 95% and 99% confidence intervals of population variance?
Sample data: $n$ = 9; average = 2.34, and $s$ = 0.25.

*95% Confidence*

$\alpha = 0.05$, therefore the 95% confidence limits are 0.025 and 0.975 @ $\nu$ = 8:

$$8 \cdot (0.25)^2/17.535 < \sigma^2 < 8(0.25)^2/2.180$$

$$0.0285 < \sigma^2 < 0.229 \qquad \text{or} \qquad 0.17 < \sigma < 0.48$$

*99% Confidence*

$\alpha = 0.01$, therefore the 99% confidence limits are 0.005 and 0.995 @ $\nu$ = 8:

$$8 \cdot (0.25)^2/21.955 < \sigma^2 < 8(0.25)^2/1.344$$

$$0.0228 < \sigma^2 < 0.372 \qquad \text{or} \qquad 0.15 < \sigma < 0.61$$

Note that the confidence interval gets larger as the confidence limits increase.

## 5.2  Determining Process Capability

Process capability is the analysis of a process to determine its quality. A single or several quality characteristics are selected, some of which might be variable or attribute. For variable characteristics, the distribution of the data collected is for normality, and the distribution average $\mu$ and standard deviation $\sigma$ are calculated. It has been shown in this and previous chapters that it takes a sample size of 30 measurements to directly obtain these two parameters and determine whether the distribution of data is normal. For low-volume production, the previous section discussed methods of determining a confidence interval for the two parameters. The confidence limits from these intervals could be used for worst-case determination of six sigma quality. For attribute processes, the defect rate is determined for parts that are manufactured in small quantities as prototypes, or from similar parts in current production. The reject rate can be translated into DPU (PPM), DPMO, FTY, Cpk, or sigma quality, as was shown in Chapters 2 and 4.

The amount of sampling required for determining process capability is also dependent on whether the process has been in production (existing) for some time or is a new process is being created. It is also desirable that once the process is operating on a regular basis, and a reasonable level of quality is achieved, the quality characteristic(s) being measured be charted for statistical control in control charts. For quality level approaching six sigma and beyond, control charting might not be required; a total quality management program to monitor individual defects per period as opposed to use the sampling methods of control charts (refer to the discussion in Chapter 3 regarding this issue) could be substituted.

### 5.2.1  Process capability for large-volume production

The following procedures are recommended when time and resources are not gating items. It is ideally suited for large-volume manufacturing, where the parts cost is low and the ease of collecting data is high. These procedures will increase the accuracy of the process capability and reduce its apparent variation with time.

1. Initial determination of process capability. Historical guidelines for variable and attribute data are given in Table 5.4. Each subgroup of data should be taken at a different point in time, preferably on different days. In this manner, day-to-day variations of the process could be integrated into the process capability calculations. There should be no allowance for process average shift in the Cpk calculations. For low volume applications, the moving range method should be used because of the low volume required. A discussion of the moving range method is given in the next section.

2. Regular updates of the process capability. The process capability should be regularly checked to determine if the process has changed. If the change is deemed significant using statistical tests, then a process quality correction project should be initiated to determine the cause of the process deviation. The amount of data required for checking the process could be less than the original data needed for initial determination. Determination of $\sigma$ can be achieved either directly from the data or through the $\overline{R}$ estimator for variable data. For large-volume production, a sample size of 30 is sufficient to perform this check of process capability for variable

**Table 5.4**  Amount of data required for process capability studies

|  | Period of time | Sample size | Total |
|---|---|---|---|
| *High Volume* | | | |
| $\overline{X}$ and $R$ charts | 1st period | 50 measurements | |
| | 2nd period | 25 measurements | |
| | 3rd period | 25 measurements | 100 measurements |
| | | | |
| *P, nP* charts | 1st period | 20–25 samples | |
| *U* and *C* charts | (50–100 units tested) | | |
| | 2nd period | 20–25 samples | |
| | | (50–100 units tested) | |
| | 3rd period | 20–25 samples | |
| | | (50–100 units tested) | 3000 min. units tested |
| *Low Volume* | | | |
| Moving range | Long period | 10 consecutive numbers | 10 |
| | Long period | 10 consecutive numbers | 10 |

data. For low-volume production, smaller sample sizes can be used and deviations tested for the probability that the average or standard deviation has shifted from the original, given a confidence interval.

3. Correction of process capability based on regular updates. Correction should only be undertaken if the manufacturing process has shifted beyond normal statistical significance of 10%, for either variable or attribute processes, and the population distribution is assumed to be normal. To check normality, many tests are available, including the graphical and $\chi^2$ (chi-square) tests discussed in Chapter 2. The distribution of the data should be symmetrical, with no skew. If not, the process should be investigated. Changes to the process capability should be tested as follows:

- Testing changes in the average $\mu$ for variable processes. The $z$ test is used for comparing sample average to the population average if the sample and population are both greater than 30. The $t$ test is used to compare sample average to population average if the sample is < 30 and the population is > 30. If both the initial process capability and the process update data are less than 30, then a compound sample standard deviation term can be calculated to compare the two samples (population $\sigma$ is either known or unknown). The purpose of this test is to determine if the average has shifted or not and, therefore, whether to recalculate six sigma process capability data for the average.

- The formulas for these tests against original population data are as follows:

$$z = \frac{\overline{X} - \mu}{\sigma/\sqrt{n}} \tag{5.10}$$

for testing a large sample $n$ with average $\overline{X}$ against a population (or large sample) of average $\mu$ and standard deviation $\sigma$;

$$t = \frac{\overline{X} - \mu}{s/\sqrt{n}} \tag{5.10}$$

for testing a small sample $n$, with average $\overline{X}$ and sample standard deviation $s$, against a population (or large sample) of average $\mu$ and an unknown standard deviation.

- The formulas for testing current samples data against original sample data, when both are < 30 and with known sample sizes, are given in Section 5.4

- For testing changes in the $\sigma$, several tests are available, depending on the size of the samples taken. If the initial variable process capability population data is greater than 30, and the capability update data is less than 30, then the $\chi^2$ test can be used. To compare current value of $\sigma$ to the initial process capability $\sigma$ when both data sets are under 30, the $F$ test should be used. $F$ tests can test for a level of significance (5% or 1%) to determine if the $\sigma$'s between the two data sets are statistically different. Depending on the results of these tests, the six sigma attributes are either retained or recalculated. The $F$ test can also be used when two or more sample data sets originate from a common population. In that case, the differences between sample variability are either due to natural variation or a deviation in the product. More details on the $F$ test are given in Chapter 7.

### 5.2.2   Determination of standard deviation $\sigma$ for process capability

There are four different methods for determining the standard deviation $\sigma$ of the population for process capability studies:

1. Total overall variation. All data is collected into one large group and treated as a single large sample with $n$ greater than 30.
2. Within-group variation. Data is collected into subgroups, and a dispersion statistic is calculated (range). All ranges of each subgroup are averaged into an $\overline{R}$. The $\sigma$ is calculated from an $\overline{R}$ estimator ($d_2$). This method is the basis for variable control chart limit calculations and discussed in Chapter 3.
3. Between-group variation. Data is collected into subgroups, and an average ($\overline{X}$) is calculated for each subgroup. The standard deviation $s$ of sample averages is calculated. The population $\sigma$ is estimated from the central limit theorem equation, $\sigma = s \cdot \sqrt{n}$. This method can be used to obtain process capability from control chart limits.
4. Moving range method. In this method, data is collected into one group of small numbers of data, over time. A range ($R$) is calculated from each two successive points. All ranges of each pair are averaged into an $\overline{R}$. The $\sigma$ is calculated from an $\overline{R}$ estimator ($d_2$) for $n$ = 2, which is equal 1.128. Method 4 is the preferred method for time series data and small data sets from low-volume manufacturing.

For processes that are in statistical control, these methods are equivalent over time. For processes not in control, only Method is 2 insensitive to process variations of the average over time. The $\sigma$ estimate is inflated or deflated with Method 1 and could be severely inflated/deflated with Method 3. An example of a process out of control is one in which one subgroup has a large sample average shift as opposed to smaller average shifts in the other subgroups. Another way to advantageously leverage Method 2 to negate the effect of average shift is to use Method 4, with the data spread over time.

### 5.2.3 Example of methods of calculating σ

**Example 5.10**

Data for a production operation was collected in 30 samples, in three subgroups, measured at different times. The four different methods of calculating σ are as follows.

| Subgroup | Measurement | Subgroup range(R) | Average | s |
|---|---|---|---|---|
| I | 4, 3, 5, 5, 4, 8, 6, 4, 4, 7 | 5 | 5 | 1.56 |
| II | 2, 4, 5, 3, 7, 5, 4, 3, 2, 5 | 5 | 4 | 1.56 |
| III | 3, 6, 7, 6, 8, 4, 5, 4, 6, 6 | 5 | 5.5 | 1.51 |
| Average of subgroups I–III | | 5 | 4.83 | 1.54 |
| For the total group | | 6 | 4.83 | 1.62 |

| Moving range for each subgroup | | Total | $\overline{R}$ | σ |
|---|---|---|---|---|
| I | 1, 2, 0, 1, 4, 2, 2, 0, 3 | 15 | 1.67 | 1.48 |
| II | 2, 1, 2, 4, 2, 1, 1, 1, 3 | 17 | 1.89 | 1.68 |
| III | 3, 1, 1, 2, 4, 1, 1, 2, 0 | 15 | 1.67 | 1.48 |
| Average moving range | | | 1.74 | 1.54 |

**Method 1. Total overall variation of 30 data points from 3 subgroups**

$$\sigma^2 = \frac{\sum_i (y_i - \bar{y})^2}{n-1} = \frac{\sum_i y_i^2 - (\sum_i y_i)^2/n}{n-1} = [777 - (145)^2/30]/29 = 2.626$$

$$\sigma = 1.62$$

**Method 2. Within-group variation; $\overline{R}$ = 5 ($n$ = 10)**

$$\sigma = \overline{R}/d_{2(n=10)} = 5/3.078 = 1.62$$

**Method 3. Between-group variation**

$$s(\overline{X}) = \sigma(5, 4, 5.5) = 0.763$$

$$\sigma = s \cdot \sqrt{n} = 0.764 \cdot \sqrt{10} = 2.415$$

## Method 4. Moving range method ($n = 2$)

For each subgroup, obtain the average range between successive numbers:

Subgroup I: $\sigma = \overline{R}/d_{2(n=2)} = 1.67/1.128 = 1.48$
Subgroup II: $\sigma = \overline{R}/d_2 = 1.89/1.128 = 1.68$
Subgroup III: $\sigma = \overline{R}/d_2 = 1.67/1.128 = 1.48$

For the total groups (I–III), $\sigma = \overline{\overline{R}}/d_2 = 1.74/1.128 = 1.54$.

As can be seen from Example 5.10, the $\sigma$ of the overall 30 numbers was 1.62 (Method 1). The 30 numbers were made of three subgroups (samples) with large shifts in sample averages. The closest indirectly calculated $\sigma$ value was obtained by Method 2, between-group variation from the $\overline{R}$ estimator of $\sigma$, because it negated the average shifts. The moving range method (Method 4) was as much as 10% off, even when using the full 30 numbers. The least accurate value was Method 3, the between-group variation, which derived $\sigma$ from a distribution of sample averages and the conversion of the sample to population $\sigma$. The number of subgroups (samples) was small and led to the largest error in $\sigma$ determination.

### 5.2.4    Process capability for low-volume production

When it is not feasible to collect the amount of data required to determine process capability because of cost or resource issues or production volume, reduced data can be used successfully to estimate process capability, provided that confidence is quantified in the data analysis. Although 30 points of data are considered statistically significant, a smaller number of data points can be taken, using predetermined error levels and confidence goals, to obtain a good estimation of process average and variability. Refer to earlier sections in this chapter for proper methods and examples.

The moving range method provides an alternate mechanism for estimating the $\sigma$ for small amounts of data, provided that data points are taken over time for both variable and attribute processes. Ten data point are required to provide an estimator for $\sigma$ with the moving range method.

### 5.2.5    Moving range (MR) methodologies for low volume: MR control charts

The moving range methodology allows for a reasonable estimate of $\sigma$ and process capability for both variable and attribute processes. It uses individual measurements or defect rates over a representative period of time. It is very useful when there is only one number to describe a particular condition or situation. It can be used to estimate production variables such as temperature, pressure, humidity, voltage, or conductivity. It can also be used for production support efforts such as costs, efficiencies, shipments, and purchasing activities. The moving range charts can also be used for attributes. Instead of counting defects, the time between defects can be counted and entered as the variable in the chart.

The moving range stands for the difference between successive pairs of numbers in a series of numbers. The absolute value of the difference is used, creating a new set of range numbers, each with two successive elements. The number of differences or "ranges" is one less than the individual numbers in the series. The chart is built up from the following:

- The centerline of the chart is the average of all the individual measurements.
- The average of the ranges of the successive numbers is called the $\overline{MR}$. The control limits are set by multiplying $\overline{MR}$ by the number 2.66. This is the result of using the factor $d_2$ for $n = 2$ (1.128) estimation of the $\sigma$ in the following equation:

$$\text{MR control limits} = \overline{\overline{X}} \pm 3 \cdot \overline{MR}/1.128 = \overline{\overline{X}} \pm \overline{MR} \cdot 2.66 \qquad (5.12)$$

Note that the conversion from the standard deviation of sample average to population $\sigma$ that is performed on $\overline{X}$, $R$ charts is not necessary here, since the moving range charts use the actual distribution of data, not those from sample distributions.

### Example 5.11
Days between defects were counted as a measure of the quality of a manufacturing process. They occurred on the following production calendar days: 23, 45, 98, 123, 154, 167, 189, 232, 287, 311, and 340. Calculate the data for the moving range chart for days between defects.

Days between defects: 22, 53, 25, 31, 13, 22, 43, 55, 24, 29; average = 31.70

Moving ranges ($R$'s): 31, 28, 6, 18, 9, 21, 12, 31, 5; $\overline{R}$ = 17.89

$MR_x$ control limits = 31.70 ± 2.66 · 17.89 = 17.89 ± 47.59 days

$R$ chart control limits: $UCL_R = D_{4(n=2)} \cdot \overline{R} = 3.27 \cdot 17.89 = 58.5$; $LCL_R = 0$

Another method to plot this defect data would be defects/month, obtained by dividing the data by 30.

**Example 5.12**

Fuses are made in a production line, with specifications of $5 \pm 2$ ohms. A sample of six fuses measurement was taken at 3, 6, 6, 4, 5, and 5 ohms. If it is desired to have an $\overline{X}, R$ control chart, what is the quality data for the fuse line?

$$\text{Moving range method data} = 3\ 0\ 2\ 1\ 0$$

$$\text{Average } \overline{X} = 4.83; \overline{MR} = 1.2$$

$$\sigma = \overline{MR}/d_2 = 1.2/1.128 = 1.0638$$

$$UCL_x = 4.83 + 2.66 \cdot \overline{MR} = 8.02$$

$$LCL_x = 4.83 - 2.66 \cdot \overline{MR} = 1.64$$

$$UCL_R = D_{4(n=2)} \cdot \overline{MR} = 3.27 \cdot 1.2 = 3.92$$

$$LCL_R = 0$$

$$Cp = 2/3 \cdot 1.0638 = 0.63; \text{Cpk} = (4.83 - 3)/3 \cdot 1.0638 = 0.57$$

$$z_1 = (3 - 4.83)/1.0638 = -1.72; f(z_1) = 0.0427$$

$$z_2 = (7 - 4.83)/1.0638 = 2.04; f(-z_2) = 0.0207$$

Defect rate $(RR) = 0.0427 + 0.0207 = 0.0634$ or 6.34% or 63,400 PPM

### 5.2.6    Process capability studies in industry

The discussions in the previous sections outlined a system for investigating and maintaining process capability for the purpose of quality planning. In the six sigma environment, process capability data will have to be maintained within one or more of the indicators that were discussed in previous chapters, including DPU (PPM), DPMO, yield, and number of sigma's quality (including six sigma). Knowing that all of these indicators are related to each other as discussed and shown by examples in previous chapters, an enterprise can decide on one of these indicators, or a combination of several, and use the indicator(s) in process capability studies. This is especially useful when the enterprise management or major customers have asked for a certain level of quality.

An example would be a factory that chose Cpk as the process capability indicator. This requires that all of the fabrication and assembly operations, as well as major part suppliers and outside manufacturing contractors, are to report on their process capabilities. For the suppliers and contractors, a supplier management team and contractual processes with quality as well as cost and delivery requirements have to be in place to indicate the need for process capability. The purpose of these activities is to inform the new product design teams of the current quality status of different operations in manufacturing and the supply chain. If the design team finds the process capability inadequate, manufacturing has to purchase better-quality equipment or select new suppliers that can meet the quality goals. The process capability data has to be updated regularly in order to keep design team abreast of quality and capability enhancements. The frequency of updates should be short enough to comfortably fit inside the new product design cycles, as well as meet yearly management goals. A typical frequency of updating process capability is every quarter.

For assembled parts, the process capability determination has to be compatible with industry standards, as well as the calculations of defect opportunities. For PCBs and their terminations, standards such as DPMO are used (see Section 4.3.3). For fabricated parts, especially those made in machine shops, the process capability determination is more difficult. The machine shop can produce parts with the desired geometry using many possible machines in the shop; some producing high-quality parts and others parts of much lower quality. The dilemma is whether a particular process should be machine dependent, especially since the machine selection is usually not included in the part or assembly documentation. If a ½″ hole needs to be drilled, there are many alternative machines in the shop to perform this operation, with varying process capabilities. So what will the design team assume for the ½″ holes defect rate?

One solution to the fabrication dilemma is to allow for an additional attribute in the six sigma methodology. This attribute would be a quality or complexity indicator. The fabrication shop could be divided into several (maximum of three) levels of complexity. As each new part is being designed, the design engineer can select from any of the three process capabilities available, depending on the level of complexity of the part.

For each process, a baseline process capability is determined, according to the sampling methods outlined in Table 5.4. Every quarter, all of the process capabilities are checked, and recalculated if they show a statistically significant shift in average or $\sigma$ using statistical comparison tests. The $z$ distribution is used to compare a large (>30) sample with the baseline population averages; the $\chi^2$ test is used to compare sample to population $\sigma$. For smaller-size samples, the sample average shift to the population average can be tested with the $t$ distribution, as shown earlier in this chapter.

Some of the process capability data can be obtained from control charts, as shown in Chapter 3, whereas others can be calculated directly by taking samples from the production line. Table 5.5 is an example of a production line of PCB assembly process capability calcula-

**Table 5.5**  Example of process capability studies for PCB assembly line

| Process | Cpk baseline | Cpk this QTR | Status | Check method | Specification limit |
|---|---|---|---|---|---|
| Lead form | 1.42 | 1.61 | Recalculate | $n = 100$ | ± 0.005 |
| Screen print | 1.41 | 1.41 | Check OK | $P$ chart | |
| Adhesive apply | 1.99 | 1.99 | Check OK | $n = 30$ | ± 0.005 |
| Place components | 1.70 | 2.66 | Recalculate | $MR = 10$ | ± 0.002 |
| Solder reflow | 1.06 | 1.06 | Check OK | $n = 30$ | ± 0.005 |
| Manual solder | 1.18 | 1.18 | Check OK | $n = 30$ | ± 0.005 |
| Connector install | 1.06 | 1.06 | Check OK | $\overline{X}, R$ | ± 0.005 |
| Hardware assembly | 1.72 | 1.72 | Check OK | $MR = 20$ | ± 0.010 |
| Conformal coat | 1.70 | 1.70 | Check OK | $\overline{X}, R$ | ± 0.005 |

tions using Cpk. It shows the baseline and the present quarter performance. The data could also be plotted versus time, with the management goals shown prominently on the graph plots.

Table 5.5 shows a process capability, measured in Cpk, for each step of the process. The process capability is checked each quarter, and the source of the check is shown. Some checks are performed by using existing control charts, including moving range (MR) charts, whereas others are checked using sampling methods. Note that two process capabilities had to be changed, since the quality performance has changed dramatically.

## 5.3  Determining Gauge Capability

An important part of capability studies when measuring the total variability in manufacturing is to account for gauge or test process variability. Variability is not limited only to the manufacturing process; the variability of the measurement system needed to test the manufactured parts should also be considered. Figure 5.6 shows typical sources of variation and error in a process and its measurement system. The majority of measurement errors, including those due to the operator (appraiser) or the equipment (gauge), can be measured and quantified through gauge reliability and reproducibility (GR&R) methodology. The use of GR&R to evaluate measurement systems quality is mandatory in achieving six sigma quality.

The following is an explanation of the terms used in Figure 5.6.

- Short and long variations in the manufacturing process are due to time-dependent parameters, such as incoming part quality changes, age of equipment, and methods for maintaining equipment. They will be discussed in the next section.

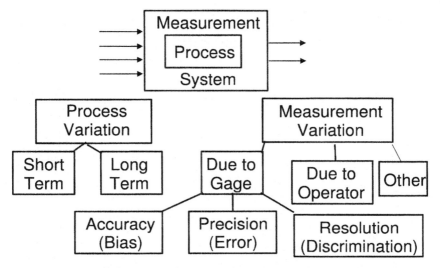

**Figure 5.6**  Sources of process variation and error.

- Precision is the relative amount of the variability of the measuring system; hence, it is an indicator of the variability of the equipment (gauges).
- Accuracy is a relative measure of achieving the measurement target. It is the difference between the true and measured values, although the true value may not be known in many cases. Accuracy is usually referred according to some standard of measurement. Hence, it is an indicator of the measurement error average of μ of the equipment (gauges). Accuracy and precision are shown in Figure 5.7, using a target analogy.
- Repeatability is a measure of the consistency of readings of the same part for a single operator (appraiser). Poor repeatability indicates measurement system problems related to equipment. Repeatability is derived from the same operator measuring different parts repeatedly using the same measurement equipment. Sometimes it is called precision or equipment variation (EV).
- Reproducibility is a measure of variation in average measurements when different operators are taking many measurements of the same part. It can be used as measure of the relative amount of training or skills for the operators. Sometimes it is called appraiser variation (AV), using the same parts and gauges and different operators.
- GR&R is the root sum of the squares (RSS) value of repeatability and reproducibility. It should be noted that the average or the

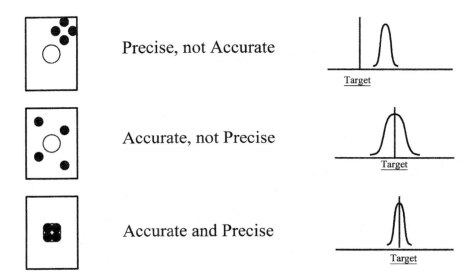

**Figure 5.7**   Accuracy and precision target example.

process and measurement errors add up algebraically, whereas the standard deviations add up in the squares, as shown in Figure 5.8

### 5.3.1   GR&R methodology

GR&R methodology consists of quantifying the measurement error due to equipment and operators. Data is collected by several operators measuring the same set of parts on the same equipment. The average ranges of the part measurements determine the equipment variability, and the differences in the measurement averages determine the operators' (appraisers) variability. The methodology for attaining the GR&R of a measurement system is as follows:

1. The parts to be used in the GR&R study should be identified. Up to 10 parts are normally used from the same production process.
2. Up to three skilled operators should be identified to make the measurements. They should be familiar with the parts and measurement equipment.
3. Each operator then measures each part on the same equipment several times; these measurements are called trials. Usually up to three trials are made by each operator.
4. The errors are thus generated by $n$ parts, which are measured and repeated $r$ times by different operators (A, B, and C). It is assumed

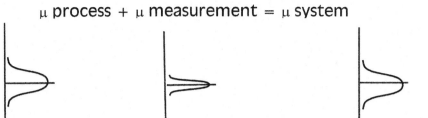

μ **process** + μ **measurement** = μ **system**

$\sigma^2$ **process** + $\sigma^2$ **measurement** = $\sigma^2$ **system**

**Figure 5.8**  Summation of averages and standard deviations.

that the GR&R measurement encompasses 99% of the normal curve variation of all measurements. This results in an error of 1% and $f(z) = 0.01/2$ or 0.005, corresponding to a $z$ value of 2.575 $\sigma$ for each side of the normal curve. A total of 5.15 $\sigma$ constitutes the total variation for the area inclusive under the curve for GR&R calculations.

5. Repeatability, or equipment variation (*EV*), is measured by 99% of the error span due to equipment. This is equivalent to 5.15 $\sigma_{EV}$, which in turn is derived from the $\bar{\bar{R}}$ = average $\bar{R}$'s of each operator:

$$\sigma_{EV} = \frac{\bar{\bar{R}}}{d_2}; \text{ and } EV = 5.15\ \sigma_{EV} \text{ or } EV = \bar{\bar{R}} \cdot K_1 \qquad (5.13)$$

EV is related directly to $\bar{R}$ by the factor $K_1$. $K_1$ is equal to 4.56 for two trials ($r = 2$) and 3.05 for three trials ($r = 3$). This is derived from the relationship introduced in the variable control chart factor $d_2$ (table 3.1) for $n = 2$: $K_1 = 5.15/d_2$ or $5.15/1.128 = 4.56$ for $r = 2$ and $5.15/1.693 = 3.05$ for $r = 3$.

6. Reproducibility, or appraise variation (*AV*), is measured by 99% of the error span due to operators. This is equivalent to 5.15 $\sigma_{AV}$ which in turn is derived from the $\overline{X_{\text{diff}}}$ = range of operator averages $\bar{X}$'s and the factor $d_2^*$ from Table 5.6. The error inherent in the equipment variation (*EV*) has to be removed from the appraiser variation (*AV*). The $\sigma_{AV}$ is based on the root sum of the squares of observed operator variation minus the normalized equipment variation, the latter divided by the number of measurements:

$$\sigma_{AV} = \sqrt{\left[\frac{\overline{X_{\text{diff}}}}{d_2^*}\right]^2 - \frac{\sigma_{EV}^2}{nr}} \text{ and } AV = 5.15\ \sigma_{AV} \qquad (5.14)$$

**Table 5.6**  $\overline{R}$ estimator of $\sigma$ for GR&R

| $n/m$ | $d_2$ | $d_2^*$ |
|:---:|:---:|:---:|
| 2 | 1.128 | 1.410 |
| 3 | 1.693 | 1.906 |
| 4 | 2.059 | 2.237 |
| 5 | 2.326 | 2.477 |
| 6 | 2.534 | 2.669 |
| 7 | 2.704 | 2.827 |
| 8 | 2.847 | 2.961 |
| 9 | 2.970 | 3.076 |
| 10 | 3.078 | 3.178 |

$d_2$ = unbiased $\overline{R}$ estimator for $\sigma$ ($n$ = sample subgroup size).
$d_2^*$ = biased $\overline{R}$ estimator based on $m$ = number of trials.
$\sigma = \overline{R}/d_2$.
$\sigma = \overline{X}_{\text{diff}}$ difference (highest to lowest trial averages)/$d_2^*$.

or

$$AV = \sqrt{\left[ (\overline{X_{\text{diff}}} \cdot K_2)^2 - \frac{EV^2}{nr} \right]} \qquad (5.15)$$

where $AV$ is related directly to $\overline{X_{\text{diff}}}$ by the factor $K_2$. $K_2$ is equal to 3.65 for $r = 2$ and 2.70 for $r = 3$. This is derived from the relationship $K_2 = 5.15/d_2^*$ or $5.15/1.410 = 3.65$ for $r = 2$ and $5.15/1.906 = 3.05$ for $r = 3$. If the result of subtraction in the $AV$ terms inside the square root term is negative, then $AV$ should be set to zero.

7. The GR&R is calculated from the RSS of $EV$ and $AV$. In most cases, it is expressed as a percentage of the total specification span.

### 5.3.2  Examples of GR&R calculations

**Example 5.13**

A process is to be analyzed for repeatability using one operator (A) measuring five parts, two times each, on one machine. The data is arranged as follows:

| Operator A | Trial | | |
|:---|:---:|:---:|:---|
| Trial # | 1 | 2 | Range |
| 1 | 1.000 | 1.010 | 0.010 |
| 2 | 1.015 | 0.995 | 0.020 |
| 3 | 0.980 | 1.015 | 0.035 |
| 4 | 0.995 | 1.010 | 0.015 |
| 5 | 0.980 | 1.025 | 0.045 |
| Total | 4.970 | 5.055 | 0.125 |
| | $\overline{X} = 1.0025$ | | $\overline{R} = 0.025$ |

$$\sigma_{EV} = \overline{R}/d_2 = 0.025/1.128 = 0.02216$$

$$EV = 5.15 \cdot \sigma_{EV} = 0.114; \text{ or alternately, } EV = \overline{R} \cdot K_1 = 0.025 \cdot 4.56 = 0.114$$

## Example 5.14

The same process in Example 5.13 is to be analyzed for repeatability and reproducibility with the addition of a second operator measuring the same set of five parts:

| Operator | A | | | B | | |
|---|---|---|---|---|---|---|
| | Trial | | | Trial | | |
| Trial # | 1 | 2 | Range | 1 | 2 | Range |
| 1 | 1.000 | 1.010 | 0.010 | 0.990 | 1.010 | 0.020 |
| 2 | 1.015 | 0.995 | 0.020 | 0.990 | 1.000 | 0.010 |
| 3 | 0.980 | 1.015 | 0.035 | 1.020 | 1.000 | 0.020 |
| 4 | 0.995 | 1.010 | 0.015 | 1.030 | 1.040 | 0.010 |
| 5 | 0.980 | 1.025 | 0.045 | 1.020 | 1.000 | 0.020 |
| Total | 4.970 | 5.055 | 0.125 | 5.050 | 5.050 | 0.080 |

$$\overline{X} = 1.0025 \; \overline{R} = 0.025 \qquad \overline{X} = 1.010 \; \overline{R} = 0.016$$
$$\overline{\overline{R}} = 0.0205 \qquad \overline{X}_{\text{diff}} = 0.0075$$

$$\sigma_{EV} = \overline{\overline{R}}/d_2 = 0.01817$$

$$EV = 5.15 \cdot \sigma_{EV} = 0.094; \text{ or alternately, } EV = \overline{\overline{R}} \cdot K_1 = 0.0205 \cdot 4.56 = 0.094$$

$$AV = \sqrt{[(0.0075 \cdot 3.65)^2 - EV^2]/nr} = \sqrt{(0.00075 - 0.094^2)/10} = \sqrt{-0} = 0$$

In this case, the $AV$ variation is smaller than the $EV$, so it is set to zero:

$$GR\&R = EV^2 + AV^2 = EV = 0.94$$

### 5.3.3  GR&R results interpretation

GR&R represents 99% of the measurement error caused by either operator or equipment. It is usually expressed as a percentage of the total variation (TV). The GR&R percentage = $GR\&R/TV$, which is the portion of the total variation consumed by the GR&R measurement error, can be derived from the following sources:

1. The specification limits have historically been used as the total variation, since it is assumed that the test of the product or part will cull out any parts outside the specifications.
2. The total variation is comprised of RSS of the GR&R and the part variation (PV). The part variation, $\sigma_P$, which is also the population variation used for six sigma calculations, can be derived from the

GR&R data by multiplying the range of part averages as measured by the operators by the constant $K_3$. $K_3$ is calculated from $d_2^*$ in Table 5.6, depending on the number of parts examined in the GR&R measurements, as follows:

$$PV = R_p \cdot K_3 \qquad (5.16)$$

$$K_3 = 5.15/d_2^* \qquad (5.17)$$

The values for $K_3$ are for number of parts examined in the GR&R:

$$K_3 = 3.65 \quad 2.70 \quad 2.30 \quad 2.08 \quad 1.93 \quad 1.82 \quad 1.74 \quad 1.67 \quad 1.62$$

$$n_p = 2 \qquad 3 \qquad 4 \qquad 5 \qquad 6 \qquad 7 \qquad 8 \qquad 9 \qquad 10$$

$$TV = \sqrt{GR\&R^2 + PV^2} \qquad (5.18)$$

3. If the process variation is known through process capability studies and is based on six sigma, then $\sigma_P$ can be derived independently from the GR&R study and used for PV and TV calculations, with $PV = 5.15 \cdot \sigma_P$.

The GR&R% of total variation can be used to determine if the measurement system is acceptable for its intended applications. General guidelines for the value of GR&R% are:

- If GR&R% < 10%, then the measurement system is acceptable
- If 10% < GR&R% < 30%, then the system may be acceptable, based on whether the part characteristic classification is not critical or from customer input
- If GR&R%0 > 30%, then the system is not acceptable. It is then desirable to seek resolution through the use of quality tools, better operator training, or the purchase of new inspection equipment.

### 5.3.4    GR&R examples

**Example 5.15**

Table 5.7 is a complete GR&R example of three operators and two trials, measuring parts with specifications ±0.500. $\bar{\bar{R}}$ is obtained from the average $\bar{R}$ of the three operators and is equal to 0.0383. $\bar{X}_{\text{diff}}$ is obtained from the difference between the highest average operator and the lowest and is equal to 0.0600.

$$EV = \overline{X_{\text{diff}}} \cdot K_1 = 0.03833 \cdot 4.56 = 0.1748$$

$$AV = \sqrt{[(0.06 \cdot 2.70)^2 - (EV^2/nr)]} = \sqrt{[0.026244 - (0.1748^2/20)]} = 0.1572$$

$$GR\&R = \sqrt{0.1748^2 + 0.1572^2} = 0.2351$$

*GR&R% from specifications.* If the specifications are given as ±0.500, then GR&R% = 0.2351/0.500 = 47%.

*GR&R% from part variation.* Taking the range of part averages from the data:

$$R_p = 1.0167 - 0.4583 = 0.55833$$

$$PV = R_p \cdot K_{3(n=10)} = 0.55833 \cdot 1.62 = 0.9045$$

$$TV = \sqrt{GR\&R^2 + TV^2} = \sqrt{0.2351^2 + 0.9045^2} = 0.9346$$

$$GR\&R\% = 100(GR\&R/TV) = 0.2351/0.9346 = 25\%$$

In this example, the measurement system is of marginal acceptance.

**Example 5.16**
An analysis of a test system with a specifications limit of 5 ± 3 consists of repeating a sample measurement three times by three operators:

| Operator | Measurements | | | $R$ | $\overline{X}$ |
|----------|---|---|---|---|---|
| 1 | 4 | 6 | 4 | 2 | 4.67 |
| 2 | 4 | 5 | 6 | 2 | 5.00 |
| 3 | 5 | 5 | 7 | 2 | 5.67 |
| | $\overline{X_{\text{diff}}} = 1$ | | Average | 2 | 5.11 |

Show quality control, six sigma, and GR&R analysis.
  For the control chart:

$$\overline{R} = 2, n = 3; UCL_R = \overline{R} \cdot D_4 = 2 \cdot 2.57 = 5.14; LCL_R = 0$$

$$\sigma_{EV} = \overline{R}/d_{2(n=3)} = 2/1.693 = 1.18133$$

$$s_{EV} = \sigma/\sqrt{n} = 1.18133/1.732 = 0.68; 3s = 2.04$$

$$\overline{\overline{X}} = \text{chart centerline} = 5.11$$

$$UCL_x = 5.11 + A_2 \cdot \overline{R} = 7.15; \text{ or } UCL_x = 5.11 + 3s = 7.15$$

$$LCL_x = 5.11 - 2.04 = 3.07$$

For six sigma calculations:

$$\text{Average shift} = 0.1111$$

$$Cp = \pm SL/3\sigma = 3/(3 \cdot 1.18133) = 0.85$$

$$Cpk = \min (8 - 5.11)/(3 \cdot 1.18133) = 0.82 \text{ or } 3.11/(3 \cdot 1.18133)$$
$$= 0.82$$

Table 5.7 GR&R example

| Operator | A | A | A | A | B | B | B | B | C | C | C | C | PART |
|---|---|---|---|---|---|---|---|---|---|---|---|---|---|
| Sample | 1st trial | 2nd trial | Average | Range | 1st trial | 2nd trial | Average | Range | 1st trial | 2nd trial | Average | Range | AVERAGE |
| 1 | 0.65 | 0.6 | 0.625 | 0.05 | 0.55 | 0.55 | 0.55 | 0 | 0.5 | 0.55 | 0.525 | 0.05 | 0.5667 |
| 2 | 1 | 1 | 1 | 0 | 1.05 | 0.95 | 1 | 0.1 | 1.05 | 1 | 1.025 | 0.05 | 1.0083 |
| 3 | 0.85 | 0.8 | 0.825 | 0.05 | 0.8 | 0.75 | 0.775 | 0.05 | 0.8 | 0.8 | 0.8 | 0 | 0.8000 |
| 4 | 0.85 | 0.95 | 0.9 | 0.1 | 0.8 | 0.75 | 0.775 | 0.05 | 0.8 | 0.8 | 0.8 | 0 | 0.8250 |
| 5 | 0.55 | 0.45 | 0.5 | 0.1 | 0.4 | 0.4 | 0.4 | 0 | 0.45 | 0.5 | 0.475 | 0.05 | 0.4583 |
| 6 | 1 | 1 | 1 | 0 | 1 | 1.05 | 1.025 | 0.05 | 1 | 1.05 | 1.025 | 0.05 | 1.0167 |
| 7 | 0.95 | 0.95 | 0.95 | 0 | 0.95 | 0.9 | 0.925 | 0.05 | 0.95 | 0.95 | 0.95 | 0.05 | 0.9417 |
| 8 | 0.85 | 0.8 | 0.825 | 0.05 | 0.75 | 0.7 | 0.725 | 0.05 | 0.8 | 0.8 | 0.8 | 0 | 0.7833 |
| 9 | 1 | 1 | 1 | 0 | 1 | 0.95 | 0.975 | 0.05 | 1.05 | 1.05 | 1.05 | 0 | 1.0083 |
| 10 | 0.6 | 0.7 | 0.65 | 0.1 | 0.55 | 0.5 | 0.525 | 0.05 | 0.85 | 0.8 | 0.825 | 0.05 | 0.6667 |
| Totals | 8.3 | 8.25 | 8.275 | 0.45 | 7.85 | 7.5 | 7.675 | 0.45 | 8.25 | 8.3 | 8.275 | 0.25 | |

$\overline{R}_a = 0.0450$

$\overline{R}_b = 0.0450$

$\overline{R}_c = 0.0250$

| $\text{Sum}_a =$ | 16.55 |
|---|---|
| $\overline{X}_a =$ | 0.8275 |

| $\text{Sum}_b =$ | 15.35 |
|---|---|
| $\overline{X}_b =$ | 0.7675 |

| $\text{Sum}_c =$ | 16.55 |
|---|---|
| $\overline{X}_c =$ | 0.8275 |

| $\overline{\overline{R}} = 0.0383$ | $\overline{X}_{\text{diff}} =$ | 0.0600 |
|---|---|---|
| Specification tolerance | | ± 0.500 |

| Rp | 0.55833 |
|---|---|
| PV | 0.9045 |
| TV | 0.93455 |

162

Test for control

$\bar{R} = 0.0383$

|  | 2 trials |
|---|---|
| UCL ($R$) | 0.1254 |
| LCL ($R$) | 0.0000 |

Measurement system gauge capability

|  | 2 trials |
|---|---|
| Equipment variation ($EV$) repeatability | 0.1748 |

|  | 3 Operators |
|---|---|
| Operator variation ($AV$) reproducibility | 0.1572 |

| Repeatability and reproducibility | 0.2351 |
|---|---|
| GR&R% from Specs = ±0.500 | 47% |

| GR&R % from $TV = \sqrt{(GRR^2 + PV^2)}$ | 25% |
|---|---|

$$z_1 = 3 \cdot \text{Cpk} = 2.45; f(-z_1) = 0.0071$$

$$z_2 = 3 \cdot \text{Cpk} = 2.63; f(-z_2) = 0.0043$$

Total error = 0.0114; or 1.14% or 11,100 PPM

$EV = 5.15 \, \sigma_{EV} = 5.15 \cdot 1.18133 = 6.08$ or $R$ double bar $\cdot K_1$
= $2 \cdot 3.05 = 6.10$

$$AV = \sqrt{(1.270)^2 - (EV^2/nr)} = \sqrt{7.29 - 1.18133^2/3} = 2.61$$

$$GR\&R = \sqrt{EV^* + AV^2} = 6.63$$

$G\&GR\%$ from specifications = $100(6.63/3) > 100\%$

Measurement system quality is unacceptable.

## 5.4   Determining Short- and Long-Term Process Capability

An important part of new product development is the development of process capabilities and specifications for new parts and products. Design engineers work with the general specification of products that are set by marketing or the customer, but these specifications do not necessarily flow down to all of the parts and to all of their attributes. It is necessary for design engineers to always question the relevance of each part specification, and whether it is too tight for its proper use in the customer's hands. It is always desirable to use tools such as quality function deployment or QFD, discussed in Chapter 1, to attempt to relate each specification for every part to the customer's wishes.

For six sigma designs of new products, process capability should be determined in the prototype stage of parts manufacturing. Some large consumer and mass product companies normally plan for large prototype runs to fully simulate the variability of the production process. This may not be feasible for many industries, due to the cost of parts or the volume of expected sales, so that process capability has to be derived from low volumes, using the techniques discussed in this chapter.

Process capability for new products can follow one of the following three scenarios:

1. The product represents an evolutionary increase in technology, and engineers build the prototypes with tight control, in special prototype shops. In this case, the process capability of the prototypes might actually be of higher quality that the early production runs.

2. For state of the art products, the part specifications are set aggressively, with the implication that the early production runs will have a poor yield. The parts in this case will attain the desired level of quality through rigorous testing against specifications. Eventually, their process capability will improve over time, thus achieving the specified first-time yield sometime after product release.

3. Using six sigma procedures for process capability implies that every purchased or manufactured part or assembly meets the six sigma requirements. Process capabilities might not be available for many of the new purchased parts and may have to be calculated from prototype purchases. For major companies, this issue is less of a problem, as they can specify the process capability or six sigma directly in the purchasing contracts for parts.

### 5.4.1  Process capability for prototype and early production parts

When prototype parts are acquired, whether through purchase or made in the company's internal factories, the following methodology is recommended for process capability calculations:

1. New parts that are very similar to current parts, or made in the same production line or process, can assume the current part process capability. Examples would be fabricated and assembled PCBs. Process capability can be derived from existing manufacturing statistical control data.

2. For parts new to the company, either purchased from the supply chain or locally manufactured, the sampling plan of Table 5.4 can be used for high-volume manufacturing.

3. For low-volume manufacturing, use smaller sample sizes, including the moving range method. Use the statistical techniques of $t$ and $\chi^2$ distributions as well as sample size determination, discussed in this chapter, to determine the ranges of population average $\mu$ and standard deviation $\sigma$. Use the confidence limits to determine the worst-case process capability.

4. To determine the specification limits, especially for six sigma design, ensure that the specifications are related to the customer wishes, and that the average and population standard deviations are within the six sigma limits of design.

5. The six sigma or the Cpk quality level target can be altered for the short versus the long term. In some cases, including prototype and early production, close attention is given to the parts and manufacturing process by the design team and manufacturing engineers in

the short term. As production ramps up, more parts are made with newer and less-skilled operators, resulting in poor quality, even if a good control system is in place. In the long term, with good use of corrective action processes and TQM, as well as increased operators' skills through the learning curve, the parts' quality levels will increase. Considering the previous arguments, it is advisable to set a higher quality level in the early production phase in order to counteract the problems when production ramps up. An example would be to set quality for early production runs to Cpk = 1.67 (five sigma), then back off to Cpk = 1.33 (four sigma) in the long term when the product matures. In Figure 5.9, the standard deviation used is the combined $\sigma$ based on the prototype and production runs.

6. The formulas for combining $s$ (small samples) or $\sigma$ (large samples) from two distinct samples with varying sample sizes ($n_1$ and $n_2$) follow. For large samples (>30) of standard deviation $\sigma_1$, $\sigma_2$ and sample sizes $n_1$, $n_2$:

$$\sigma_{\text{combined}} = \sqrt{\frac{\sigma_1^2}{n_1} + \frac{\sigma_2^2}{n_2}} \tag{5.19}$$

For small samples (<30) of standard deviation $s_1$, $s_2$ and sample sizes $n_1$, $n_2$:

$$s_{\text{combined}} = \sqrt{\frac{s_1^2(n_1 - 1) + s_2^2(n_2 - 1)}{n_1 + n_2 - 2}} \tag{5.20}$$

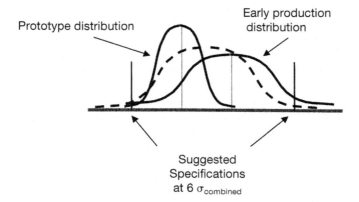

**Figure 5.9**  Distributions of prototype and early production of parts.

To compare large samples to see if the differences between sample averages are significant, a test statistic $z$ is generated:

$$z = (\overline{X}_1 - \overline{X}_2)/\sigma_{\text{combined}} = (\overline{X}_1 - \overline{X}_2)/\sqrt{\frac{\sigma_1^2}{n_1} + \frac{\sigma_2^2}{n_2}} \tag{5.21}$$

Repeating the above for differences of small sample averages, $t$ is calculated with $n_1 + n_2 - 2$ degrees of freedom (DOF):

$$t = (\overline{X}_1 - \overline{X}_2)/\left(S_{\text{combined}} \cdot \sqrt{\frac{1}{n_1} + \frac{1}{n_2}}\right) \tag{5.22}$$

$$t = (\overline{X}_1 - \overline{X}_2) \cdot \sqrt{\frac{n_1 n_2 (n_1 + n_2 - 2)}{n_1 + n_2}} / \sqrt{(n_1 - 1)s_1^2 + (n_2 - 1)s_2^2} \tag{5.23}$$

The use of the combined standard deviation can then be expanded to the confidence limits based on the combined degrees of freedom of $n_1 + n_2 - 2$.

### Example 5.17
Two equal samples were measured, from two presumably equal variances that are normally distributed, one for the original process capability study and the other for a later check performed three months later: $n_1 = n_2 = 10, \overline{X}_1 = 108, s_1^2 = 211, \overline{X}_2 = 100, s_2^2 = 86$. Should the difference in the samples necessitate recalculating the process capability?

From Equation 5.23:

$$t = (108 - 100) \cdot \sqrt{10 \cdot 10 \cdot (10 + 10 - 2)/10 + 10}/\sqrt{(9 \cdot 211) + (9 \cdot 86)} = 1.47$$

$$\text{DOF} = 18$$

From Table 5.1 and with DOF = 20 (which is close to DOF = 18 in this example), the $t_{0.05,20}$ is 1.725 for 95% confidence. Based on this probability, the differences in the sample process capabilities is small and should not be calculated.

### Example 5.18
Two large samples—$n_1 = 30, \overline{X}_1 = 9.9, \sigma_1 = 4.9$, and $n_2 = 35, \overline{X}_2 = 16.7, \sigma_2 = 7$—were taken, one for the original process capability study and the other for a later check performed three months later:

From equation 5.19:

$$z = (9.9 - 16.7)/\sqrt{(4.9^2/30) + (7^2/35)} = -4.58$$

The $z$ corresponds to a probability of value less than 4.5 $\sigma$, which is 0.0000034. The samples are indeed different and the process capability should be recalculated.

### 5.4.2 Corrective action for process capability problems

The previous section described a methodology for calculating process capability for new parts. If a process capability study was done with existing parts, and it was found to be unacceptable, the following suggestions might be followed to bring the process capabilities in compliance with six sigma or Cpk targets:

- Can specifications be amended (enlarged) and still meet system requirements?
- Can increased training, corrective action processes, design of experiments, or other quality improvement tools be used to increase process capability?
- If current processes remains not capable, can new equipment or outside suppliers be investigated?

## 5.5   Conclusions

This chapter showed how to handle the common problem of applying six sigma quality methodology to small as well as large production volumes. Statistical tools such as moving range and the $z$, $t$, $f$, and $\chi^2$ distributions can be used to quantify the attributes of the population distribution for average and standard deviations based on samples taken. Many examples were given to demonstrate sampling techniques and their relationship to populations. Process capability as well as gauge capability were also demonstrated with formulas, examples, and case studies. Finally, the process capability applications in short- versus long-term production were also shown, with examples and strategies for handling process capability in the prototype as well as long-term production.

## 5.6   References and Bibliography

Burr, I. *Engineering Statistics and Quality Control.* New York: McGraw Hill, 1953.

Bronshtein, I. and Semendyayev, K. *Handbook of Mathematics.* Leipzig: Verlag Press, 1985.

Ducan, A. J. *Quality Control and Industrial Statistics,* 4th ed. Homewood, IL: Richard D. Irwin. 1995.

Johnson, R., *Probability and Statistics for Engineers,* 5th ed. Englewood Cliffs, NJ: Prentice-Hall, 1994.

Walpole R. and Myers, R. *Probability and Statistics for Engineers and Scientists.* New York: Macmillan, 1993.

# Six Sigma Quality and Manufacturing Costs of Electronics Products

In this chapter, the need for accurate estimates of cost and quality will be shown, especially for mature technology products. In addition, expected cost and quality levels can be used as design guidelines for product functional partition, design quality assessment, and material and process selection in manufacturing. Developing an accurate quality and cost model for new electronic products, specifically for printed circuit boards (PCBs) is important, since PCBs represent the major part of cost, especially for assembly and test requirements. The model should be used as early as possible during the design stage, and is based on the design and manufacture of the PCB assembly operations as well as the manufacturing line equipment selection and layout. The following aspects of the relationship between quality and cost will be explored:

1. The overall electronic product life cycle cost model. In Section 6.1, the generalized product life cycle is reviewed, outlining the different phases that products and technologies go through, and the relationship of cost and quality to each phase. The elements that make up each electronic product cost are outlined, and techniques for monitoring and controlling costs are shown. These techniques include developing cost models especially for the primary cost factors, which are the PCBs.

2. The quality and cost relationship. The relationship of quality and cost are explored in Section 6.2 through the quality loss function (QLF). Formulations and examples of this system are given, and its use in estimating the relative value of making products to target or reducing variability explored. In addition, the use of this function to set factory process targets is shown to be a trade-off of defect removal either in the manufacturing plant or at the customer site.

3. Electronic products cost estimating systems for PCB fabrication. In Section 6.3, the technologies used for PCB fabrication and assembly are reviewed and their costs are quantified based on their manufacturing operations and complexity factors. A cost model for PCB fabrication is presented with a case study. The cost and quality assessment has to be tempered by other factors such as design time and new product introduction impact.

4. Electronic products cost estimating systems for PCB assembly. In Section 6.4, several systems are examined for determining the cost of PCB assembly. These systems vary in their complexity, from simple PCB components' material-cost-based systems to the more complex quality-based cost models, including a cost and quality model to examine the tradeoffs in design and manufacturing. Defects generated by alternative design, manufacturing and test strategies can be examined and a decision made for the lowest-cost alternative. Each system is discussed with examples and case studies.

## 6.1    The Overall Electronic Product Life Cycle Cost Model

The manufacturing costs of products are highly dependent on life cycle stage, as shown in see Figure 6.1. The first stage is called start-up or market development. During this stage, emphasis is on the performance of the product. Features such as speed, capacity, response time, and other "bells and whistles" dominate the product cycles. At this stage, the benefits of the product to the customer are perceived to be very high in increased productivity or personal comfort and satisfaction. The number of competitors is large, since entry into the market is wide open, and a new company can establish a niche in the marketplace for a relatively low investment. Product development during the start-up stage is marked by the intense drive to arrive at the market as early as possible, with minimum concern over manufacturing cost. A good indicator of this stage is the number of wire cuts and changes to printed circuit boards (PCBs) in new products. The quality

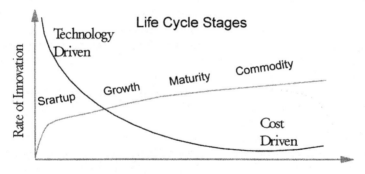

**Figure 6.1**  Product life cycle stages.

and reliability of the new product in manufacturing is achieved by extensive inspection and testing.

The second stage is the growth stage. As the marketplace is expanded and general acceptance of the product is assured, the number of competitors drops and the rate of market development begins to slow. The issue is not the acceptance of the technology or the particular use of the product, but the differentiating aspects of the manufacturers. Elements of the long-term cost of ownership of the product such as the quality and field support of the product, the commitment of the manufacturers to the particular business segment, and the growth of ancillary products and services supporting the product and its technology are emphasized. In addition, there is increasing customer confidence in the evolution of the product technology.

Product development during the growth stage is characterized by the focus on introducing the manufacturing guidelines of capabilities and constraints to the new product, and beginning to concentrate on manufacturing as a strategic weapon to achieve low cost and high quality. Coordination with suppliers is increased by the introduction of just-in-time (JIT) schedules into the manufacturing process.

The third stage is the maturity period. This phase is characterized by the emergence of a dominant technology or technique for the product design. At the same time, the relative growth of the market is slowed, being only proportional to the growth of the population or the customer base, as the product saturates the market. The number of manufacturers continues to decrease, as they either go out of business or get bought out by larger companies. The competitive emphasis in this stage is on price and quality, as the dominant technology does not allow too much variation on the basic design of the product.

Product development in the maturity phase is focused on continued improvement in manufacturing processes, such as a stronger emphasis on quality through the tools of control charts, continuous quality improvement, and robust processes. Variability reduction through implementing the techniques of design of experiments (DoE) and heavy emphasis on automating part or all of the manufacturing processes is increased. The suppliers for the product are involved early and often, and the design process is made more robust through the use of analysis and simulation tools.

The last stage can take one of two forms: either the product will decline as the need for it is overwhelmed by new technology (as was the case for 8-track cassette players and electric typewriters), or the product will develop into a commodity. In either case, the number of product manufacturers will decline to a select few big companies, and entry into this market will become very expensive and risky. The emergence of standards of use, manufacture, interconnection, and quality will make price the only competitive factor. The products will essentially be interchangeable from one manufacturer to another, with high customer expectations of quality and reliability. The revenue per unit decreases rapidly, as manufacturing techniques become the major factor in ensuring the long-term survival of the product's manufacturing company. Follow-on products will be evolutionary, with a market leader establishing a very careful trend that locks on his customer base and provides a definite upgrade path for the new generation of products. The attributes of each stage in the product development life cycle are shown in Table 6.1.

The product development emphasis in the commodity stage is on reducing manufacturing cost while maintaining the high quality expected by the customer. There is a much higher level of automation, as manufacturing knowledge and the stability of the design are increased. Few companies can enter into a market at the commodity stage since costs of recruiting personnel with the required knowledge

**Table 6.1**  Product development life cycle stages attributes

|  | Startup | Growth | Maturity | Commodity |
|---|---|---|---|---|
| Product variety | Great variety | Standardization | Dominant design | Mature standards |
| Volume | Low | Increasing | High | Very high |
| Industry structure | Many companies | Consolidation | Few companies | Survivors |
| Competition basis | Options | Delivery | Quality | Price |
| Critical processes | Innovation | Speed | Project management | Cost management |

or developing the internal learning curve for the necessary expertise can be prohibitively high.

The electronics industry has followed many other industries into this pattern. The automobile industry is a prime example. In the early part of the last century, there were hundreds of auto manufacturers, and any of the competing technologies could have become dominant: electric, steam, or internal combustion. The computer industry has gone through the stages discussed above for various products. Mainframes have all but disappeared, the personal computers have become a commodity industry, with exchangeable software programs and plug-in PCBs and modules.

This chapter is mainly focused on electronic products in the maturity or commodity stages, since the emphasis is on quality and cost. Maintaining a good level of cost accuracy during the development stage is important in the success of later stages of the life cycle of technological products.

### 6.1.1  The use of the quality and cost model to achieve world-class cost and quality

The cost and quality model developed in this chapter can be used at the earliest possible time in design to develop an accurate estimate of quality and cost of new products and to help design and manufacturing engineers make tradeoffs in material and manufacturing equipment acquisition and selection.

The design of new electronic products can be partitioned effectively into modules, each comprising units or collections of PCBs, mechanical parts and assemblies, software, and special requirements such as hybrid integrated circuits. As described in earlier chapters, a quality assessment of the design of each part up to the completed product can be undertaken to determine the quality of the design and the proposed manufacturing plan. The results of this process will input into the quality and cost model.

The model can also be interconnected to a simulation of the current manufacturing process as it exists in equipment and work flow. The results of adding the new product to the factory can be shown clearly through the model. The manufacturing equipment can be reorganized for better work flow or new machines can be added and their impact on cost and quality shown. In addition, a cost-effective test strategy can be developed from the quality attributes of design and manufacturing, as well as a strategy to most efficiently remove defects by using the various test equipment available, as was discussed in Chapter 4.

The cost and quality model can help company management keep abreast of how the new product is meeting its initial goals. This will guide the engineers in making the necessary adjustments in order to keep quality and cost at competitive levels.

### 6.1.2   Developing the background information cost estimating for electronic products

An accurate cost estimate for electronic products is dependent on many factors:

- Development schedule realization. The cost estimate should improve as a new product moves closer to production. In addition, the timing of the product introduction might influence the sales forecast, especially if there is new technology incorporated in the design. The cost estimate plans should include provisions for aggressive (50%) as well as standard new product introduction schedules (90% probability of realization).

- The sales forecast should be as accurate as possible. The marketing department should include up and down sales potential, competitive analysis, and price performance curve strategies. These help in selecting the optimum manufacturing strategy in equipment and tooling and hence determine the appropriate cost structure of the product.

- Nonrecoverable expenses (NRE) should be quantified, including tooling and capital equipment costs. A determination should be made whether some of those costs could be shared with other products or resources in the form of a cost center that allocates an overhead or burden rate to other products that use the NRE tools and equipment. A depreciation schedule and methodology, whether straight line (SL) or sum of the years digits (SOYD) should be agreed upon. Typically, 3–5 years and SL are used.

- The bill of materials (BOM) should be as complete and up-to-date as possible. It should include provisions for options, raw materials, and identified suppliers. Nonidentified suppliers should be investigated and estimates of material costs as well as reliability studies initiated. In addition, there should be a material cost reduction program for developing lower-cost material alternatives to the current BOM. These materials may be substituted when newer technology is available or when lower-specification materials might offer comparable performance in the product. A good target for such a program is 3–5% cost reduction per quarter after release to manufacturing. Material volume discount schedules should be available and readily incorporated with the forecast into the cost structure.

- The product routing scheme should be reasonably developed. The routing includes all of the manufacturing operations or steps necessary to fabricate, assemble, inspect and test the product. A determination should be made whether intermediate steps in the product assembly should be treated as line fabrication items with no inventory control points or as subassemblies. It is always desirable to have the minimum level of assembly to reduce assembly time and cost as well as lower inventory requirements.

- The direct labor needed to produce, assemble, inspect, and test the product should be accumulated for each manufacturing step. The amount of labor needed is dependent on other factors such as outsourcing, which turns in-house labor into purchased materials, the use of tooling, level of automation, and production volume based on the marketing forecast.

- The overhead rate for the product and whether it is different than the typical overhead rates for the product family should be determined. The overhead should include provisions for equipment and workspace allocations, special requirements due to energy and environmental considerations, and special skills needed to manufacture and technically support the product. As materials might contribute significantly to product cost, and because of the increasing trend toward outsourcing, several overhead rates can be applied, including one for material and another for labor. Material overhead should include costs for material warehousing, obsolescence, purchasing, and inventory control.

- The quality plan for the product, including the quality goals (six sigma or a certain level of Cpk), costs of expected yield, rework, scrap, inspection, and testing. The defects imparted by the raw materials suppliers should be added to the defects inherent in the design as well as those incurred in production. A test strategy is then developed for the optimum removal of these defects.

- General and administrative costs, including royalties paid to corporate R&D investments, profit margins, and provisions for taxes and reinvestment.

- Startup costs. These should include costs for design revisions, equipment and tooling debug, and support costs for additional technical and material support during the prototype and beta production phases of the product.

- A typical cost distribution of an electronic product is given in Figure 6.2. The cost estimates are regularly updated during the different phases of product development, due to increased clarity about the selection of components and manufacturing processes, and the resulting fallout in the costs of material, labor, overhead, deprecia-

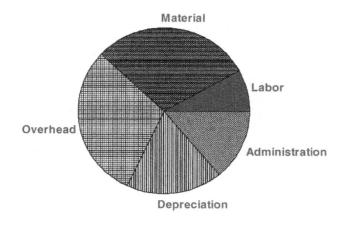

Figure 6.2    Typical cost distribution of an electronic product.

tion, administration, and yield of the new product. The cost information collected during these stages can help in understanding the impact of design decisions made. This effort might be spearheaded by a representative from the financial part of the organization temporarily assigned to the design team.

### 6.1.3   Determination of costs and tracking tools for electronics products

After collecting background information, several tracking tools can be used in order to make proper product marketing or financial decisions affecting the cost of the product. Some of these tools are as follows.

- The return factor of the product, which is the total profit (sales revenues minus manufacturing costs) returned by the product during its life cycle (up to 3–5 years), divided by the development costs. This return factor should be compatible with the historical trends of the product family and its competitors, expressed in return on investment (ROI) terms, which is determined by the time-adjusted present worth of the return factor. It should be in the range of 12–18% for typical electronic products. Obviously, this factor is dependent on the expected volume of the product. The volume will change the percentage of each element discussed in Section 6.1.2. In addition, this volume will determine where the product will be

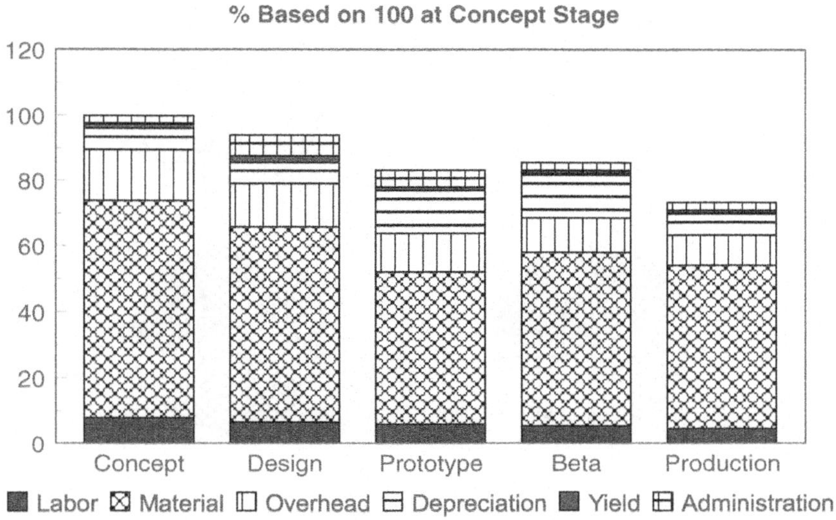

**Figure 6.3**   Cost history of an electronic product based on the concept stage.

manufactured, either in the company's own facilities or in the global supply chain.

- Cost history of the product. The costs of the product can be identified in terms of labor, material, overhead, depreciation on capital, NRE tooling, quality, and administration costs. These costs can be tracked over the design as well as the production phases of the product to show impact of design changes and investment in automation. An example of the cost history of an electronic product based on the concept stage is given in Figure 6.3

- Volume sensitivity of the product. Depending on forecast accuracy and upside potential, several levels of automation and manufacturing strategies can be used to estimate product costs. An example of the volume sensitivity in the typical cost percentages of a consumer electronic product is given in Figure 6.4.

## 6.2   The Quality and Cost Relationship

The impact of using quality metrics such as six sigma is that they develop a good accounting of defect causes in the product but do not show the impact of the cost to the company. Several attempts to link the two elements of quality and cost were developed. The quality loss function (QLF) is one of the tools attempting to link quality and cost.

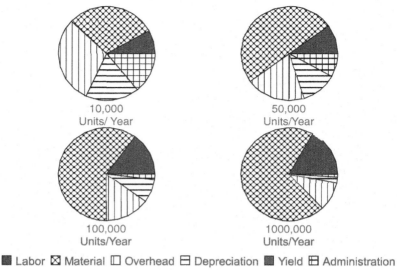

10,000
Units/ Year

50,000
Units/Year

100,000
Units/Year

1000,000
Units/Year

■ Labor ⊠ Material ☐ Overhead ⊟ Depreciation ■ Yield ⊞ Administration

**Figure 6.4**   Volume sensitivity of the cost of an electronic product.

### 6.2.1   The quality loss function (QLF)

The quality loss function was defined by Genishi Taguchi, its major author, as "the financial loss to society imparted by the product due to deviation of the product's functional characteristic from its desired target value." It is a negative definition of quality, which totals up the quality loss after the product is shipped. This loss is not widely used by product designers since the data required to calculate it are not readily available in the early part of the design of the product. The loss could be tangible as in-service and warranty costs that companies have to pay to repair the product. There are other costs that cannot be measured quantitatively: loss of market share, customer dissatisfaction, and lost future sales.

Quality loss function is a quadratic expression estimating the cost of a product quality characteristic not meeting its target. This deviation from target can be measured by the average shift from target and by the standard deviation of the quality characteristic. Even when a product leaves the factory within its specifications, it carries with it the inherent loss due to not exactly meeting its target. The cost is proportional to the loss to society due to a product defect, as measured in monetary loss due to repair as well as the loss of customer satisfaction. This could lead to lost future sales and to the company loosing its market share.

The loss function $L$ indicates a monetary measure for the product characteristic average versus its target value and the distribution around the average. Generally, it is expressed in terms of the cost of each failure divided by the square of the deviation from the average at which the failure occurs:

$$L(y) = \frac{A}{\Delta^2}(y - m)^2 \tag{6.1}$$

where
$L$ = loss function
$y$ = design characteristic
$m$ = target value or specification nominal
$A$ = cost of repair or replacement of the product
$\Delta$ = functional limit of the product, where customer dissatisfaction occurs. This could be wider than the product specifications.

Rewriting the formula by using the fact that $(y - m)^2$ is similar to the expression for mean square deviation (MSD) or the variance for the product characteristics:

$$L = \frac{A}{\Delta^2} \cdot MSD \tag{6.2}$$

The loss formula can be translated into familiar statistical terms of actual product characteristic average $\mu$ and the standard deviation $\sigma$. The $\sigma$ term is based on the $n$ divisor of the standard deviation formula and not $n - 1$ for the sample deviation:

$$L = \frac{A}{\Delta^2}[(\mu - m)^2 + \sigma^2] \tag{6.3}$$

### 6.2.2 Quality loss function example

An example of the quality problems that occur in the fabrication of printed circuit boards (PCBs) is the fit of a PCB edge male connector into the product housing female connector or "card cage." If the variability of the edge connector size is large, the fit is difficult or impossible to achieve, which could result in scrapping the PCB.

Assume that the tolerance for acceptable fit is ±6 mm, the cost of removing a defect in the PCB at the fabrication shop is $100, and the cost of removing a defect at the customer site after the PCB has been assembled is $500. A typical lot of 18 PCBs from the PCB fabricator was measured. The following shows the calculations of the loss function due to the variability of the edge connector and estimation of the savings incurred by either adjusting the average to target or reducing variability of the PCB edge connector.

Assuming actual deviations from the target value of a set of 18 PCBs at fabrication shop: 0, 0, –3, 0, 0, 1, 0, –5, –2, –2, 3, –5, –1, 0, –4, 3, 0, 1. Then

$$L = \frac{A}{\Delta^2} \cdot MSD, \quad MSD = (Y - M)^2$$

$$MSD = \frac{1}{N}(Y_1^2 + Y_2^2 + Y_3^2 + \ldots + Y_n^2)$$

where $n$ is the number of $Y$ deviations.

$$MSD = \frac{1}{18}(0^2 + 0^2 + \ldots + 1.0^2) = 5.778 \text{ mm}$$

$$L = \frac{A}{\Delta^2} \cdot MSD = \frac{\$500}{36} \cdot 5.778 = \$80.25/\text{PCB}$$

or

$$\sigma_n = 2.274; \quad \text{average deviation from target} = -0.778$$

$$L = \frac{A}{\Delta^2}[(\mu - m)^2 + \sigma^2] = \frac{\$500}{36} \cdot (0.778^2 + 2.274^2) = \$80.25/\text{PCB}$$

There are two ways to improve quality: set the average to target, or reduce variability. It can be readily seen that the second alternative results in the greatest quality cost improvement:

$$L_{\text{Average}} = \frac{A}{\Delta^2} \cdot \sigma^2 = \frac{\$500}{36} \cdot (-0.778)^2 = \$8.40/\text{PCB}$$

$$L_{\text{Variability}} = \frac{A}{\Delta^2} \cdot (\mu - m)^2 = \frac{\$500}{36} \cdot (2.274)^2 = \$71.84/\text{PCB}$$

The importance of the loss function is that it gives a monetary value to the state of the output of the process, both in terms of the process average not meeting the specification nominal and the process deviation. In the example outlined above, the average for all 18 measurement was –0.78 mm and the standard deviation was 2.274. Note that in this case the $\sigma_n$, which is 2.274, is different than the $\sigma_{n-1}$, which is 2.34. The maximum loss function for an assembled PCB that causes customer dissatisfaction is set at $500, and if it does not cause dissatisfaction, there is no loss. Using the formula, the loss due to the process average not being equal to target is calculated to be $8.40, whereas the loss due to variability around the average is $71.84. Taguchi used this technique to compare two Sony television factories in Tokyo and San Diego, CA in 1973.

The quality loss function can also be used to find an optimum level of quality at which the target factory quality can be balanced by the customer dissatisfaction of escaping potential defects. This would imply balancing the product shipping tolerance at $100 per defect removal at the factory versus the advertised specifications (±6mm) with a defect removal of $500 at the customer site. This can be shown mathematically as follows:

$$L_{\text{factory}} = L_{\text{customer}} = \frac{A}{\Delta^2_{\text{factory}}} \cdot MSD = \frac{A}{\Delta^2_{\text{customer}}} \cdot MSD \qquad (6.4)$$

$$\frac{100}{\Delta^2_{\text{factory}}} = \frac{500}{6^2}$$

$$\Delta_{\text{factory}} = \sqrt{3600/500} = \pm 2.68 \text{ mm shipping tolerance}$$

The above calculations indicate that the factory should set the tolerance of the manufacturing process at ±2.68 mm with a $100 cost per defect in order to balance the customer tolerance of ±6 mm and $500 cost per defect.

It can be seen that this methodology can provide an alternate approach to six sigma is setting product specifications based on the trade-offs of removing defects at various points in the product life cycle. This analysis is similar to the one performed for testing strategy in Chapter 4. Obviously, the quality loss function methodology is difficult to quantify, especially since the customer defect cost, as expressed in terms of loss to society, is difficult to ascertain.

### 6.2.3 A practical quality and cost approach

Both six sigma and the quality loss function discussed above are useful tools that can be used to achieve an assessment of product quality in design and manufacturing and relate it to the cost of the product.

For six sigma, the connectivity to cost is that the desired quality target of 3.4 PPM is required by customers to maintain a high level of growth enjoyed by electronics companies such as Motorola. There is no volume adjustment to the six sigma philosophy, so that the quality level is expected to be the same for mass-produced items such as cellular phones and pagers as for low-volume products such as those used by aerospace and the military.

The quality loss function (QLF) can be used as comparative estimate of the loss to the product incurred because of its process average shift versus target or its variability. It can also be used to measure the trade-off of quality between the factory and the customer, as shown in the example above (Section 6.2.2). The cost of a potential defect at the customer is estimated by a monetary value of the expected level of customer dissatisfaction with that defect. The strategy is to allow for a shipping tolerance at the factory narrower than the advertised specifications.

One of the obvious difficulties of the QLF strategy is the monetary estimate of customer dissatisfaction. It is larger that the cost of repairing or replacing a defect at the customer, since it includes the cost of removing the defective unit as well the loss of the use of the product and customer dissatisfaction.

A practical quality and cost approach is to use six sigma and its associated tools to calculate the potential number of defects in design and manufacturing. The result will be added to a cost model as follows:

- The quality level will be used to estimate the number of defects to be found in the product based on its current configuration.
- The defective parts will be replaced and the replacement cost added to the manufacturing operation cost.
- The defects generated will have to be removed through testing and inspection, and an estimate of the removal cost will be added to the model depending on the type of test performed.
- The model can be used to monitor the cost trade-offs in the selection of alternate design methodologies, materials, and manufacturing processes, as well as different test methodologies.

## 6.3    Electronic Products Cost Estimating Systems

Typically, PCBs account for 90% of the total material cost of an electronic product. Developing PCB cost models can vary depending on the accuracy level needed. Consumer products are sensitive to cost variation, whereas new technology products are less sensitive.

The electronic design cycle and its implementation in PCBs is divided into several steps. For most current electronic design activities, computer aided engineering (CAE) is used to document the design and provide the basis for electronic analysis and iterations of the design. Its function is also to physically partition the design into distinct electronic groupings or models that are then incorporated into each PCB. It also acts as a data source for further steps in the cycle. Figure 6.5 shows the steps involved in the PCB design cycle which are:

- The logical design phase of matching the product specification requirements by completing the electronic circuits design, selecting the components, and documenting the circuit connectivity.

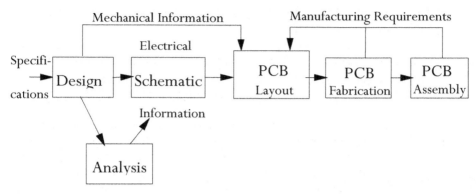

**Figure 6.5**  Electronic design implementation in PCBs.

- The analysis phase, in which the design is checked out to produce the optimum performance in terms of minimizing errors in connectivity, loading, and race conditions, optimizing testability and conformance to specification. This is usually performed using analysis tools for analog and digital simulation and modeling to verify the functionality of the electronic design. In addition, the design review concept at this phase is important to ensure both the technical validity of the PCB design, its connectivity to other PCBs in the product, and its suitability for manufacturing. The design review is a good alternative in the absence of effective analysis tools, especially in today's complex design environments.
- The PCB layout phase uses computer aided design (CAD) techniques to physically place the components and their interconnections to each other and to the outside world. This function determines the tooling and manufacturing environments for the PCBs and their future cost.
- The supporting and follow-on processes, which include activities such as device library creation, prototype PCB fabrication, assembly, and testing.

The alternatives in the design and layout processes include the selection of process factors for the components, layout, fabrication, assembly, and testing technologies. These factors affect the overall product cost and quality differently, as follows.

*Component technology* affects the component count directly and hence the PCB layout space required, the assembly production rate, and the reliability estimates of the product. These technologies include the following:

1. Through-hole (TH) components, which have leaded terminals to attach them to holes drilled in the PCBs.
2. Surface mount technology (SMT) components, which are leadless or have low-profile leads to attach them to the surface of the PCBs.
3. Printed circuit materials, which can include single and multilayer plated-through PCBs as well as one sided, nonplated holes.

These components have different footprints (spacing), production rates, assembly equipment investment, and required support.

*PCB layout* offers a clear choice of faster development time versus fabrication costs. Two layers or several levels of multilayer fabrication technology are some of the alternatives presented in the PCB layout phase. As the layer count decreases, there is a proportional effect on the cost and reject rate of PCB fabrication, but an inverse relationship to the time required to completely lay out a complex electronic design.

*Fabrication strategy* is dependent on the desired physical and electrical characteristics of the PCBs, as well as the maturity of the design and the time required for completion. Multiple alternatives are available such as PCB materials, layer count, hole and line specifications, and construction technologies. Many design engineers are not aware of these choices and do not fully understand the cost–benefit ratios of each.

*PCB assembly strategy* is influenced by the selection of the component technology in the design phase and the machine complement on the production floor. The chosen technology dictates a particular set of assembly operations. Several levels of manual versus automatic production processes can be used, depending on the physical electronic components chosen for the design.

*Test strategy* allows for logical and physical interconnection between the PCBs and the test systems. Additional target test points and test circuits influence both the layout timing and the physical constraints of the design.

### 6.3.1  Relating quality data to manufacturing six sigma or Cpk levels

There are many steps in the manufacture of printed circuit boards (PCBs). They include the preparation of the components and the fabricated PCB, the placement of the components or their insertion into predetermined locations on the PCBs, and the attachment of the components to the PCBs through the application of solder joints.

In order to control the quality of manufacturing the PCBs, some of these steps have their own recommended specifications from the equipment and material suppliers. However, a direct relationship of these specifications and the defects occurring during the PCB manufacturing steps is not readily discernible. This has sometimes led to a manufacturing process having a high-quality Cpk for the process meeting its individual specifications, yet having a very poor effective PCB assembly yield. This could result in a loss of credibility in the Cpk values in manufacturing.

An example of such a problem is in the SMT assembly operation in PCBs. The assembly consists of applying solder paste onto PCB component pads through a thin metal stencil in a screening machine, then placing the components onto the pads using an automatic placement machine. The components remain on the PCBs because of the tackiness of the solder paste. The final operation consists of passing the PCB through a conveyer oven to reflow the solder. The solder paste suppliers recommend a particular paste volume and height of the solder deposited on the pads and a particular temperature profile for the reflow oven. A Cpk of the solder paste and reflow operations can easily be obtained from control chart or process capability data.

High Cpk levels in solder deposition, oven profiles, and other indirect measurements of quality do not necessarily lead to high yields in PCB assembly. This has resulted in the need to develop composite Cpk analysis based on direct defect analysis for each step of the PCB assembly operations. These will be discussed in Chapter 8.

### 6.3.2    Printed circuit board (PCB) fabrication technologies

Conventional PCB fabrication (raw PCBS) utilizes subtractive copper etching to produce circuitry. This process is generally carried out in a number of steps, as shown in Figure 6.6, where different metals are

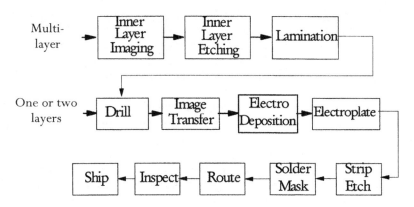

**Figure 6.6**    PCB fabrication steps.

plated on the raw PCB only to be removed at subsequent processes. This includes the bulk of the copper plated on the original laminate. Additive elements of the PCB have typically been restricted to metal plating operations and outside layer dielectric solder mask. Acid copper and electroless plating operations used in PCB fabrication utilize large amounts of wet chemistry. This chemistry is costly to the fabricator due to both the process control and effluent waste treatment required.

A fully additive process capable of achieving the desired electrical properties of raw PCBs without the use of wet chemistries could potentially reduce the PCB fabrication cost by reducing the costs associated with extra processing and waste management.

Additive processes will continue to become more prevalent in the future of the PCB fabrication industry. Currently, standard etching operations continue to decrease line width and space between the lines down to 2 mils. The critical limitation of the lines and spaces is the metal etching operation, which removes material in an isotropic manner.

Electronic component designs continue to become smaller with additional electrical performance requirements. This is driving designers and PCB fabricators to incorporate many electrical elements of the finished assembly into the raw PCB itself, either on the surface or buried within the layers. The passive components, such as resistors and capacitors, are beginning to be designed as buried elements within the PCB.

An additive technology that appears to hold great promise as a significant contributor to the future of raw PCB fabrication is polymer thick film (PTF). PTF is an attractive technology because it is fully additive and produces very little waste. The technology is also very easily processed. Typical applications use simple screen printing operations followed by a drying step. The drying step is generally done at temperatures comparable to that of current solder mask cures. The PCB fabrication industry is already very familiar with screen printing operations for both solder masking and legend printing. Most fabricators already possess both the equipment and general background to process PTF technology.

PTF technology can be used for applications that include both fully additive circuits and configurations that use a mixture of etched or plated copper in conjunction with PTF PCB features. PTF is already widely used for applications that include jumper circuits for product revision changes, precious metal plating replacements for switch applications, electrical shielding, and both surface and buried resistor configurations. Silver through-hole and via connections using PTF are very popular practices that are used in Europe and are gaining use in the United States. Fabricators of simpler raw PCBs can deposit switching configurations and through-hole connections in a single printing pass. Inks are being used as part of filled via hole configurations to add hole wall integrity to the via as well as remove heat from high-density packages such as ball grid arrays (BGAs). These filled via connections can be used as part of buried via configurations for sequential and nonsequential configurations that utilize tag drilling.

Although fully additive circuits are currently possible using PTF inks, the application has generally been limited to low-power switching operations. The finished circuit trace consists of metal flakes encapsulated in a polymer binder. A nonhomogenous conductor is created by contact between adjacent flake material. Unfortunately, even the most conductive PTFs currently have line resistance values that are considerably higher than an etched circuit. The line resistance tends to attenuate electronic signals and produce line resistance.

Dimensional limitations have also hampered the widespread use of PTF products for fully additive circuitry. Typical PTF line widths and spaces are 10–20 mils. Current etched cooper line widths and spaces are significantly tighter, generally in the 4–6 mil range and as low as 2 mils.

In its current state, PTF additive circuits cannot become a replacement for high-power etched copper raw PCBs. If the line resistance and dimensional problems can be overcome, fabricators could benefit from the savings accorded to a "greener" method of manufacturing printed circuit boards. In the interim, PTF's versatility for specialized applications should continue to provide PCB designers with good solutions to today's complex electronic circuit designs.

### 6.3.3  Printed circuit board (PCB) design, fabrication, cost, and quality issues

The PCB manufacturing process is not standard and is heavily dependent on design parameters. A low-component-density PCB could be implemented in a two-sided (two-layer) fabricated PCB, which does not require inner-layer processing or lamination. Similarly, many fabricators use inner-layer inspection only for very dense or controlled impedance designs. The type of solder mask selected (screened, dry film, or liquid photoimageable) is dependent on the density of signal lines and the types of assembly requirements for the PCBs.

The operating cost of a manufacturing facility will include overhead for administration and marketing. However, these are usually applied evenly across all products and are therefore not design-dependent.

By examining the list of cost factors, the PCB features that affect the cost of the design can be identified. The number of layers will determine how many times the innerlayer processing rate is applied, and the material cost of copper-clad dielectric and prepreg material for adhesion of the layers. Since the drill cost rate is often expressed per hole, this cost will depend on the stack height, which is in turn based on the PCB thickness, hole diameter, and the required accuracy of the location of the drilled hole. Too tight a hole tolerance, which affects the solder mask operation, will result in a lower drilled stack, and therefore higher drilling costs but less misregistration defects. Most fabricators process full panels (such as the 18" × 24" size), so the more images or PCBs that can be designed to fit on a single panel, the smaller the PCB cost.

The cost of a fabricated PCB should provide a conversion process from PCB design features to manufacturing cost. An important part of this process is the designer's understanding of the manufacturing process, capabilities, and constraints. For example, the dry film solder mask is the most expensive, yet offers the best quality in terms of soldering defects in PCB assembly. The inverse is true of the screened solder mask.

To calculate the effect of different design alternatives, a manufacturing engineer must provide information about each cost parameter in the fabrication process that would influence the final cost. The cost ($C_i$) for each PCB parameter is:

$$C_i = P_i \cdot F_i \tag{6.5}$$

Where $P_i$ is the PCB specific number applied for each cost parameter and $F_i$ is the respective cost factor for the fabricator.

The cost factor, $F_i$, is derived from the fabricator cost rates based on their actual material, labor overhead, and support costs. For example, the relative cost factor $F_{ilp}$ of inner layer processing is calculated from the following formula:

$$F_{ilp} = (C_{ll} + C_{lo} + C_{ls})/N_l \tag{6.6}$$

Where $C_{ll}$ and $C_{lo}$ are the inner layer imaging and etching direct labor and overhead expended from the last financial period reported, respectively, and $C_{ls}$ is the department support and maintenance costs such as percentage of utilities, maintenance, and general management costs incurred to support the inner layer department. $N_l$ is the number of layers consumed by the fabricator during that period. Obviously, this cost factor system necessitates alternative accounting procedures by which costs are accumulated based on the cost factor structure.

Cost factor systems provide a standard measure of the contribution of individual design features to overall PCB manufacturing cost. They allow manufacturing engineers to compare multiple design alternatives in order to select the optimum design.

Yield prediction for PCB fabrication is an important element of the cost equation, yet it has proven to be difficult to estimate. Examples given above suggest that there is a trade-off between the higher cost of materials and processes and the resultant yield from them.

Historically, PCB tooling departments provided manufacturability reviews of new PCB designs prior to production release. These reviews are successful in identifying major errors such as spacing violations or missing features. However, in most cases, the yield has already been determined by decisions made far upstream and it is too late to significantly alter the design. Although factors that contribute to yield loss are well known (including high layer count, fine lines, and small holes), higher performance unavoidably requires selecting features that create less-manufacturable PCBs.

Yield prediction is required to evaluate the effect of feature selection on yield at the early stages of the design process, thereby minimizing the PCB cost for a set of performance requirements. One possibility is to express the complexity of a design technology set with a single standard metric containing values of the significant design elements. The yield prediction model then becomes a functional relationship between fabrication yield and the complexity metric. This method allows several different design alternatives to be quantitatively compared to determine the yield (or cost) improvements associated with selected design changes. An example of complexity-based process DPUs from a typical PCB fabrication shop is shown in Table 6.2.

The development of a complexity metric should be guided by a study of the influence of design elements on fabrication yield. First, each printed circuit manufacturing process should be examined to uncover possible sources of yield loss. Then the most common fatal defects observed in manufacturing are investigated to determine proba-

**Table 6.2**  Complexity-based process DPUs from a typical PCB fabrication shop

|  | Complexity | | |
| Process | Low | Medium | High |
| --- | --- | --- | --- |
| Image transfer | 0.02 | 0.03 | 0.04 |
| Copper etching | 0.01 | 0.02 | 0.02 |
| Lamination | 0.005 | 0.01 | 0.01 |
| Drilling | 0.004 | 0.01 | 0.01 |
| Metallization | 0.04 | 0.05 | 0.05 |
| Solder mask | 0.02 | 0.03 | 0.03 |
| Total DPUs/PCB | 0.099 | 0.15 | 0.16 |
| Process yield | 91% | 86% | 85% |
| Cpk | 0.55 | 0.48 | 0.47 |

ble design and process-related causes. This complexity factor is based on two elements: the geometry of the PCBs as well as any special electrical requirements that can lead to material and process considerations in plating, solder coating, or solder mask selection.

The geometry of the PCBs is based on the components to be used, their pad sizes, and the line widths and spaces connecting them. Small rectangular pads for SMT components are replacing the round insertion pads of TH technology. The old standard of 0.060" insertion component pads on 0.100" centers has been replaced with rectangular pads that are only 0.030" wide on 0.050" centers, or even 0.005" wide pads on 0.010" centers for TAB configurations. Added to these factors are the increasing number of interconnecting holes or vias and their associated small-diameter pads. Many SMT device component leads require an attached via pad for electrical testing as well as interconnection to inner layers. This connectivity is an important part of the complexity metric.

The most common expression of connectivity is inches of wiring per square inch of circuit board (in/in$^2$). One method of measuring this factor is the total line lengths necessary to implement the PCB design, based on the theoretical optimum placement of the components on the PCB. In the PCB industry, the line length is measured by track (number of traces between grid points) and layer count. Unfortunately, this metric can only be known after the layout is completed, and therefore cannot be used to describe the necessary connectivity required to select an alternative.

Most designers currently make an estimate of track and signal layer count from prior experience. The CAD software autorouter is set up with "standard" feature dimensions (usually from a design specification) and allowed to work away. After the autorouting cycle is completed, the layout is checked for remaining disconnects. If a significant number of signals are still unconnected, the usual procedure is to add another pair of signal layers.

A possible design alternative is to increase track by using smaller trace widths, spacing and/or pad sizes, and thus reducing the total layer count and PCB costs. However, this increased track could be offset by the greater use of fabrication technology.

A geometry performance model should use the information available to the PCB designer prior to the start of layout phase. Several industry studies have demonstrated that wiring demand can be calculated from the number of input/output connections (I/Os) per component, the number of components, and the approximate spacing between components. This can be expressed in terms of equivalent integrated circuits (EICs). In addition, the choice of the pad, hole, and line widths and spaces should be determined prior to layout. If the capability of the CAD system and its operator can be expressed as an efficiency factor, then wiring capacity and PCB technology sets can be estimated from standard EIC densities.

The overall fabrication yield of a printed circuit board is limited by the maximum capabilities and normal process variations of the individual steps. The key processes that introduce yield loss are:

Image transfer
Copper etching
Lamination
Drilling
Metallization
Solder mask

An analysis of PCBs scrapped at major fabrication shops revealed that the majority of fatal defects fell into three categories: electrical opens, electrical shorts, and solder mask defects (e.g., cracking, flaking, or loss of adhesion). A list of possible design-related causes of these defects is given in Table 6.3.

In order to verify the apparent effect of these features on yield and obtain an estimate of their relative significance, actual production yield data should be collected for current production part numbers.

### 6.3.4   PCB fabrication cost and quality alternative example

An example of the cost and quality alternatives for PCB fabrication is the choices made when laying out a PCB, depending on the quality

**Table 6.3**   Design-related causes of PCB defects

| Defect | Critical design feature |
| --- | --- |
| Electrical open | Trace width |
|  | Line length |
|  | Board area |
|  | Layer count |
|  | Hole count |
|  | Board thickness |
|  | Hole diameter |
| Electrical short | Spacing |
|  | Line length |
|  | Board area |
|  | Layer count |
|  | Hole count |
| Solder mask | Solder mask clearance |

and complexity of the alternatives selected. An example would be a design with the following attributes that could be completed with three types of alternative layouts, requiring different levels of complexity factors in the PCB fabrication process:

| Alternative #                | 1       | 2       | 3       |
| ---------------------------- | ------- | ------- | ------- |
| Complexity factor            | Low     | Medium  | High    |
| Number of layers             | 6       | 4       | 4       |
| Number of vias               | 130     | 270     | 234     |
| Total line length            | 411.7"  | 466.5"  | 411.6"  |
| Geometry of lines and spaces | 12/12   | 8/7     | 6/6     |
| Total number of holes        | 791     | 931     | 895     |

If the complexity factor yields shown in Table 6.2 are applied to the alternatives in the example above, then other factors can also be evaluated in reaching the most optimal alternative:

| Alternative #      | 1    | 2    | 3    |
| ------------------ | ---- | ---- | ---- |
| Estimated yield    | 91%  | 86%  | 85%  |
| Cpk                | 0.55 | 0.48 | 0.47 |
| Relative cost      | 17   | 15   | 15   |
| Layout time (days) | 4    | 10   | 8    |

In this case, the savings of the material realized by reducing the layer count from 6 to 4 outweighs the slightly lower quality due to narrow line widths. Obviously, the best decision is the one that balances the effect on the speedup of the layout time based on the choice of complexity and hence faster new product introduction, versus lower complexity process selection and hence lower cost of the PCB fabrication. In this example, the highest quality and relative cost alternative 1 (at Cpk = 0.55 and 17 relative cost) provided the fastest layout time, and hence greater profit from faster new product introduction.

## 6.4  PCB Assembly Cost Estimating Systems

The PCB assembly process is usually a complement of equipment with different capabilities and constraints in terms of automation, speed, component technology, and quality. Table 6.4 contains a listing from the classifications for the different types of PCB assemblies. This listing was developed by the Institute for Interconnecting and Packaging of Electronic Circuits (IPC).

It can readily be seen that the task of assembling a PCB can be achieved through many different alternatives of equipment and

**Table 6.4**    Classifications for different types of PCB assemblies

| | |
|---|---|
| Type 1. | Components (mounted) on only one side of PCBs |
| Type 2 | Components (mounted) on both sides of the PCBs |
| Class A | Through-hole (TH) component mounting only |
| Class B | Surface mount technology (SMT) components only |
| Class C | Mixed assembly of TH and SMT |

processes, some of which are overlapping in function, yet different in quality and productivity.

A typical approach to printed circuit board assembly is presented in Figure 6.7. The process presented shows a two-sided mixed printed circuit board (PCB) assembly, including all the steps required to populate the boards with components on both sides. Using a technology mix of TH and SMT, components can be loaded by machine or by hand, depending on the geometry of the components and the volume and number of component types that the manufacturing facility processes. Automatic sequencing and insertion equipment are limited by the number of heads available and geometry of the parts.

SMT components can also have different material and manufacturing options. Components can be placed by hand or machine, depending on the constraints of geometry and package variety. In addition, some of the smaller-size components might be assembled by more accurate placement machines, such as robots, requiring their own set of processes in parallel to the regular-size SMT components.

Three levels of PCB costs systems are in common use in the electronics industry, depending on accuracy and resources available to manage the cost system.

### 6.4.1  Material-based PCB assembly cost system

Since material costs account for the majority of PCB costs, all other costs are calculated as a percentage of the material used in the PCB BOM. Cost parameters are shown in Table 6.5 for typical communication PCBs of analog and digital designs. The ranges are based on querying four different PCB manufacturers. The NRE expenses are based on the tooling necessary for making stencils for applying the circuit images on the PCBs and the test setup. Test costs are based on several iterations (revisions) of the PCB design.

### 6.4.2  The technology cost driver system

In this system, component technology and process methodology selection is used in calculating the costs of the PCBs. A single cost driver is assigned to each PCB production or assembly step, including a driver

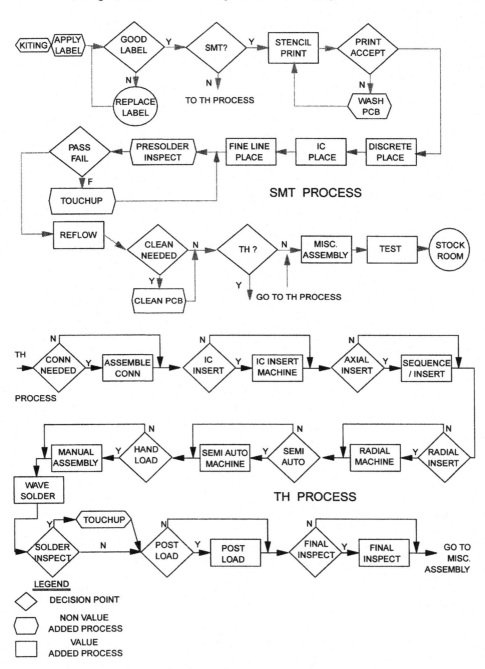

**Figure 6.7**   A typical approach to printed circuit board (PCB) assembly.

**Table 6.5**   Material-based cost model, NRE and test costs

| Direct costs | Digital | Analog | Range (4 companies) |
|---|---|---|---|
| Raw material | $500 | $1,000 | ± 5% |
| Mark-up (material overhead) | $50 | $100 | 5–20% of material |
| Labor | $50 | $100 | 10% of material |
| General/administration | $50 | $100 | 8–10% of material |
| Earnings before taxes | $50 | $100 | 7–15% of material |
| Total direct costs | $700 | $1,400 | |
| | | | |
| Nonrecoverable expenses (NRE) | | | |
| Stencils | $500 (etch) | $800 (laser) | 1 each side |
| Programming/documentation | $1,000 | $1,000 | Laser for fine pitch |
| Total NRE | $2,000 | $2,600 | |

| Test hardware costs (All PCBs) | # Revisions | Cost/revisions | Total |
|---|---|---|---|
| In-circuit test (ICT) fixture | 2 | $10,000 | $20,000 ($6/test pin) |
| Setup electrical test | 1.5 | $3,000 | $4,500 |
| Test tooling NRE | | $5,000 | $5,000 |
| Test hardware subtotal | | | $29,500 |

| Test support costs | # of weeks | Cost/engineer/week | Total |
|---|---|---|---|
| Engineering test and debug | 6 | $2,000 | $12,000 |
| Quality assurance test | 6 | $2,000 | $12,000 |
| Test engineering and QA subtotal | | | $24,000 |
| Total test costs | | | $53,500 |

for each type of component technology used. Through-hole (TH) and surface mount technology (SMT) are the most commonly used. The driver is a representation of the cost of the assembly function. The dollar per driver rate is based on the total costs for each manufacturing step for a certain period of time, typically a quarter, divided by the number of driver units processed by the manufacturing department for that period. All costs specific to the assembly step should be used in determining the driver rate, including costs for factory space, consumables, energy, supervisory and technical support, and equipment maintenance and calibration. The number of cost drivers used to determine the PCB costs will vary with the accuracy needed. A 15% accurate model can be constructed with 4 drivers, whereas a high-accuracy system of 5% error should be made with at least 15 drivers. To be very effective in reducing costs of new products, the cost per driver rates should be recalculated periodically (quarterly) at a faster rate than the new product development cycle.

In order to simplify the determination of cost and quality for PCBs in the technology model, several extraneous factors have to be considered:

- Capital equipment purchases of new machines and modifications or enhancement costs of existing ones should be part of the capital budget of the plant, and hence not considered in the cost model. This equipment might be used for other products and its useful life might extend beyond the new product life cycle. In a typical multiproduct company, equipment purchases are financed from retained earnings or borrowed from banks. Customers ultimately pay for these purchases through interest expenses applied against the balance sheet of the company.
- Utilization rates, consumables, and energy costs are dependent on the volume of the new product, its impact on the sales of current products, and the model mix on the factory floor. These rates can only be determined through the use of factory floor simulation runs.

An example of cost rate calculations for one of the PCB assembly process steps, which is the automatic insertion of machine-loaded through-hole (TH) components, is given in Table 6.6. The table shows a step-by-step procedure for calculating the manufacturing cost based on a typical PCB, including the number of parts and volume produced.

Table 6.6 starts with the advertised productivity information of the TH component insertion machine rate of 3000 components per hour (c/hour). A typical PCB, containing an average of 100 TH components, can be theoretically machine loaded in $(100/3000) \cdot 60$ or 2 minutes. However, this rate is reduced by the stoppage of the machine due to misloaded components. The stoppage of the machine occurs every fifth PCB with a 7.5 minutes average time to clear the machine. This represents 1.5 minutes per PCB for a total machine run time of 3.5 minutes per PCB. Run time per component is calculated using the 100 components per PCB at $(3.5 \cdot 60)/100 = 2.1$ seconds per TH component. This results in an effective insertion rate of $3600/2.1 = 1714$ components/hour, down from the advertised machine rate of 3000 components/hour.

The second part of Table 6.6 calculates the cost driver rate for machine-loaded TH components. This rate is calculated from the 2.1 seconds per component, multiplied by the labor ($10 per hour) and the process overhead rate (100%). The cost driver is thus calculated at $(\$20/3600) \cdot 2.1 = \$0.0117$ per machine-loaded TH component. This

**Table 6.6**   Cost rate calculations for machine-loaded TH components (machine type: Universal Instrument Company)

| Productivity information | Machine rate | Rate/PCB/ Component/Job |
|---|---|---|
| TH component insertion rate | 3000 | c/hour |
| Average TH components inserted/PCB | 100 c | |
| Theoretical machine load time | | 2 minutes/PCB |
| Average machine stoppage/PCB | 1/5 PCB | |
| Stoppage time (to clear machine) | 7.5 minutes | 1.5 min/PCB |
| Total run time/PCB | | 3.5 min/PCB |
| Run time/TH Component | $(3.5 \cdot 60)/100 = 3600/2.1$ | 2.1 seconds/component |
| Effective TH insertion rate | | 1714 components/hour |
| **Cost information** | | |
| Average TH component cost | | $0.15/component |
| Average operator pay | | $10/hour |
| Process overhead rate | | 100% |
| Effective labor rate | $20/hr | $0.00556/sec |
| Cost driver rate/axial insertion | $0.00556 × 2.1 | $0.0117/component |
| **Modifier information** | | |
| Machine load time/PCB | 25 seconds | |
| Setup modifier $h$ | $(25/3600) \cdot \$20$ | $0.14/PCB |
| Average changeover time/job | 25 minutes | |
| Batch modifier $b$ | $(25/60) \cdot \$20$ | $8.33/job |
| Average misloaded components | 2000 PPM | 1/500 components |
| Average cost of TH component | $0.15 | |
| Quality modifier ($I$) | $0.15/500 = | $0.0003/component |

rate can then be used as part of a PCB cost model containing most of the PCB processes shown in Table 6.7.

### 6.4.3   PCB assembly cost modifiers in the technology cost model

The manufacturing plan for printed circuit boards can change frequently. Some of the changes could be the batch run sizes, the work holder sizes, or the quality of the assembly process steps. This can have a significant effect on the calculations shown in Table 6.6. A more accurate methodology is required to make the cost calculation more reflective of the changes in the manufacturing cycle. Hence, the use of cost modifiers can make the cost model more flexible and responsive to changes in the manufacturing plans.

The modifiers represent the impact of changing manufacturing components and parts volume due to the new products being intro-

**Table 6.7**   Technology cost model with modifiers for PCB assembly

| Process function | Driver units | Rate $ | Modifiers | | |
| --- | --- | --- | --- | --- | --- |
| | | | Setup $/h$ | Batch $/b$ | Quality $/l$ |
| Component procurement | Material $ | 0.032 | | | 0.00045 |
| Material storage & delivery | # of part numbers | 0.022 | | | 0.0005 |
| Machine load through hole | # of insertions | 0.012 | 0.14 | 8.33 | 0.0003 |
| Hand load TH postsolder | # of insertions | 0.24 | | | 0.0005 |
| Machine load passive SMT | # of placements | 0.024 | 0.08 | 3.15 | 0.0002 |
| High-accuracy active SMT | # of placements | 0.075 | 0.10 | 5.00 | 0.0002 |
| Hand load through Hole | # of insertions | 0.120 | | | 0.0005 |
| Hand load SMT | # of placements | 0.036 | | | 0.005 |
| Solder and wash for TH | # of board/holder | 3.56 | | | |
| Solder paste and reflow | # of boards/holder | 10.00 | | | |
| Assembly operations | # of parts | 0.360 | | | 0.0010 |
| Test and repair | # of components | 0.035 | | 1.00 | |
| Visual inspection | # of components | 0.030 | | | |

*$h$ = PCBs per work holder modifier; $b$ = batch size modifier; $l$ = quality loss/component modifier.

duced. This might require additional purchase or upgrade of equipment, new factory layout and material flow, as well as updated manufacturing methods. Since machine setup and product batching account for a significant portion of manufacturing resources, these effects should be taken into consideration when estimating the costs of new products. Some of the modifiers that should be included are:

1. Batch run setup modifier. This modifier represents the setup requirements for each product part number. They include computer-integrated manufacturing (CIM) information transfer, the setup of the machine parameters, and first piece inspection for each batch. In PCB assembly, the setup times could be quite lengthy. For example, a third of the production time allocated for automatic insertion could be spent on properly preparing the components for sequencing prior to insertion. Many different techniques have been used to reduce this setup time, including permanent component allocation to loading heads as well as off-line loading of heads. These modifiers are in effect only at those operations where batch specific tasks are to be performed.

2. Assembly holder modifier. This modifier represents the load/unload time for each machine-based process step. It is highly dependent on the machine setup procedures and the type of work holder to be used. In PCB assembly, this modifier represents the allocation to the number of PCBs that can be processed in a single work holder at different process steps. In some cases, this holder can be the

solder frame, while in other cases the PCB panel can be considered as the carrier. These modifiers are in effect only at those operations where positioning for loading and unloading is required.

3. Standard process modifiers. These are standard process steps that every PCB goes through without any special attention to the type of material or function of the board. They include universal process steps such as oven curing, soldering, or cleaning. To implement the use of the modifiers in PCB assembly, they could be calculated in terms of unit cost drivers per component, as shown in Table 6.7, for a typical PCB assembly of 100 components. This will allow for simple calculations based on the driver units of each process step to determine the total cost per PCB.

The procedure outlined above can be used to calculate all of the rates for individual manufacturing process driver units outlined in Table 6.7. The $0.0117 rate, calculated for axial automatic insertion in Table 6.6, is entered as the machine-loaded through-hole rate of $0.012 in Table 6.7. The cost driver unit for that process is the number of component insertions for the PCB. The setup modifier $h$ for machine-loaded through-hole insertion is based on the number of PCBs that can be accommodated in the machine work holder. It takes 25 seconds to load each PCB into the machine. At the $20/hour effective labor rate, this amounts to $(25/3600) \cdot 20 = \$0.14$ additional cost for every PCB loaded on the machine. Since a PCB contains an average of 100 components, this cost represents an addition of $0.0014/component. This cost is halved when the work holder used can accommodate two PCBs at the same time.

The batch modifier $b$ is based on the batch size or the number of PCBs in each job. For each new job, the machine has to be emptied of the previous job's components and then loaded with the programming information and the components of the new job. From Table 6.6, axial insertion machine changeover time is 25 minutes. At the $20/hour rate, this amounts to $(25/60) \cdot 20 = \$8.33$ additional cost for every PCB job loaded on the machine. For a typical job of 100 PCBs and 100 components to be machine loaded, this modifier represents an additional cost of $0.00083 per component. This cost is inversely proportional to batch size.

4. The material loss quality modifier ($l$) is based on the defect rate of the assembly process. From Table 6.6, the typical misloaded component rate for axial insertion is 2000 PPM, or 1 in 500 components. At the $0.15 typical cost of a through-hole component, the quality loss is $0.15/500 = \$0.0003/component$. When six sigma is achieved for all operations, the quality modifier could be deleted and the cost of quality can be determined at the next-highest level.

The rates for the various PCB operations in Table 6.7 are based on calculations similar to the ones for TH components, as shown in Table 6.6, including material, labor, overhead, and reject rates. The values of the modifiers $h$, $b$, and $I$ depend on the current levels of tooling, scheduling, and quality of the manufacturing process.

The use of this cost method for PCB assembly will be rendered more effective in conjunction with a simulation of the PCB assembly area, which will determine the optimum levels for the setup and batch factors $h$ and $b$. Typically, the batch size $b$ in production is determined by the material acquisition and inventory policies of the manufacturing operations, while the work holder factor $h$ is determined by the relevant manufacturing equipment work area and the size of the PCB. These modifiers can be best evaluated through many iterations of the simulation to determine an optimum operating methodology. For example, altering the material technology by switching to more SMT components, which are smaller, will help in reducing the PCB size. This will result in more PCB's per work holder and an increase in the value of $h$, and hence a decrease the cost of the PCB assembly.

A composite technology cost model using several manufacturers in the defense, instruments, and communications industries is shown in Table 6.7. For very high volume companies, the batch $b$ and work holder $h$ modifiers are not used, since their PCB manufacturing includes a high level of automation in the setup as well as production, and therefore the setup time is reduced to zero.

Similar technology cost models could be built for other manufacturing processes, incorporating the yields of each step of the process into the model. Table 6.8 is an example of a list of cost model drivers for sheet metal fabrication.

**Table 6.8**  Cost model drivers example for sheet metal fabrication

 1. Total # of pieces/lot
 2. Material thickness and type
 3. Length in X direction
 4. Length in Y direction
 5. # of nonstandard tools used
 6. # of hits/piece
 7. # of fold angles
 8. # of folds/piece
 9. # of pems and/or rivets per piece
10. # paint yes/no
11. # type of paint finish
12. # class of paint finish
13. # of silkscreen colors

### 6.4.4    Quality based product cost models

The test and repair cost driver in Table 6.7 is a historical estimate in the technology-based cost model, and is based on previous or similar products' FTY yields. In the quality-based cost model, test costs are derived from FTY defect prediction based on quality analysis of the product design and each manufacturing step. For all cost drivers, parts per million (PPM) defect rates are calculated based on actual machine operations, determined from periodically dividing defects generated by all of the driver unit output for each process. The defects generated are then summed up to the product level and a test strategy developed for the most efficient removal of these defects.

An example of creating an overall design cost model for PCBs by combining all of these principles discussed earlier is shown in Table 6.9. All of the possible process steps needed to produce a typical electronic product are included in the cost model, including the various technologies of PCB assembly. The model produces a tally of the material, manufacturing, and support costs necessary to manufacture

**Table 6.9**  PCB quality-based technology cost model

| Process | Driver | # | % PCB cost | Cost Process | Cost Driver | Quality # Defects | Quality Cost | Adjusted process |
|---------|--------|---|-----------|--------------|-------------|-------------------|--------------|------------------|
| Auto Insertion | | | 22.79% | $349.89 | | | | $349.90 |
| | # of axial parts | 600 | parts | | $152.84 | 0.046725 | $0.01 | |
| | # of axial PNs | 22 | PNs | | $18.58 | | | |
| | # of ICs | 20 | parts | | $172.91 | 0.000267 | $0.00 | |
| | # of IC PNs | 5 | PNs | | $5.56 | | | |
| SMT | | | 42.07% | $645.88 | | | | $646.35 |
| | # of passive parts | 4000 | parts | | $283.21 | 0.93091 | $0.44 | |
| | # of active parts | 400 | parts | | $362.66 | 0.046546 | $0.03 | |
| Hand load | | | 0.96% | $14.80 | | | | $14.80 |
| | # of parts | 40 | parts | | $14.80 | 0.018639 | $0.00 | |
| Wave and | | | 0.12% | $1.88 | | | | $1.88 |
| or wash | # of PCBs per frame | 2 | PCBs | | $0.00 | | | |
| | Normal wave | 1 | times | | $0.94 | 0.000346 | $0.00 | |
| | PCB Wash | 1 | times | | $0.94 | 0.000346 | $0.00 | |
| In circuit | | | 17.59% | $270.09 | | | | $270.09 |
| test | # of defects | 1.6 | defects | | $27.59 | | | |
| | # of parts | 1 | parts | | $0.01 | | | |
| | # of fixture heads | 1 | heads | | $242.50 | | | |
| | Volume per month | 1 | units/mo | | | | | |
| Assembly | | | 0.34% | $5.15 | | | | $5.15 |
| | # of parts types | 20 | parts | | $5.15 | | | |
| Functional | | 1 | 16.12% | $247.44 | | | | $247.44 |
| test | # of defects | 3 | defects | | $247.28 | | | |
| | # of connectors | 1 | each | | $0.16 | | | |
| Total | | | | $1,535.14 | | | | $1,535.62 |

and ship the products. Each manufacturing process is broken down to its smallest discernible step, and a cost driver is assigned to each step. The process cost is then summed up and presented as a percentage of the total cost of the product.

These steps include the following:

- Automatic insertion of TH parts costs, including the cost of the type of part to be inserted: axial or integrated circuit (IC). The setup cost per axial or IC part number (PN) is also included in the cost model.
- Automatic placement of SMT components costs, shown by different types: passive or active SMT parts
- Manual (hand) solder costs, which are dependent on the number of parts to be hand loaded.
- Solder and wash operations, which are dependent on the number of PCB boards per solder frame.
- Mechanical assembly, which has a cost for each part to be assembled.
- In-circuit test (ICT), which is an inspection function to remove all manufacturing defects created by the previous operations. There is a normal cost associated with running the test as well as a cost per defect to be removed.
- Functional test, which is required to remove all design-caused defects, as well as those not removed by ICT.

The cost column by driver is shown as a baseline. It is listed by process, and then the number of parts or drivers to arrive at the baseline cost multiplies each process cost. The quality cost is automatically connected to a Cpk calculator, which determines the number of defects, based on an analysis similar to that set out in previous chapters, and provides for the number of defects and the cost of removing those defects. The cost of replacing defective parts is shown in the quality column of the cost model. This cost is added to each process step and a resultant adjusted cost is shown. The test labor required to identify the defective parts is shown as part of the in-circuit test (ICT) for assembly defects and functional test (FT) for removing design-caused defects and those not removed by ICT.

Trade-offs are determined by calculating the capital equipment and tooling investments versus the labor expended and their impact on the cost of the product. The assembly method selection is dependent on the investment in automation or robotics versus the use of manual assembly methods. The mix of component material and attachment technology selection is also important in determining the cost of placing and soldering those components on the PCBs.

In addition, the design engineers can work with manufacturing to plan additional production equipment procurement based on the projected sales volume, especially for those machines that can become the bottleneck in production capacity.

A good understanding is needed to evaluate the relationship of electronic product cost to other factors such as marketing and forecasting, material and process selection, the level of automation of manufacturing equipment, and designing process methods and flows. Cost determination, estimation, tracking, and control varies depending on the life cycle phase of the product and the market it competes in. Several types of PCBs cost systems are developed, as shown in this chapter, since PCBs are the major cost components of electronic products.

## 6.5   Conclusion

Accurate estimation of new product cost and quality is becoming very important for electronic enterprises. A good understanding of how cost and quality affect major decisions in material and process selection is needed to determine the manufacturing equipment, material flows, and manufacturing processes required to build the new product and compete globally for customers. Current levels of world-class performance has resulted in requiring that the new product cost and quality estimation process become very accurate, given the variability of products and companies making them. Accurate and flexible cost and quality estimation models are thus needed for electronic products and especially for printed circuit boards, since they represent the principal electronic product cost.

In this chapter, several cost and quality models were developed to help in estimating the quality of design and manufacturing, calculating an accurate cost of the assembly of the product, and developing a test strategy for removing the defects generated in the design and manufacture of the product.

## 6.6   References and Bibliography

Banks S. "After ISO there is ABC." *Surface Mount Technology,* August 1993, p. 25.

Dunlop R. "Design for profit." In *Proceedings of the National Electronic Packaging and Production Conference.* Anaheim, CA, 1992, pp. 139–147.

IPC. *Guidelines for Printed Board Component Mounting.* Lincolnwood, IL: IPC, 1993, IPC-CM-770.

Manko H. *Soldering Handbook for Printed Circuits and Surface Mounting.* New York: Van Nostrand Reinhold, 1993.

Saigal A. and Shina S. "An Algorithm for selecting the electronic design implementation in printed circuit board fabrication based on cost factors." In *Proceedings of ASME Winter Annual Meeting,* Atlanta, GA, 1996, pp. 757–769.

Saigal A. and Shina S. "A design quality based cost model for new electronic systems and products." *Journal of Materials,* April 1998, pp. 29–33.

Saigal A. and Shina S. "Technology Cost modeling for the manufacture of printed circuit boards in new electronic products." *Journal of Manufacturing Science and Engineering,* May 1998, 368–375.

Wheelwright S. "Operation as strategy: Lessons from Japan." *Journal of Stanford Graduate Business School, 59,* 1, 58–67, 1981.

# Six Sigma and Design of Experiments (DoE)

The concept of the design of experiments (DoE), alternatively known as "robust design" or "variability reduction," has been used to reduce some of the sources of manufacturing variation or manipulate a design toward its intended performance. These attributes make DoE one of the most effective tools for reaching six sigma in design and manufacturing.

DoE influences both ends of the six sigma ratio: manufacturing operations produce parts with defects, either because of tight design specifications or manufacturing process variability. Using DoE, the need to have narrow specification limits can be eliminated, and the product can operate satisfactorily within wide production process variability. Most of the applications of DoE have been made in the production or process development phases of new products, because the use of DoE is most beneficial in multidisciplinary applications, where traditional engineering analysis, simulation, and verification are difficult to achieve.

In design applications and new product development, DoE is very effective in systems design when there is considerable interaction among the system components in achieving system performance. Six sigma tools can be effectively used in the selection of the quality characteristic in DoE experiments because they can point to where the most benefit can be extracted. This chapter will address the issues of using DoE methods in design and manufacturing of new and current products striving to achieve six sigma. The topics to be discussed in this chapter are:

1. DoE definitions and expectations. In Section 7.1, the definition of DoE is given, as well as the expectations of proactive improvement of the product and process design. The reasons for using DoEs are discussed, including the effects of noise and other external and internal conditions that contribute to the variability of products and processes.

2. Design of experiments (DoE) techniques. These techniques are introduced in Section 7.2 with an algorithm for conducting a DoE project, and selection of the quality characteristic.

3. The DoE analysis tool set. These tools are presented in Section 7.3 with case studies for each. They include graphical and statistical analysis of the average and the variance of the quality characteristic.

4. Using DoE methods in six sigma design and manufacturing. DoE design techniques have been used mostly to reduce manufacturing variability. Section 7.4 addresses the use of DoE methods for design engineering applications as well as optimizing manufacturing.

## 7.1   DoE Definitions and Expectations

Design of experiments (DoE) is a systematic method for determining the effect of factors and their possible interactions in a design or a process toward achieving a particular output of the quality characteristic(s). It is used in order to quantify the source and resolution of variation and the magnitude of the error when comparing the average of the quality characteristic to the target. Figure 7.1 is an example of how these elements are arranged.

Using DoE techniques, a design or a process can be manipulated to provide a target or minimal/maximum performance of the quality

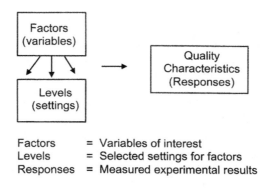

Factors     = Variables of interest
Levels      = Selected settings for factors
Responses   = Measured experimental results

**Figure 7.1**   Basic elements of DoE.

characteristic average or reducing its variability or both. This is accomplished by setting factors that affect the quality characteristic to predetermined levels and analyzing the output sets of factorial, partial factorial, or orthogonal experimental arrays.

Reducing production variability is one of the most commonly used methods to increase the process capability index and attain six sigma quality. Variability can be addressed by using a combination of two strategies:

1. On-line control. Here the focus is on maintaining the current production processes within a specified area of variability through control charts, optimal maintenance, and calibration of production processes and equipment. This is the traditional method of maintaining quality, and was discussed in Chapter 3.
2. Off-line control. Here a proactive effort is aimed at reducing the process variability or increasing design robustness through defect analysis and design of experiments (DoE). This allows for achieving six sigma through targeting of specific process operations or design elements. This effort can be guided by many of the tools of TQM and corrective action processes discussed in Chapter 3, as well as this chapter.

On-line control methods should be instituted before attempting off-line control projects. No amount of design of experiments and defect analysis can rectify a poor-quality operation that is out of control. In that case, the benefits of off-line control improvement can only be felt temporarily, before being negated by a manufacturing operation that is out of control, where the production factors, materials, and processes are constantly changing. The sources of defects, as outlined earlier in this book, are due to the interaction between product specifications and process variability. This interaction originates from one of two sources: either the process is not centered (the process output average does not equal the target value), or the product and process variability, as measured by the standard deviation of the manufactured product characteristics, is too high. Either one or a combination of both can influence the product defect level.

It is much easier to identify, collect data, and rectify the first situation: a process average not equal to the target. Incoming materials, equipment settings, and performance can be measured, and if not equal to target, can be corrected by strict adherence to specifications. Materials properties such as geometrical tolerances, density, tensile strength, hardness, etc., can easily be measured and rectified by working with production personnel and suppliers. Equipment factors such as temperature, pressure, speed, and feed and motion accuracy can be measured by calibration gauges against original purchase specifications, and readjusted as necessary.

The calibration of production equipment is usually achieved by using an instrument or gauge that is inherently more accurate that the equipment to be calibrated. In addition, the instrument's accuracy has to be certified through traceability to the National Institute of Standards and Technology (NIST). It is common to use calibration equipment whose accuracy and resolution are at least one-tenth that of the equipment being calibrated, as was shown by the gauge capability (GR&R) section in Chapter 5 of this book.

The maintenance of production variability and keeping the production average equal to the target is best accomplished by using control charts, discussed in detail in Chapter 3. Variable control charts show control of the quality characteristic average in the $\overline{X}$ chart, and its variability in the $R$ chart. Attribute charts do not make a distinction as to defect source between the average deviation versus the variability of the process, and therefore it is more difficult to ascertain the causes of the defects.

Electronic production operations, such as those producing PCB assemblies, that are in good control and operating within six sigma quality generate a small amount of defects, normally in the range of 1–20 PPM, amounting to a few defects per working day. Individual defects can thus be analyzed using the tools of TQM and corrective action process improvements presented in Chapter 3: brainstorming, cause and effect, pareto diagrams, data collection, and sampling, etc. These tools can be used to determine the most probable cause for each defect. If a deviation of the production process was found to be the cause, the process can be adjusted accordingly.

Reducing the variability of the production process is more difficult and requires a thorough examination of the sources of variability. Some of these causes are uncontrolled factors or noise. They can be generated from the following:

- External conditions, imposed by the environment under which the product is manufactured or used, such as temperature, humidity, dirt, dust, shock, vibration, human error, etc. These conditions are beyond the control of the design and manufacturing process planners. They are difficult to predict, and it is expensive to design specific characteristics to satisfy all of the possible conditions under which the product is expected to operate.
- Internal conditions under which the product is stored or used, such as friction, fatigue, creep, rust, corrosion, thermal stress, etc. These conditions have to be specified correctly within the normal use of

the product. However many customers will overuse the product, and expect that it will continue to operate even beyond its normal range. Therefore, the design has to be made more robust to ensure proper operation beyond advertised specifications.

DoE is focused on improving the robustness of product functionality in external and internal conditions of operation. It seeks to determine the best set of process materials and factors in order to ensure that the product characteristics average is equal to the specified nominal, and the variability of the product characteristics is as small as possible. A set of designed experiments can be performed to find such an optimum level of factors influencing the operation or manufacture of the product.

### 7.1.1  DoE objectives and expectations

The objectives of DoE are to adjust the quality characteristics (or design or process output) to the optimum performance by properly choosing the best combination of factors and levels, as shown in Figure 7.1. This is accomplished by collecting maximum information from the DoE experiment results using minimum resources. The factors can be categorized to determine which factors effect the average, variability, both average and variability, or have no effect on quality characteristics. Figure 7.2 shows these possible effects. The results of a DoE experiment can be one of the following:

1. Identify the most important factors that influence the quality characteristic
2. Determine factor levels for the important factors that optimize desired quality characteristics (output responses)

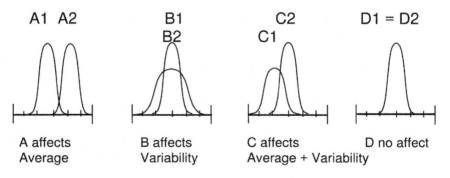

Figure 7.2  Possible effects of different factors.

3. Determine the best or most economic setting for factors that are not important
4. Validate (confirm) responses and implement in production or design

The success of an experiment is not determined solely by just achieving the desired quality level. Important information about the design or the manufacturing process can be gleaned from any experiment. This information can be put to use in future experiments or through using more traditional quality improvement processes such as TQM. Information gained from DoE can be listed as follows:

• The factors that are significant for influencing the quality characteristic average, reducing variability, or both, and which factors are not significant. If none of the factors are found to be significant, then the design of the experiment has to be repeated to include factors or levels not previously considered.

• The proper balance between average shift from target versus variability reduction by choosing the proper factor levels. The choices of certain factor levels can shift the average, whereas others can reduce the variability, or both. Although good results can be obtained by moving the average to the maximum or minimum possible or achieve a target for the design, this action can be tempered by selecting alternate factors and levels to achieve the greatest robustness in reducing variability. The quality loss function discussed in Chapter 6 can be used to make decisions based on economic considerations.

• The predicted experiment outcome can be determined when the design or production factors are set to the specified levels. Confidence intervals and the expected error can also be shown for the predicted outcome.

• The goodness of the experiment design and the proper selection of factors and levels can be evaluated by statistical analysis.

## 7.2   Design of Experiments (DoE) Techniques

DoE is best characterized as making several assumptions about the design or the process being studied, quantifying these assumptions by the choice of factors and levels, and then running experiments to determine if these assumptions are valid. It a mix of several tools that has been developed to optimize performance, based on statistical analysis, significance tests, and error calculations.

Improving the process capability requires the concurrent efforts of both product and process designers. Product designers should increase the allowable tolerance to the maximum that will still permit the successful functioning of the product. Process designers should center the process to meet the specification target and minimize the variability of the process. DoE is a tool that can help with both of these goals.

DoE uses statistical experimental methods to develop the best factor and level settings to optimize a process or a design. Some of the statistical methods have been simplified by the use of specialized software analysis packages. In many cases, the engineers responsible for the process or design can perform the necessary steps to conduct the experiments from knowledge presented in this or other books on DoE, or after taking minimum training in the techniques of DoE, perhaps with the assistance of a statistician. In addition, the technical knowledge of the basic science or technology necessary for optimizing a design is not critical. Neophytes can optimize product and process design just as well as experienced engineers using DoE techniques.

### 7.2.1  Steps in conducting a successful DoE experiment

Conducting a DoE experiment involves using many of the tools of six sigma quality that were outlined in previous chapters. It is always advantageous to form a team to perform the tasks of designing the experiments and interpreting the results. Teams have shared experiences in the design, and can achieve broad consensus on different approaches to the DoE and the problem being analyzed.

The success of a DoE project is dependent on selecting the proper team members, identifying the correct factors and levels, focusing on optimizing and measuring the quality characteristics, and analyzing the results. Steps in performing a successful robust design of experiments are as follows:

- Problem definition. The first task in performing a DoE project is to outline the goals of the project and to define the quality characteristic(s) of process or the design to be optimized. Although only one characteristic can be optimized at a time, many characteristics can be measured from the same experiment matrix while performing the experiments and analyzed separately. The final-level selection can be a mix of the recommended factor and level settings, depending on the compromise of the different objectives of each quality characteristic.

- Design space. Creating the boundary of the product or design to be optimized is important. The experiment should not be constrained to a small part of the design and hence not provide the opportunity to study the interactions between the different parts of the total design. On the other hand, the experiment should not be all-encompassing in an attempt to optimize a wide span of product design steps or processes. Ideally, the total design should be analyzed, and a compromise made in developing a plan for a succession of DoEs, each providing additional information about the design to be optimized.

- Team creation and dynamics. A project team should be selected to conduct the experiments and perform the analysis. The team should be composed of those knowledgeable in the product and process, and should solicit inputs from all parties involved in the design to be optimized. It is not necessary to have an in-depth technical understanding of the science or technology of the problem, but the team members should have experience in similar or previous designs. Knowledge in statistical methods, and in particular DoE techniques, should be available within the team, either through a statistician or someone having received training or experience in DoE.

- Factor and level selections. DoEs can be performed using two approaches. One method is to select a large number of factors and use a screening experiment, usually a saturated design (to be explained later), to narrow down the factor selections. Then a follow-on experiment, preferably a full factorial experiment, is used to complete the selection of the optimal factors and their levels. The second method is to have the team members consider this DoE project as a single opportunity to try out as many possible factors, levels, and combinations of both, because of the lack of time or resources available. In this case, partial factorial experiments are used, with some assumptions as to the relationships of factors, in order to maximize the benefits, resources, and time spent on a single experiment.

- Brainstorming techniques should be used to select the number of factors, and the different levels for each factor. The selection process should outline factors that are as independent as possible from other factors, and hence are additive in controlling the quality characteristic(s) to be optimized. This is important in reducing the interactions of factors, which are difficult to quantify statistically.

An example of selecting independent factors and reducing their interactions is the case of an infrared conveyorized oven for the reflowing of surface mount technology printed circuit boards (SMT PCBs). The reflow process is characterized by three factors: the ramp-up of temperature to the solder melting stage, the maximum temperature level reached during reflow, and the time during which the temperature remains above the solder liquid state, usually called time above liquidus (TAL). There are several heater zones in the reflow oven, and the oven temperature can be controlled by setting the zones on the top and bottom of the reflow oven to predetermined levels, as well as varying the conveyor speed. Choosing temperature zones and the conveyor speed as the factors for a reflow experiment would result in strong interaction between the factors. The proper choice of factors would be the ramp-up temperature rate, the TAL, and the maximum reflow temperature. The factor levels selected should be achieved by actually experimenting with the temperature zones and conveyer speed to reach the desired levels in the experiment.

- Level selection. Proper selection of the levels for each factor used in the experiments is important in achieving the proper design space. Levels that are either too close together or too far apart in value should not be selected, because they do not represent a continuum of the impact of the factor on the measured characteristic. Level selection should follow these guidelines:

  1. Three level designs could be chosen if the project team is confident that the current design is performing adequately but needs to be improved. The current level should be in the center of a 20% span represented by the other two levels. In this manner, the DoE can help in finding a more optimized operating set of factors levels in the design space.

  2. Two levels could be selected if there is little confidence in the adequacy of the current design, based on the collective judgment of the team. By choosing two levels, more factors can be tested within a small number of experiments, as will be demonstrated later. In addition, the direction of better design performance can be ascertained for future DoEs

  3. Multiple level factors should be chosen for survey experiments. In these DoEs, a team can select many new technologies or materials within one factor to identify which one can perform best in the design. The number of multiple levels should be close to squares of two or three levels, such as four, eight, or nine levels. They are easier to perform since they fit easily into the set of predetermined experiment arrays.

  4. The selected levels should be well within the operating range of a working characteristic within the design space. In the soldering reflow experiment mentioned above, the combination of temperature factors and levels should not result in having components soldered beyond their maximum temperature and time exposure specifications.

- Experiment arrays. Most DoE experiments use a set of standard orthogonal arrays available to conduct the experiment, with two or three levels. There are only certain combinations of factors and their levels available in order to perform the experiment. Compromise might be necessary to achieve economy in DoEs by selecting a given number of factors and levels that can fit within one of the orthogonal arrays. There are only a small number of these arrays of two and three levels, and their size increases geometrically with the number of factors selected.

- Conducting the experiments is based on the selected orthogonal arrays. The arrays are arranged in terms of the number of experiments, factors, and levels. The experiments should be conducted in a random order from the array matrix. The measurements of the characteristic to be optimized could be repeated using various scenarios, depending, on the variability considerations of the design (see the later section on variability reduction).

- Data analysis. Once the experiments are performed, the data can be analyzed graphically to determine the optimal settings of levels of significant factors. In addition, statistical analysis can be performed in order to determine the significance of each factor's effect on the quality characteristic, through the use of analysis of variance (ANOVA). Important factors can be set to the proper level, and least significant factors can be ignored, or set to the most economic conditions.

- Graphical analysis of the data is sufficient to determine the best factor setting to adjust the design average to target and reduce design variability. The statistical analysis provides more details on the probability of the effect of each factor on the characteristic measurement. In addition, statistical analysis can quantify the usefulness of the DoE project: low significance of the total experiment usually results from the lack of significant factors. In this case, the experiment is not providing useful guidance to the design team and it should be repeated with additional or different factors and levels.

- Prediction and confirming experiments. Once the graphical and statistical analysis is completed, the characteristic value can be predicted based on the choice of factor levels. These choices could be a compromise between setting the design characteristic average to the target value versus reducing variability. A recommended factor level might cause variability to be reduced, yet at the same time the process average will be shifted from target. Another case is when multiple characteristics are to be optimized using one experiment with many separate output measurements and data analysis.

For example, a robust design experiment could be performed to design a new plastic material to be injected molded. The material and process design can have several desired characteristics including modulus of elasticity, density, amount of flash after injection, gel time, flow rate, and free rise density. A DoE experiment could be designed using an orthogonal array that determines what ratios and composition of raw materials are to be used, as well as the injection molding machine parameters. Measurement of all the desired characteristics will be performed, then the data analyzed to determine the best set of raw material ratios for each characteristic. A compromise of all recommended factor levels will have to be made in order to achieve the best overall plastic product.

- Confirmation experiment. Once all the choices and predictions of the DoE experiment have been agreed upon, a confirmation experiment run should be made before final adoption of the design decision, to verify the analysis outcome. This confirmation will test the entire robust design process before full implementation takes place. In manufacturing, the newly adjusted process should continue to be monitored through statistical quality control methods for a six month minimum time period, before any attempts are made to further increase the robustness of the process by launching another DoE.

### 7.2.2 Types of DoE experiments using orthogonal arrays

The arrays most commonly used in the design of experiments are the orthogonal arrays. These arrays are balanced: there are an equal number of levels for each factor in the experiments. The behavior of each factor level can be studied while other factors are changing their levels. This technique results in an array matrix with $n$ columns and $n + 1$ experiments with two level factors.

Orthogonal arrays are different from the one-factor-at-a-time experiments, as shown in Table 7.1. In this design, shown with four factors and two levels, the first experiment consists of all of the factors set to level 1. Additional experiments are added in which the factor levels are varied to the second level individually, while the rest of the factors are kept constant. There are many deficiencies in this technique: it does not allow for measuring the effect of varying the other factors at the same time as the one factor being changed. This mathematical relationship of factors, called factor interaction, is very important, and needs to be analyzed to take full advantage of DoEs. Orthogonal arrays can measure interactions through several techniques:

**Table 7.1**    "One factor at a time" experiments

| Experiment number | Factors | | | |
|---|---|---|---|---|
| | *A* | *B* | *C* | *D* |
| 1 | 1 | 1 | 1 | 1 |
| 2 | 2 | 1 | 1 | 1 |
| 3 | 1 | 2 | 1 | 1 |
| 4 | 1 | 1 | 2 | 1 |
| 5 | 1 | 1 | 1 | 2 |

1. Full factorial DoEs are used to evaluate the effects of all factors and their interactions. For every number of factor columns $n$, there are at least $n - 1$ interactions column to be considered; and for $m$ levels, there are $m^n$ experiments. For example, If four factors are considered ($A$, $B$, $C$, and $D$) with two levels, there are 11 interactions such as $AB$, $AC$, $AD$, $BC$, $BD$, $CD$, $ABC$, $ABD$, $ACD$, $BCD$, and $ABCD$, and $16(2^4)$ experiments. The levels in the interaction columns are derived from the multiplication of the levels of the originating factors, using an exclusive OR (XOR) logical formula shown in Table 7.2. For full factorial designs, the number of experiments increases geometrically as the number of factors increases.

2. Fractional factorial DoEs provide a cost-effective way of determining the significance of selected factor interactions. A fractional factorial DoE uses a portion of the full factorial columns to estimate main factor effects and their interactions. The unused interaction columns are then assigned to other factors. This results in the condition called "confounding," in which the assigned factor could be confounded with the interaction that is normally found in the column. By selectively choosing where to confound, a fractional factorial DoE could be used to study more factors with less experiments. Good planning of DoEs could minimize the confounding problems. There are several levels of confounding called "resolutions," including:

**Table 7.2**    XOR logic table for interaction level determinations

| Levels | | Resulting level |
|---|---|---|
| *A* | *B* | *AB* |
| 1 | 1 | 1 |
| 1 | 2 | 2 |
| 2 | 1 | 2 |
| 2 | 2 | 1 |

A. Resolution III. No interactions are considered. Main factors are used for each column in the orthogonal arrays. All interactions are confounded with main factors.

B. Resolution IV. Two factor interactions are confounded with two other factor interactions only.

C. Resolution V. Two factor interactions are confounded with three other factor interactions only.

3. Saturated Design DoEs are Resolution III designs that allow all of the columns in the OA to be assigned to different factors. They represent a minimum set of experiments for the number of factors considered. They are called "screening designs" because they are commonly used to whittle down the number of factors quickly through smaller DoEs, then full factorial DoEs can be performed on the remaining factors. The assumption in saturated designs is that interactions are small and can be ignored compared to the main factor effects.

### 7.2.3 Two-level orthogonal arrays

The most commonly used two-level orthogonal array is the L8. It is an eight experiment array, sometimes referred to as $2^7$, having seven columns to be used as factors (1 through 7 or $A$ through $G$), and each factor is to be considered at two levels (1 and 2), as shown in Table 7.3. The symbols for the factors are given for the top two rows as satu-

**Table 7.3**  L8 orthogonal array

|  | Factor symbols | | | | | | |
|---|---|---|---|---|---|---|---|
|  | 1 | 2 | 3 | 4 | 5 | 6 | 7 |
|  | $A$ | $B$ | $C$ | $D$ | $E$ | $F$ | $G$ |
| Experiment | $A$ | $B$ | $AB$ | $D$ | $AD$ | $BD$ | $ABD$ |
| number | $A$ | $B$ | $AB$ | $D$ | $AD$ | $BD$ | $G$ |
| 1 | 1 | 1 | 1 | 1 | 1 | 1 | 1 |
| 2 | 1 | 1 | 1 | 2 | 2 | 2 | 2 |
| 3 | 1 | 2 | 2 | 1 | 1 | 2 | 2 |
| 4 | 1 | 2 | 2 | 2 | 2 | 1 | 1 |
| 5 | 2 | 1 | 2 | 1 | 2 | 1 | 2 |
| 6 | 2 | 1 | 2 | 2 | 1 | 2 | 1 |
| 7 | 2 | 2 | 1 | 1 | 2 | 2 | 1 |
| 8 | 2 | 2 | 1 | 2 | 1 | 1 | 2 |

Number of levels

Number of experiments $\quad\downarrow\quad$ Number of factors

L8 $(2 \times 7)$

rated design, for the next row for full factorial, and the bottom row for partial factorial design using Resolution IV. It can be noted there are three uses for the L8:

1. Use as Full Factorial array to check three factors (A, B, and D) at two levels and four interactions [C(AB), E(AD), F(BD), and G(ABD)].
2. Use as a saturated (screening) design for up to seven factors at two levels and no interactions. When using this array for saturated designs, factors should be assigned according to potential significance as follows. The most important factors should be the assigned to the primary columns A, B, and D. Column G, which confounds with the three-way interaction ABD, should be assigned next. For the last three factors to be assigned, use the columns C, E, and F, which confound with two-way interactions of the primary factors.
3. Use the L8 as a partial factorial Resolution IV design with three primary factors assigned to columns A, B, and D, a fourth factor G, and interactions C, E, and F. There are several confounding and missing interactions in this application of L8. Factor G confounds with the three-way interactions ABD and two-way interaction of the three primary factors A, B, and D with factor G are missing and assumed to be insignificant. Factor G is assumed to be independent of the previous three factors.

All eight experiment lines in an L8 can be repeated as necessary to establish average and variability analysis of the quality characteristic(s), and to obtain a statistically relevant sample. A simple rule is to use a minimum of 30 values for assuming a population distribution. In this case, the L8 should be repeated four times. For large processes with many different factors, such as an IC manufacturing line, the process can be divided into segments and each segment can be optimized individually with a DoE. It is much easier to conduct two L8s than a single larger experiments such as an L32.

The use of the L8 as a saturated design with seven independent factors contrasts with their full factorial use. A full factorial design with seven factors would require a DoE with an L128 ($2^7$) experiments. What is gained by much less experiments in the saturated design (L8) is offset by its inability to calculate interactions as in the full factorial designs (L128). A pictorial presentation of the eight versus 128 experiments is given in Figure 7.3, where the eight experiments are shown as blackened squares in the 128 experiments' matrix.

The balance of the orthogonal arrays can be shown with the L8. For each particular level in a column, all of the other levels in the other

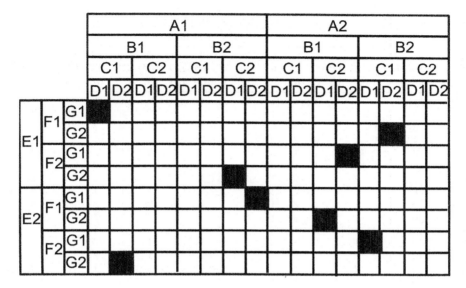

**Figure 7.3**  The use of an L8 as full factorial versus saturated design.

columns are rotated through their values. Experiments 1 through 4 have column A with level 1 only, whereas the levels in columns B through G contain both levels 1 and 2, in equal numbers of 2 each. The balance of the orthogonal arrays allows for a simple solution for the values of the factors in the experiments using Cramer's rule. In each array, there are $n$ unknown variables that can be solved in $n - 1$ equations. The L8 can be represented with seven unknowns and eight simultaneous equations, and thus each variable (factor) can be solved for.

The next higher two-level array is an L16, shown in Table 7.4. The L16 includes 16 experiments and 15 factors (several symbols are shown for each factor—top row for saturated design, middle row for full factorial, and bottom row for partial factorial with Resolution IV design) at two levels. L16 can be thought of as two L8s stacked on top of each other, with an additional column used for an extra factor and its interactions. It is not necessary to choose all available factors to be included in the experiment: an L16 experiment can be performed with 10 factors in saturated design, the other factors (array columns) can be left unassigned. This does not jeopardize the utility of the experiment, since the analysis of the effect of the 10 factors on the output characteristic is valid. The remaining factors could be used for calculating some of the interactions, according to the assignment of the main factors.

**Table 7.4**  L16 orthogonal array

| | Factor symbols | | | | | | | | | | | | | | |
|---|---|---|---|---|---|---|---|---|---|---|---|---|---|---|---|
| | 1 | 2 | 3 | 4 | 5 | 6 | 7 | 8 | 9 | 10 | 11 | 12 | 13 | 14 | 15 |
| Experiment | *A* | *B* | *AB* | *C* | *AC* | *BC* | *ABC* | *D* | *AD* | *BD* | *ABD* | *CD* | *ACD* | *BDC* | *ABCD* |
| number | *A* | *B* | *AB* | *C* | *AC* | *BC* | *DE* | *6* | *AD* | *BD* | *CE* | *CD* | *BE* | *AE* | *E* |
| 1 | 1 | 1 | 1 | 1 | 1 | 1 | 1 | 1 | 1 | 1 | 1 | 1 | 1 | 1 | 1 |
| 2 | 1 | 1 | 1 | 1 | 1 | 1 | 1 | 2 | 2 | 2 | 2 | 2 | 2 | 2 | 2 |
| 3 | 1 | 1 | 1 | 2 | 2 | 2 | 2 | 1 | 1 | 1 | 1 | 2 | 2 | 2 | 2 |
| 4 | 1 | 1 | 1 | 2 | 2 | 2 | 2 | 2 | 2 | 2 | 2 | 1 | 1 | 1 | 1 |
| 5 | 1 | 2 | 2 | 1 | 1 | 2 | 2 | 1 | 1 | 2 | 2 | 1 | 1 | 2 | 2 |
| 6 | 1 | 2 | 2 | 1 | 1 | 2 | 2 | 2 | 2 | 1 | 1 | 2 | 2 | 1 | 1 |
| 7 | 1 | 2 | 2 | 2 | 2 | 1 | 1 | 1 | 1 | 2 | 2 | 2 | 2 | 1 | 1 |
| 8 | 1 | 2 | 2 | 2 | 2 | 1 | 1 | 2 | 2 | 1 | 1 | 1 | 1 | 2 | 2 |
| 9 | 2 | 1 | 2 | 1 | 2 | 1 | 2 | 1 | 2 | 1 | 2 | 1 | 2 | 1 | 2 |
| 10 | 2 | 1 | 2 | 1 | 2 | 1 | 2 | 2 | 1 | 2 | 1 | 2 | 1 | 2 | 1 |
| 11 | 2 | 1 | 2 | 2 | 1 | 2 | 1 | 1 | 2 | 1 | 2 | 2 | 1 | 2 | 1 |
| 12 | 2 | 1 | 2 | 2 | 1 | 2 | 1 | 2 | 1 | 2 | 1 | 1 | 2 | 1 | 2 |
| 13 | 2 | 2 | 1 | 1 | 2 | 2 | 1 | 1 | 2 | 2 | 1 | 1 | 2 | 2 | 1 |
| 14 | 2 | 2 | 1 | 1 | 2 | 2 | 1 | 2 | 1 | 1 | 2 | 2 | 1 | 1 | 2 |
| 15 | 2 | 2 | 1 | 2 | 1 | 1 | 2 | 1 | 2 | 2 | 1 | 2 | 1 | 1 | 2 |
| 16 | 2 | 2 | 1 | 2 | 1 | 1 | 2 | 2 | 1 | 1 | 2 | 1 | 2 | 2 | 1 |

### 7.2.4   Three-level orthogonal arrays

Three-level orthogonal arrays are popular in manufacturing. Most current operations can be improved with DoEs using the current process value as the middle level, and then extending 10–20% above and below the current value for the other two levels. Three-level graphical analysis of the relationship between the factors and the quality characteristic(s) can be plotted using three points; hence any curvature of the data can be shown, as opposed to the straight line of the two-level experiments. Three-level columns have two two-way interactions, so that factors A and B have two interactions—*AB* and *BA*.

The smallest three-level orthogonal array that can be used is the L9, shown in Table 7.5. The top two rows are two different factor symbols commonly used for saturated design, and the bottom row is the factor symbol for full factorial design. The three-level orthogonal arrays, such as L9, have two uses similar to the L8, they are:

1. Use as full factorial design to check two factors (*A* and *B*) at three levels and all their interactions [one two-way interaction with two columns *C*(*AB*) and *D*(*BA*)]

2. Use as saturated (screening) design to check up to four factors (*A*, *B*, *C*, and *D* or 1, 2, 3, and 4) at three levels. The current process is cho-

**Table 7.5** L9 orthogonal array

| | Factor symbols | | | | |
|---|---|---|---|---|---|
| | A | B | C | D | |
| Experiment | 1 | 2 | 3 | 4 | |
| number | A | B | AB | BA | Results |
| 1 | 1 | 1 | 1 | 1 | Y1 |
| 2 | 1 | 2 | 2 | 2 | Y2 |
| 3 | 1 | 3 | 3 | 3 | Y3 |
| 4 | 2 | 1 | 2 | 3 | Y4 |
| 5 | 2 | 2 | 3 | 1 | Y5 |
| 6 | 2 | 3 | 1 | 2 | Y6 |
| 7 | 3 | 1 | 3 | 2 | Y7 |
| 8 | 3 | 2 | 1 | 3 | Y8 |
| 9 | 3 | 3 | 2 | 1 | Y9 |

sen as the mid-level, with ±20% variations from the current process as the other two levels.

A partial factorial design is not possible in an L9. All nine experiment lines are repeated as necessary to establish average and variability analysis of the quality characteristic(s). For large processes, they can be divided into major segments with a DoE for each. It is much easier to conduct two L9s than larger experiments such as L81. The use of the L9 as a saturated design of four factors and nine experiments can be contrasted with the full factorial design of four factors and 81 ($3^4$) experiments.

### 7.2.5 Interaction and linear graphs

Interaction occurs when one factor modifies the conditions of another. If this is deemed significant, the interaction should be derived from its own column in the array, and no factor should be assigned to this column. Table 7.6 is an example of an L4, a two-factor, two-level ar-

**Table 7.6** Interaction example using L4 orthogonal array

| | Factor symbols | | | Experiment results with interaction | |
|---|---|---|---|---|---|
| | 1 | 2 | 3 | | |
| Experiment | A | B | C | | |
| number | A | B | AB | None | Large |
| 1 | 1 | 1 | 1 | 74 | 74 |
| 2 | 1 | 2 | 2 | 80 | 80 |
| 3 | 2 | 1 | 2 | 78 | 78 |
| 4 | 2 | 2 | 1 | 84 | 72 |

ray. Results are shown for two different experiments—one with no interaction and another with a large interaction. The four points of the experiment results are plotted in Figure 7.4. The left part of the figure, representing the no-interaction condition, shows the two lines formed by the pair of points $A1B1$, $A1B2$ and $A2B1$, $A2B2$, are parallel. The right part, formed by the same four points, shows that the two lines are intersecting, indicating large interactions between factors $A$ and $B$. It should be noted that the contribution value of the interaction, that is, the difference between the condition of level 1 versus level 2 of factor $AB$, is equal to zero when there is no interaction. In the opposite case, a large interaction indicates that the main factors should be considered as one single combined factor, and graphically analyzed using the methodology in Figure 7.4

Although there is only one interaction column in orthogonal array L4, there are four interaction columns in orthogonal array L8. The interaction of columns 1 and 2 can be found in column 3, forming an exclusive OR relationship in the levels for column 3. These relationships can be grouped into primary factors and interaction factors. For example, in the partial factorial design with resolution IV, columns 1, 2, 4, and 7 are primary factors in array L8, and the remaining columns are due to the interactions of the first three primary factors. Two possible scenarios of assigning interactions are as shown in Table 7.7.

These scenarios can also be shown graphically through linear graphs, which are provided in Figure 7.5. All L8 factors and their interactions are apportioned in one of two ways: scenario I shows that primary factors 1, 2, and 4 are equal in importance and their interactions are available for analysis, since no factors were confounded with the interaction columns 3, 5, and 6. It is also assumed that factor 7 is an independent factor with no interactions with the other three factors (1, 2, and 4). Scenario 2 shows that factor 1 predominates and all of the interactions of other factors (2, 4 and 7) with it (3, 5, and 6) are available for analysis. However, the two-way interactions of the other factors (2, 4, and 7) with each other are not available. In the case of L8, either scenario can be analyzed once the data is recorded for the experiments. This is not true in higher-order arrays: the assignment

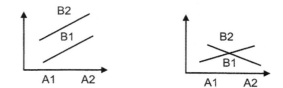

**Figure 7.4**   The plot of interactions of the example in Table 7.6.

**Table 7.7**  Interaction scenarios for L8 with partial factorial Resolution IV design

| Scenario I (factors 1, 2, equal and interacting) | | | Scenario II (factor 1 dominant) | | |
|---|---|---|---|---|---|
| Primary columns | Iteration columns | Missing interactions | Primary columns | Iteration columns | Missing interactions |
| 1, 2, 4, 7 | $3 = 1 \times 2$ | $1 \times 7$ | 1, 2, 4, 7 | $3 = 1 \times 2$ | $2 \times 4$ |
| | $5 = 1 \times 4$ | $2 \times 7$ | | $5 = 1 \times 4$ | $2 \times 7$ |
| | $6 = 2 \times 4$ | $4 \times 7$ | | $6 = 1 \times 7$ | $4 \times 7$ |

scenarios of factors and their interactions cannot be changed once the experiment is designed and then carried out.

As the array size increases, the number of choices in the factor selections increases. Table 7.8 shows some of the scenarios available for the orthogonal array L16. The choice of interactions and factor grouping relationship depends on the DoE team visualization of the design to be optimized. Scenario I, which is a Resolution IV design, represents an equal relationship and importance of the first five primary factors, with 10 two-way interactions available for analysis. No higher order interactions analysis are available for scenario I. Scenario II, which is a Resolution V design, assumes that primary factor 1 is the most important, with all other primary factors interacting with it. Two other scenarios, III and IV, are given, which are a combination of the first two scenarios. Whichever of the four scenarios the team selects, the experiment data analysis proceeds on that assumption, and the other scenario results cannot be calculated. The team should be very careful when selecting the interaction scenario, and should spend an adequate amount of time brainstorming this issue.

Interactions have caused much confusion for DoE teams. If an interaction is to be considered, less primary factors can be used, which reduces the utility and economy of orthogonal arrays. For example,

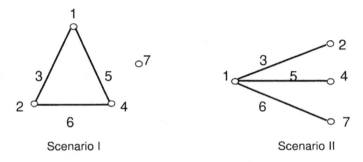

Scenario I                              Scenario II

**Figure 7.5**  Linear graphs for the interactions of L8 shown in Table 7.7.

**Table 7.8** Interaction scenarios for L16 partial factorial with confounding

| Scenario I (5 factors equal) | | Scenario II (factor 1 dominant) | |
|---|---|---|---|
| Primary columns | Iteration columns | Primary columns | Iteration columns |
| 1, 2, 4, 8, 15 | 3 = 1 × 2 | 1, 2, 4, 6, 8, 10, 12, 14 | 3 = 1 × 2 |
| | 5 = 1 × 4 | | 5 = 1 × 4 |
| | 6 = 2 × 4 | | 7 = 1 × 6 |
| | 7 = 8 × 15 | | 9 = 1 × 8 |
| | 9 = 1 × 8 | | 11 = 1 × 10 |
| | 10 = 2 × 8 | | 13 = 1 × 12 |
| | 11 = 4 × 15 | | 15 = 1 × 14 |
| | 12 = 4 × 8 | | |
| | 13 = 2 × 15 | | |
| | 14 = 1 × 15 | | |
| Scenario III | | Scenario VI | |
| Primary columns | Iteration columns | Primary columns | Iteration columns |
| 1, 2, 4, 8, 10, 12, 15 | 3 = 1 × 2 | 1, 2, 4, 5, 6, 8, 11, 12 | 3 = 1 × 2 |
| | 5 = 1 × 4 | | 7 = 1 × 6 |
| | 6 = 2 × 4 | | 9 = 1 × 8 |
| | 7 = 8 × 15 | | 10 = 2 × 8 |
| | 9 = 1 × 8 | | 13 = 1 × 12 |
| | 11 = 1 × 10 | | 14 = 5 × 11 |
| | 13 = 1 × 12 | | 15 = 4 × 11 |
| | 14 = 1 × 15 | | |

L27, which is 13 columns at three levels, can be used as a full factorial array with three columns. Using Resolution IV design, the number of factors can increase to seven primary factors and three two-way interactions (two columns for each interaction). If more than seven factors are to be used, then they should each be assigned to a column where the interaction is considered insignificant.

Interaction represents a mathematical value of the effect of one factor on others. If an assigned factor confounds an interaction column, then the analysis of the effect of that factor could be either negated or amplified by the interaction effect. In actuality, the effect of interactions is usually much smaller than expected. For DoE teams concerned with the confusion of interactions, noninteracting orthogonal arrays such as L12, L18, or L36 could be used. Table 7.9 shows the noninteracting array L12, otherwise known as the Plackett and Burman design. In these arrays, any third column does not confound the interaction of any two columns. L12 is a two-level array, whereas L18 is a combination of two- and three-level factors, popular in manufacturing DoEs, as shown in Table 7.10.

**Table 7.9**  Plackett and Burman L12 orthogonal array

| Experiment number | Factors | | | | | | | | | | |
|---|---|---|---|---|---|---|---|---|---|---|---|
| | 1 | 2 | 3 | 4 | 5 | 6 | 7 | 8 | 9 | 10 | 11 |
| 1 | 1 | 1 | 1 | 1 | 1 | 1 | 1 | 1 | 1 | 1 | 1 |
| 2 | 1 | 1 | 1 | 1 | 1 | 2 | 2 | 2 | 2 | 2 | 2 |
| 3 | 1 | 1 | 2 | 2 | 2 | 1 | 1 | 1 | 2 | 2 | 2 |
| 4 | 1 | 2 | 1 | 2 | 2 | 1 | 2 | 2 | 1 | 1 | 2 |
| 5 | 1 | 2 | 2 | 1 | 2 | 2 | 1 | 2 | 1 | 2 | 1 |
| 6 | 1 | 2 | 2 | 2 | 1 | 2 | 2 | 1 | 2 | 1 | 1 |
| 7 | 2 | 1 | 2 | 2 | 1 | 1 | 2 | 2 | 1 | 2 | 1 |
| 8 | 2 | 1 | 2 | 1 | 2 | 2 | 2 | 1 | 1 | 1 | 2 |
| 9 | 2 | 1 | 1 | 2 | 2 | 2 | 1 | 2 | 2 | 1 | 1 |
| 10 | 2 | 2 | 2 | 1 | 1 | 1 | 1 | 2 | 2 | 1 | 2 |
| 11 | 2 | 2 | 1 | 2 | 1 | 2 | 1 | 1 | 1 | 2 | 2 |
| 12 | 2 | 2 | 1 | 1 | 2 | 1 | 2 | 1 | 2 | 2 | 1 |

## 7.2.6  Multilevel arrangements and combination designs

The techniques for DoE designs using the orthogonal arrays for more than two or three levels are explored in this section. Multilevel arrangements can be made when columns representing factors are combined to form a new column with multiple levels. For example,

**Table 7.10**  L18 orthogonal array

| Experiment number | Factors | | | | | | | |
|---|---|---|---|---|---|---|---|---|
| | 1 | 2 | 3 | 4 | 5 | 6 | 7 | 8 |
| 1 | 1 | 1 | 1 | 1 | 1 | 1 | 1 | 1 |
| 2 | 1 | 1 | 2 | 2 | 2 | 2 | 2 | 2 |
| 3 | 1 | 1 | 3 | 3 | 3 | 3 | 3 | 3 |
| 4 | 1 | 2 | 1 | 1 | 2 | 2 | 3 | 3 |
| 5 | 1 | 2 | 2 | 2 | 3 | 3 | 1 | 1 |
| 6 | 1 | 2 | 3 | 3 | 1 | 1 | 2 | 2 |
| 7 | 1 | 3 | 1 | 2 | 1 | 3 | 2 | 3 |
| 8 | 1 | 3 | 2 | 3 | 2 | 1 | 3 | 1 |
| 9 | 1 | 3 | 3 | 1 | 3 | 2 | 1 | 2 |
| 10 | 2 | 1 | 1 | 3 | 3 | 2 | 2 | 1 |
| 11 | 2 | 1 | 2 | 1 | 1 | 3 | 3 | 2 |
| 12 | 2 | 1 | 3 | 2 | 2 | 1 | 1 | 3 |
| 13 | 2 | 2 | 1 | 2 | 3 | 1 | 3 | 2 |
| 14 | 2 | 2 | 2 | 3 | 1 | 2 | 1 | 3 |
| 15 | 2 | 2 | 3 | 1 | 2 | 3 | 2 | 1 |
| 16 | 2 | 3 | 1 | 3 | 2 | 3 | 1 | 2 |
| 17 | 2 | 3 | 2 | 1 | 3 | 1 | 2 | 3 |
| 18 | 2 | 3 | 3 | 2 | 1 | 2 | 3 | 1 |

combining factors 1 and 2 and their interaction column 3 in an L8 would allow for creating a substitute factor of four levels, as shown in Table 7.11. It is important to maintain the degrees of freedom (DoF) in this arrangement. DoF is the number of levels in the column minus 1. For an L8 with a four-level column, the column has DoF = 3. This is made up from three columns (1, 2, and 3) of two levels each (DoF = 1 for each two-level column). In an L16, the combination of columns 1, 2, and 4 and their interactions 3(12), 5(14), 6(24), and 7(124), shown in Table 7.4, can be combined to form a new column with eight levels and seven degrees of freedom.

If it is desired to use less than the four or eight multilevel arrangement, then only the desired number of levels are used, and some levels are repeated until the end level is reached. For example, if five levels are desired, then an L16 can be used with a combined column of the first seven columns, which has eight levels. The levels used in the combined column could be 1, 2, 3, 4, 5, 1, 2, 3. If one factor level is deemed important, then it can be multiply assigned such as 1, 2, 3, 4, 5, 4, 4, 4.

Combination designs can also be used for the insertion of two-level factors into a three-level orthogonal array, resulting in the ability to analyze more factors than originally available in the array. For example, a column representing two two-level factors could substitute one of the columns in an L9. In this case, the two columns $X$ and $Y$ of two levels each could be changed to a combined column of three levels

**Table 7.11**  Multilevel designs with L8 orthogonal arrays

| Experiment number | L8 original Factors | | | | | | | L8 multilevel factors | | | | |
|---|---|---|---|---|---|---|---|---|---|---|---|---|
| | 1 | 2 | 3 | 4 | 5 | 6 | 7 | A | 4 | 5 | 6 | 7 |
| 1 | 1 | 1 | 1 | 1 | 1 | 1 | 1 | 1 | 1 | 1 | 1 | 1 |
| 2 | 1 | 1 | 1 | 2 | 2 | 2 | 2 | 1 | 2 | 2 | 2 | 2 |
| 3 | 1 | 2 | 2 | 1 | 1 | 2 | 2 | 2 | 1 | 1 | 2 | 2 |
| 4 | 1 | 2 | 2 | 2 | 2 | 1 | 1 | 2 | 2 | 2 | 1 | 1 |
| 5 | 2 | 1 | 2 | 1 | 2 | 1 | 2 | 3 | 1 | 2 | 1 | 2 |
| 6 | 2 | 1 | 2 | 2 | 1 | 2 | 1 | 3 | 2 | 1 | 2 | 1 |
| 7 | 2 | 2 | 1 | 1 | 2 | 2 | 1 | 4 | 1 | 2 | 2 | 1 |
| 8 | 2 | 2 | 1 | 2 | 1 | 1 | 2 | 4 | 2 | 1 | 1 | 2 |

Conversion table

| Columns | | New column |
|---|---|---|
| 1 | 2 | A |
| 1 | 1 | 1 |
| 1 | 2 | 2 |
| 2 | 1 | 3 |
| 2 | 2 | 4 |

($X1Y1$, $X1Y2$ and $X2Y1$). The analysis of the L9 would be performed, resulting in determining the three levels of the two factors $X$ and $Y$. Individual factor effects could be further calculated as follows:

The main effect of $X$ at constant $Y$ level = $X2Y1 - X1Y1$
The main effect of $Y$ at constant $X$ level = $X1Y2 - X1Y1$

### 7.2.7  The Taguchi contribution to DoE

One of the pioneers in using DoE for new product design and manufacturing is Dr. Genishi Taguchi. Sometimes his name was synonymous with DoE, as in the "Taguchi methods." He is credited with transforming DoE from the realm of statisticians to be generally used by engineers and even production operators. His important contributions reduce the experiment design complexity and introduce new terminology to illustrate and simplify DoE concepts. These include the following:

* Ignore three-way interactions and above as in Resolution IV designs
* Use linear graphs to visualize interactions instead of interaction tables
* Use the signal to noise (S/N), to be discussed later, as a method to combine average and variability analysis
* Use $p\%$ contribution as a method to quantify the $F$ test in the ANOVA table
* Use quality loss function (QLF) as a methodology to optimize quality as discussed in Chapter 6 of this book

### 7.3  The DoE Analysis Tool Set

The DoE analysis tool set consists of using graphical as well as statistical analysis to determine which individual factors are significant, and how to set the quality characteristic to its design goal or reduce its variability.

The graphical analysis takes advantage of the Cramer's rule of the solution of simultaneous equations to solve for each value of factor levels. In the L9 orthogonal array in Table 7.5, it takes nine experiments to perform a solution of four factors at three-level saturated design. The average of the results of the first three experiments, $Y1$, $Y2$, and $Y3$, is the average performance of the product or process due to selecting level 1 of factor $A$, whereas the other factors negate themselves by averaging out their levels. The average of $Y2$, $Y5$, and $Y8$ is the effect of selecting level two of factor $B$. In this manner, the average of all 12 possible combinations (factors $A$, $B$, $C$, and $D$ and their levels 1, 2, and 3) is examined in terms of attaining the best result for the product or process specifications. For an L9 with $n$ repetitions, the level values for factor $A$ can be calculated as follows:

$$A1 = \frac{\sum_n Y1 + Y2 + Y3}{n \cdot 3}; A2 = \frac{\sum_n Y3 + Y4 + Y5}{n \cdot 3}; A3 = \frac{\sum_n Y6 + Y7 + Y8}{n \cdot 3}$$

$$(7.1)$$

The data can be plotted graphically so the intended results of either maximum, minimum, or targeted quality characteristic values can be used to manipulate the design to the intended or "expected" values.

The expected value (EV) of the DoE output is the result of applying all of the recommended levels. This is constructed from the overall experiment average, then the contribution of each recommended level is added to the EV. The contribution is the recommended level value minus the experiment average. The contribution of interactions can be calculated from the selected levels of primary factors.

The $EV$ value is usually calculated for significant factors only. The significant factors are determined by performing the $F$ test using the ANOVA analysis in the next section. The contribution of nonsignificant factors could be lost within the error of the experiment (the confidence interval of EV). If the selected factor levels are within the experiment design as one of the experiment lines, the expected value should equal the value attained by the experiment line, and no calculations are necessary. All expected values are bounded by the confidence interval of the error, as mentioned in Chapter 5.

### 7.3.1   Orthogonal array L9   saturated design example: Bonding process optimization

In this example, the specification for the peel strength for a mechanical bonding assembly was increased using an L9 DoE. An RTV adhesive was selected as the bonding agent (glue). Parts were cleaned prior to bonding in an ultrasonic bath filled with a cleaning chemical and a measured volume of glue was applied to both halves of the parts. Parts were then cured in an oven after bonding. The levels for cleaning time and chemical as well as curing temperature were arbitrarily selected in the design stage of the product. The product was not performing adequately in the field, as several parts separated during customer use. A DoE experiment was designed to increase the bonding process for maximum peel strength. It was decided to measure the peel strength by a special spring force tool, which is commonly used to determine the maximum outside force necessary to cause the parts to separate.

A DoE team was formed and the team decided to use a three-level L9 orthogonal array with four factors in order to maintain the current process settings as the middle level. The levels of the factors were then varied up and down in order to observe their effects. Four factors were considered in the L9 array: the cure temperature at 30°C (room temperature), 50°C (the current mild oven bake), and 70°C (a higher level of oven bake). The effect of ultrasonic cleaning, which was used in cleaning the parts, was to be tested after immersion in a chemical for 1, 3, and 5 minutes, with 3 minutes being the current time. The volume of RTV dispensed, using different dispensing heads, was varied around the current volume of 1.7 cc. The values were 1.2 cc, 1.7 cc, and 2.5 cc, respectively, which corresponded to commercially available dispensing heads. Finally, the soak chemical used in the ultrasonic bath was varied from the current methylene (MET) to other cleaners such as methyl ethyl ketone (MEK) and plain water ($H_2O$). Although other factors were considered, such as different bonding materials or the humidity of the bonding environment, the team did not elect to use these factors either because of cost of the material or resources needed to change the production environment.

The experiment was designed as shown in Table 7.12. There were nine experiments; each experiment was a unique combination of factor levels selected prior to the its running. For example, in experiment number 3, 30°C was the cure temperature in the oven, the RTV volume was 2.5 cc, the ultrasonic cleaner was MEK, and the cleaning time in the ultrasonic bath was 5 minutes. The experiment resulted in an average of 24.11 pounds of pressure applied before separating the assembly into two halves.

The graphical data analysis for the experiment only requires the use of a four-function calculator and is shown in Table 7.12 and Figure 7.6. The effects of each level were added, then averaged as to the contribution of each level of each factor. For improving the process, the highest average output for each factor level can be selected: A3 (70°C), B2 (3 minutes), C3 (2.5 cc RTV volume), and D1 (water). The following conclusions can be drawn from this experiment:

- The graphical analysis chart can be completed easily. Mathematical mistakes in the analysis table are minimized because all factor averages must add up to the same number (217 in this case).
- The most important factor is the cure temperature (factor A), since it causes the most change in output.

**Table 7.12**  Bonding process DoE

| Factors selected | Levels of each factor | | | |
| --- | --- | --- | --- | --- |
| $A$ = cure temperature | 30 | 50 | 70 | Degrees |
| $B$ = ultrasonic cleaning | 1 | 3 | 5 | Minutes |
| $C$ = RTV volume | 1.2 | 1.7 | 2.5 | CC |
| $D$ = soak chemical | $H_2O$ | MET | MEK | |

| Experiment number | L9 (3 × 4) Orthogonal array saturated design | | | | | | | | |
| --- | --- | --- | --- | --- | --- | --- | --- | --- | --- |
| | $A$ | $B$ | $C$ | $D$ | $A$ | $B$ | $C$ | $D$ | Peel force |
| 1 | 1 | 1 | 1 | 1 | 30 | 1 | 1.2 | $H_2O$ | 11.5 (lbs.) |
| 2 | 1 | 2 | 2 | 2 | 30 | 3 | 1.7 | MET | 22.7 |
| 3 | 1 | 3 | 3 | 3 | 30 | 5 | 2.5 | MEK | 22.6 |
| 4 | 2 | 1 | 2 | 3 | 50 | 1 | 1.7 | MEK | 19.0 |
| 5 | 2 | 2 | 3 | 1 | 50 | 3 | 2.5 | $H_2O$ | 28.5 |
| 6 | 2 | 3 | 1 | 2 | 50 | 5 | 1.2 | MET | 24.0 |
| 7 | 3 | 1 | 3 | 2 | 70 | 1 | 2.5 | MET | 25.1 |
| 8 | 3 | 2 | 1 | 3 | 70 | 3 | 1.2 | MEK | 30.3 |
| 9 | 3 | 3 | 2 | 1 | 70 | 5 | 1.7 | $H_2O$ | 33.3 |

Summing all experiments with the same factor levels:

| Factor | $A$ | $B$ | $C$ | $D$ |
| --- | --- | --- | --- | --- |
| Level 1 | 56.8 | 55.6 | 65.8 | 73.3 |
| Level 2 | 71.5 | 81.5 | 75.0 | 71.8 |
| Level 3 | 88.7 | 79.9 | 76.2 | 71.9 |
| Total | 217.0 | 217.0 | 217.0 | 217.0 |

Averaging all experiments with the same factor levels:

| Factor | $A$ | $B$ | $C$ | $D$ |
| --- | --- | --- | --- | --- |
| Level 1 | 18.9 | 18.5 | 21.9 | 24.4 |
| Level 2 | 23.8 | 27.2 | 25.0 | 23.9 |
| Level 3 | 29.6 | 26.6 | 25.4 | 24.0 |
| Average | 24.1 | 24.1 | 24.1 | 24.1 |

Set parameters to maximum value each level: $A3$, $B2$, $C3$, and $D1$

Contribution is additive yielding expected value (EV):
EV = Experiments average + ($A3$ + $B2$ + $C3$ + $D1$) contribution
EV= 24.1 + (29.6 - 24.1) + (27.2 - 24.1) + (25.4 - 24.1) + (24.4 - 24.1)
EV= 24.1 + 5.5 + 3.1 + 1.3 + 0.3        = 34.3 ± confidence interval

- The least important factor is the soak chemical (factor $D$), since it hardly made a difference.
- The expected value (EV) of the peel strength obtained by using the factor levels selected can be estimated by adding up the contributions of the four factors ($A3$ + $B2$ + $C3$ + $D1$): 24.1 + 5.5 + 3.1 + 0.3 = 34.3 lbs. This represents approximately a 50% increase over the average of all experiments (217/9 = 24.11).

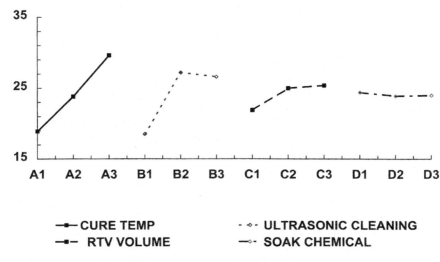

**Figure 7.6**  Bonding process DoE graphical analysis.

- The combination of the selected levels for the four factors (*A*3, *B*2, *C*3, and *D*1) is not within the L9 array table (it is within the 81 experiments of the full factorial array). By using saturated design, it was only necessary to perform nine experiments instead of 81 in order to find the optimum set of factors levels. However, the interaction of the factors cannot be calculated, and the confounding of the factors might render some of the factor effects incorrect.
- The selection of the factors appears to be appropriate, since the goal of increasing the peel force was successfully achieved.

### 7.3.2  Graphical analysis conclusions

As can be seen by the example above, Design of experiments can optimize a process or a product easily and quickly by using very simple mathematical techniques. It is also not necessary to have an in-depth understanding of the physics or the chemistry of the process or product to be optimized.

This particular example illustrates how the process average can be shifted to the desired level, in this case to the maximum possible. A similar method can be applied to reduce the variability, with several replications for each experiment line. Four replications are preferable for more than 30 points of analysis, approximating the population distribution of bonding. A mathematical transformation can convert the four numbers into a single number indicating variability. The graphical analysis for variability can be performed on the transformed number. The two analyses for average and variability can be contrasted and factor level selected for the most efficient process improvement through trade-offs of average and variability, if any.

There are two important terms used in DoEs. One is the design space, which is the limit of the investigation of the factors, as bounded by the selection of the levels for each factor. The other is the "direction of steepest ascent." This is direction of increasing or decreasing the amount of factor level values when expanding the current DoE analysis results in future DoEs.

In the design space for the bonding DoE, the selection of the levels for factor $B$, the time for ultrasonic cleaning, was optimal, as shown by Figure 7.6. The best-level position was in the middle of the three levels. The maximum point can be calculated by drawing a best fit curve through the three points and thus can be determined accurately, rather than declaring that 3 minutes (level 2) are better than 1 or 5 minutes (levels 1 and 3) of cleaning. A second-order equation can be fitted through the three points and the maximum point can be determined by setting the derivative to zero.

For factor 1, the oven temperature, the design space is not optimal. It can be seen from Figure 7.6 that the level 3 temperature (70°C) results in the highest peel force. But what happens if the oven temperature is higher than 70°C? The current design space does not allow for any conclusions regarding higher temperatures than 70°C. If more information is desired regarding the bonding process, then a second DoE could be performed. Some factors could be expanded in the direction of steepest ascent such as having higher temperature levels, while other factors could be dropped from the experiment (such as chemical used) in favor of partial or full factorial analysis.

### 7.3.3 Analysis of DoE data with interactions: Electrical hipot test L8 partial factorial Resolution IV example

In this example, a DoE was used to increase the specification tolerance for an electronic design by investigating different design methods and material selections. An electronic box with a display monitor has performed poorly in high potential or "hipot" testing. In this test, a high-voltage probe is allowed to make contact with the box and the monitor was observed for degradation (flickering) of the screen pattern. A L8 DoE was used in order to increase the voltage at which the monitor flickers when exposed to the high-voltage probe, and thus improve the performance of the design against noise conditions of high voltage and sparking caused by the customer or the product use environment.

The L8 experiment was performed with four primary factors being considered, in a partial factorial design with Resolution IV. The remaining three columns were used to measure the interaction of the selected factors, using scenario I in Table 7.7. The four primary factors consisted of using two different connectors ($X$ or $Y$) to connect the box to the monitor, two connection methods (spring or screw) for the connectors, whether to use a metal shim to seal the box cover (0 or 1 shims), and whether to paint the inside of the box with a conductive paint to ground out the high voltage. In the last factor, the two levels considered were "yes" or "no" paint.

The selection of the factors and the DoE design and layout are given in Table 7.13. The probe voltage value causing the screen to flicker

**Table 7.13**  Hipot DoE experiment

| Factors selected | Levels of each factor | | |
|---|---|---|---|
| $A$ = Connector type | $X$ | or | $Y$ |
| $B$ = Different contact methods | Spring | or | Screw |
| $D$ = Conductive paint the box | Yes | or | No |
| $G$ = Number of shims | 0 | or | 1 |

| Experiment number | $A$ $A$ | $B$ $B$ | $C$ $AB$ | $D$ $D$ | $E$ $AD$ | $F$ $BD$ | $G$ $G$ | DoE Factors | | | | |
|---|---|---|---|---|---|---|---|---|---|---|---|---|
| | | | | | | | | Connector | Contact | Paint | No. of shims | Probe Kvolts |
| 1 | 1 | 1 | 1 | 1 | 1 | 1 | 1 | $X$ | Spring | Yes | 0 | 18.5 |
| 2 | 1 | 1 | 1 | 2 | 2 | 2 | 2 | $X$ | Spring | No | 1 | 14 |
| 3 | 1 | 2 | 2 | 1 | 1 | 2 | 2 | $X$ | Screw | Yes | 1 | 18.5 |
| 4 | 1 | 2 | 2 | 2 | 2 | 1 | 1 | $X$ | Screw | No | 0 | 12.5 |
| 5 | 2 | 1 | 2 | 1 | 2 | 1 | 2 | $Y$ | Spring | Yes | 1 | 18.5 |
| 6 | 2 | 1 | 2 | 2 | 1 | 2 | 1 | $Y$ | Spring | No | 0 | 13 |
| 7 | 2 | 2 | 1 | 1 | 1 | 2 | 1 | $Y$ | Screw | Yes | 0 | 9.5 |
| 8 | 2 | 2 | 1 | 2 | 2 | 1 | 2 | $Y$ | Screw | No | 1 | 8 |

Averaging all experiments:

| Factor | Connector | Contact | Paint | Shims | Conn · Contact | Conn · Paint | Contact · Paint |
|---|---|---|---|---|---|---|---|
| | $A$ | $B$ | $D$ | $G$ | $C = AB$ | $E = AD$ | $F = BD$ |
| Level 1 | 15.875 | 16.00 | 16.250 | 13.375 | 12.500 | 14.500 | 14.375 |
| Level 2 | 12.250 | 12.125 | 11.875 | 14.750 | 15.625 | 13.625 | 13.750 |
| Contribution | +1.8125 | +1.9375 | +2.1875 | −0.6875 | −1.5625 | +0.4375 | +0.3125 |
| Average | 14.0625 | 14.0625 | 14.0625 | 14.0625 | 14.0625 | 14.0625 | 14.0625 |

Set factors to $A1$ (connector $X$), $B1$ (spring), $D1$ (paint), and $G2$ (1 shim) to improve hipot specification.

Contribution is additive yielding expected value (EV):

EV = Experiments average + ($A1 + B1 + D1 + G2$) primary contributions + interaction contributions ($C2 + E1 + F1$)

EV = 14.0625 + (1.8125 + 1.9375 + 2.1875 + 0.6875) + (−1.5625 + 0.4375 + 0.3125) = 19.8750 Kvolts was recorded in Kvolts. The average and expected value analysis was performed using the primary factors and their interactions. The resulting expected value, at 19.875 Kvolts is almost 50% greater than the experiment average of 14.06 Kvolts. Production units will more readily achieve six sigma quality with a 41% wider specification.

The graphical analysis of data is given in Figure 7.7. The first three primary factors (connector, contact, and paint) have a much greater effect on the design than the fourth factor (number of shims). In addition, only one interaction was larger than the rest, the interaction of connector type and contact method. This strong interaction indicates that the connector type and the connection method should be treated as one combination. The statistical analysis to be explored in the next section could determine which of these factors or interactions are significant.

### 7.3.4   Statistical analysis of DoEs

Statistical analysis of DoEs is based on the analysis of variance (ANOVA), which is a method of determining the significance of each factor in terms of its effects on the output quality characteristic(s). The ANOVA analysis apportions the total effect of the output characteristic average and variability to each factor in the orthogonal array. The significance test is based on the $F$ distribution, which is a ratio of the degrees of freedom for the factor divided by the degrees of freedom for the error. The least significant factors are lumped together as the error of the experiment, since they are not important in affecting the output characteristic.

The terms for determining the ANOVA table for $n$ total values in the DoE experiment are given as follows:

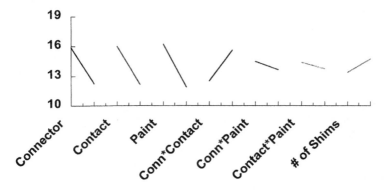

**Figure 7.7**   Hipot design DoE graphical analysis.

Total sum of the squares $(SS_T) = \Sigma(Y_i - Y_{\text{averge}})^2 = \Sigma Y_i^2 - (\Sigma Y_i)^2/n$    (7.2)

$(\Sigma Y_i)^2/n$ is sometimes called the correction factor.

Sum of squares for each factor $(SS_F) = (\Sigma Y_{\text{level 1}})^2/n_{\text{level 1}}$
    $+ (\Sigma Y_{\text{level 2}})^2/n_{\text{level 2}} + \ldots -(\Sigma Y_i)^2/n$ (repeat for as many levels)    (7.3)

Degrees of freedom (DoF):

DoF Total = number of data points – 1

DoF Orthogonal Array = number of experiments – 1

DoF Factor = number of levels – 1    (7.4)

DoF Error = Total DoF – DoF of significant factors and interactions

DoF Interaction = product of the DoF of each factor

= 1 for two-level factor interactions

= 2 for three-level factor interactions

Variance $(V) = SS/\text{DoF}$

(also called mean square deviation or MSD). Also,

$$V_T = = \sigma^2_{\text{total experiment}}$$    (7.5)

$$F \text{ ratio for each factor} = V_F/V_{\text{error}}$$    (7.6)

Modified sum of squares for each factor $(SS_F') = SS_F - V_{\text{error}} \cdot \text{DoF}_F$
    (7.7)

$$\text{Percentage contribution } (p\%) = SS_F'/SS_T$$    (7.8)

The $F$ test values are given in Table 7.14 for a confidence level of 95% and the DoF of the factor versus the DoF of the error. The $F$ test is used to determine the significance of the calculated variances. It is a ratio of the factor variance over the error variance. The error of the DoE experiment could be obtained from either of the following:

1. Replicate the whole experiment, generating error due to repetition.
2. For single repetition DoE results data, the smallest factors or interactions (with the smallest $SS_F$) can be used as the error, especially higher-order interactions.
3. Replicate the center point of the design space of the experiment.
4. Replicate some points of the experiments, such as the endpoints of the design space.

For a given confidence level, the $F$ test determines whether the effect of a factor is due to chance or due to the factor itself (the factor is deemed significant). If a factor's $F$ ratio value is less than the value in

**Table 7.14**  *F* table value for 95% confidence or 0.05 significance

| DoF error | \multicolumn{6}{c}{DoF factors} | | | | | |
|---|---|---|---|---|---|---|
|  | 1 | 2 | 3 | 4 | 5 | 6 |
| 1 | 161.4 | 199.5 | 215.7 | 224.6 | 230.2 | 234.0 |
| 2 | 18.51 | 19.00 | 19.16 | 19.25 | 19.30 | 19.33 |
| 3 | 10.13 | 9.55 | 9.28 | 9.12 | 9.01 | 8.94 |
| 4 | 7.71 | 6.94 | 6.59 | 6.39 | 6.26 | 6.16 |
| 5 | 6.61 | 5.79 | 5.41 | 5.19 | 5.05 | 4.95 |
| 6 | 5.99 | 5.14 | 4.76 | 4.53 | 4.39 | 4.28 |
| 7 | 5.59 | 4.74 | 4.35 | 4.12 | 3.97 | 3.87 |
| 8 | 5.32 | 4.46 | 4.07 | 3.84 | 3.69 | 3.58 |
| 9 | 5.12 | 4.26 | 3.86 | 3.63 | 3.48 | 3.37 |
| 10 | 4.96 | 4.10 | 3.71 | 3.48 | 3.33 | 3.22 |
| 15 | 4.54 | 3.68 | 3.29 | 3.06 | 2.90 | 2.79 |
| 20 | 4.35 | 3.49 | 3.10 | 2.87 | 2.71 | 2.60 |
| 25 | 4.24 | 3.39 | 2.99 | 2.76 | 2.60 | 2.49 |
| 30 | 4.17 | 3.32 | 2.92 | 2.69 | 2.53 | 2.42 |
| 60 | 4.00 | 3.15 | 2.76 | 2.53 | 2.37 | 2.25 |
| $\infty$ | 3.84 | 3.00 | 2.60 | 2.37 | 2.21 | 2.10 |

the *F* table given the DoF of factor and error, then it is deemed not significant and can be pooled into the error. The *F* ratios are then recalculated and the *F* test redone on the remaining factors. When a factor is significant to less than 0.05 (or the confidence is greater than 95%), then the probability of this factor affecting the experiment happens 5% by chance or once every 20 times. Since this is remote in nature, the factor must be significant, and hence it affects the experiment outcome.

The last two terms (7.7 and 7.8) in the ANOVA table were developed by Taguchi to simplify the pooling process. Instead of using the significance based on the *F* table as the source of pooling, Taguchi suggested pooling a factor if its percent contribution is less than 5%.

### 7.3.5  Statistical analysis of the hipot experiment

For the hipot experiment, the initial ANOVA table is constructed in Table 7.15. An example is given at the top of how to calculate the sum of the squares for factor $A$. In order to calculate the *F* ratio, each variance must be compared against the error variance. Since all columns are used, and there is no repetition of the experiment, the factor with the smallest $SS_F$ is used as the source of error. This is the interaction $B \times D$, or contact method × paint, with a sum of the squares ($SS_{B \times D}$) of 0.79. When the *F* ratios are calculated for the remaining factors, not a single factor was more than 95% significant. Therefore, pooling is necessary to increase significance.

**Table 7.15**   Hipot design ANOVA statistical analysis

$SS_A = (\Sigma Y^2_{\text{level 1}})/n_{\text{level 1}} + (\Sigma Y^2_{\text{level 2}})/n_{\text{level 2}} - (\Sigma Y)^2/N$

$SS_A = 1/4 \, [(18.5 + 14 + 18.5 + 12)^2 + (18.5 + 13 + 9.5 + 8)2] - 112.5^2/8 = 26.28$

| Source | DoF | Sum of squares | Mean of squares | $F$ value* |
|---|---|---|---|---|
| $A$ | 1 | 26.28 | 26.28 | 33.27 |
| $B$ | 1 | 30.03 | 30.03 | 38.01 |
| $A \times B$ | 1 | 19.53 | 19.53 | 24.72 |
| $D$ | 1 | 38.28 | 38.28 | 48.45 |
| $A \times D$ | 1 | 1.53 | 1.53 | 1.94 |
| $G$ | 1 | 3.78 | 3.78 | 4.78 |
| Error ($B \times D$) | 1 | 0.79 | 0.79 | |
| Total | 7 | 120.22 | 17.17 | |

*No factors is better than 95% confidence level

Pooling starts with the smallest remaining sum of the squares ($SS_F$) being added to the error $SS$ to see if significance is achieved for the experiment. The process is continued until no greater significance is achieved. The insignificant factors are combined with the error to obtain the pooled error. In this manner, $G$ and $A \times D$ are pooled with error $B \times D$. This implies that these factors, consisting of shims, and the two interactions connector × paint and contact × paint are not significant, showing that the paint operation is independent of the rest of the factors. Only the four factors shown in Table 7.16 are significant: connector type, contact method, their interaction, and painting the box. This clearly matches the observed values in the graphical plot of factors in Figure 7.7. The factors can be ranked in importance according to the percent contribution: paint, contact method, connector, and the interaction of connector × contact method. The total percent contribution of the error is less than 12%, indicating good confidence in the experiment. If the error percent is greater than 30%, the significance of the total DoE experiment is lessened.

**Table 7.16**   Hipot design ANOVA statistical analysis with pooled error

| Source | DoF | Sum of squares | Mean of squares | $F$ value | $SS'$ | $p\%$ |
|---|---|---|---|---|---|---|
| $A$ | 1 | 26.28 | 26.28 | 12.95 | 24.25 | 20.2 |
| $B$ | 1 | 30.03 | 30.03 | 14.79 | 28.00 | 23.3 |
| $A \times B$ | 1 | 19.53 | 19.53 | 9.62 | 17.5 | 14.6 |
| $D$ | 1 | 38.28 | 38.28 | 18.86 | 36.25 | 30.1 |
| Pooled error | 3 | 6.10 | 2.03 | | 14.21 | 11.8 |
| Total | 7 | 120.22 | 17.17 | | 120.22 | 100% |

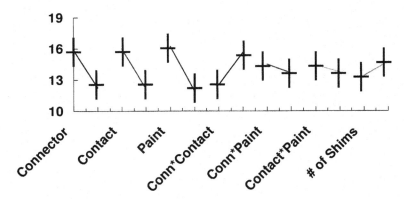

**Figure 7.8**   Visualizing the error of the hipot experiment.

An interesting method for visualizing the error of the hipot experiment is shown in Figure 7.8. The confidence interval of the error, as measured by the $3\sigma_{error}$ from the error variance, is shown superimposed on the graphical plot of the factors and levels. It can be seen that in factors that are not significant, the error span does not allow for distinguishing between the two levels of these factors. Given this pooling and significance information, the expected value should be recalculated as follows:

$EV$ = experiments average + contributions of $A1$, $B1$, $D1$, and $C2$

$EV = 14.0625 + 1.8125 + 1.9375 + 2.1875 + 1.5625 = 18.4375$ Kvolts

### 7.4   Variability Reduction Using DoE

Variability reduction, which is an important goal of six sigma quality, can also be successfully achieved by DoE. The techniques for reducing variability are the same as discussed in the DoE analysis section for the hipot experiment, with the exception of one additional step: the experiment array has to be replicated several times in order to quantify variance of data for each line in the experiment. A technique is needed to convert the repetition of each experiment line into a single number signifying variability. Once the conversion has been achieved, then the analysis can proceed similarly to the examples mentioned above. Several conversion schemes are available:

1. Signal-to-noise techniques (S/N). This technique was introduced by Taguchi, and uses a combination formula depending on whether the quality characteristic is to be made equal to a nominal target,

as small as possible with zero as the target, and as large as possible with infinity as the target. The conversion formulas are dependent on the three conditions mentioned and the number of repetitions $n$. Note that in all cases, the desired level in all the formulas is the one with the largest positive value. They are:

$$S/N = -10 \cdot \log_{10}\left(\frac{1}{n}\sum_{i=1}^{n} y_i^2\right), \text{ for smaller is better} \tag{7.9}$$

$$S/N = -10 \cdot \log_{10}\left(\frac{1}{n}\sum_{i=1}^{n}\frac{1}{y_i^2}\right), \text{ for larger is better} \tag{7.10}$$

$$S/N = 10 \cdot \log_{10}\frac{\mu^2}{s^2}; \mu = \frac{1}{n}\sum_{i=1}^{n} y_i; s^2 = \frac{1}{n-1}\sum_{i=1}^{n}(y_i - \mu)^2, \text{ for nominal} \tag{7.11}$$

2. Coefficient variation squared (CVS). This is similar to the S/N formula for the nominal (7.11). It is based on the relationship of the average versus the standard deviations.

3. Log variance conversion. In this case, the formula is equal to $-10 \cdot \log_{10}$(variance) or $-10 \cdot \log_{10}\sigma^2$. When the target is zero or minimum, this formula is equal to the S/N smaller is better equation (7.9). In the case of the repetitions having the same value for the quality characteristic in an experimental line, the $\sigma$ will be equal to zero, and Equation 7.9 will become infinite due to the logarithm term. In that case, the two numbers should be made slightly different so the calculations can proceed.

4. Mean square error (MSE). In this case, the distance from the result to the target ($T$) is used to minimize shift. The formula is

$$-10 \log\left[\frac{1}{n}\sum(y_i - T)^2\right] \tag{7.12}$$

Any of these conversion methods can be used to either reduce variability, independent of average or using a common average and variability formula such as S/N or CVS. In the first case, two mathematical analyses have to be performed, one for the average and the other for variability. In the second case, only one analysis is sufficient. However, most engineers prefer to perform both analyses so they can examine the level selection independently of improving the average or the variability. The necessary trade-offs can then be made by the engineers in choosing the proper factor levels.

The result of using these conversion formulas is to express in a single number the variability of the output quality characteristic(s). This number can then be treated in a manner similar to the analysis for improving the average, as was done in the peel strength and hipot DoE examples. In the peel strength example, repeating the experiments four times would produce four sets of results for each line of experiments. The signal-to-noise (S/N) ratio for each experiment line would be calculated from the formula for S/N for the four repetitions.

The number of repetitions is dependent on the external conditions to be simulated by the DoE, called noise factors. Unit-to-unit variability can be simulated by several repetitions of the experiment. Specific noise conditions and their levels can determine the number of repetitions to be performed. Three noise factors, with two levels each, will mean six repetitions of the experiments. As was indicated in the DoE methodology, an orthogonal array can be used in the outer array to reduce the number of repetitions: from 6 to 4 using an L4 array. An example of using variability reduction will be given in the next chapter.

DoEs can be used for the tolerance analysis of all the factors. An orthogonal array can be repeated three times for each tolerance set of each factor (nominal, USL, and LSL). If there are four factors, the experiments will have to be repeated 12 times. Although this technique is rather lengthy, it could indicate whether some of the tolerances are significant or not, and therefore could be altered accordingly.

## 7.5   Using DoE Methods in Six Sigma Design and Manufacturing Projects

One of the most important consequences of implementing six sigma has been the increased use of DoEs by the design engineering community. DoEs can be used effectively to augment the traditional design engineering methods of computer simulation and analysis of worst-case design and materials selection. The DoE techniques outlined in this chapter can be used effectively for new product quality improvement as well as manufacturing process variability reduction. Several opportunities for using DoE for design engineering are:

- Worst-case study is the method by which engineers analyze designs using a combination of the worst case of the individual parts or materials specification limits. Design engineers might overspecify parts to tighter tolerances to ensure that they meet worst-case conditions. DoE methods can be used to analyze design tolerances, resulting in the proper specification of parts. Expensive tight tolerance parts should be used only when actually needed for the design to meet the specifications.

- DoE methods can be used in computer simulation of the design to obtain optimal results. The orthogonal array experiment conditions can be inputted into the simulation. The results could then be analyzed as to the optimal design.

- DoEs can be used in new products to solve some of the "black magic" type of problems specific to electronic products, including the successful completion of environmental and transportation tests. Examples are reductions in electrical noise and radio frequency interference (RFI), and product mounting, shipping, and packaging techniques

- DoEs can be used effectively by multidisciplinary teams that need to work together to achieve performance to specifications for new products through trade-offs in design disciplines. A thermal printer case study is used to illustrate this use of DoE for new products in the next chapter.

- DoEs can be used for robust designs to achieve a linear region of performance of the factors for the quality characteristic. By selecting this linear region, the design is less sensitive to small factor changes, and hence less rigorous specifications can be used for the factors.

## 7.6   Conclusions

It has been shown, through several examples, that DoE is an excellent tool for optimizing designs by shifting the average characteristic(s) of the design to target and reducing variability. Both of these actions are very important in achieving six sigma quality. The mathematical background for DoE is a mix of tools of orthogonal arrays, designed experiments, and analysis of variance. There are several techniques in DoEs that should be thought out well in advance: the definition of the characteristics to be optimized, the selection of factors and levels, the treatment of factor interactions, the selection of experiment arrays, and how to simulate and measure variability and error.

An initial DoE project should be selected carefully to optimize a design that is relevant but not too complex. Careful hand calculations should be made to complete the analysis. Only after initial successes should software-based methods of analysis be attempted.

## 7.7   References and Bibliography

Box, G., Bisgaard, S., and Fung, C. *An Explanation and Critique of Taguchi's Contribution to Quality Improvement.* Report from Center for Quality and Productivity Improvement, University of Wisconsin, 1987.

Box, G. "Studies in Quality Improvement: Signal to Noise Ratio, Performance

Criteria and Transformation." Report No. 26, Center for Quality and Productivity Improvement, University of Wisconsin, 1987.

Cochran, W. and Cox, G. *Experimental Designs, 2nd ed.* New York: Wiley, 1981.

Diamond, W. J. *Practical Experiment Design.* New York: Van Nostrand Reinhold, 1981.

Ealy, L. "Taguchi Basics." *Quality Journal,* November 1988, 26–30.

Guenther W. *Concepts of Statistical Interference.* New York: McGraw-Hill, 1973.

Hicks, C. *Fundamental Concepts in the Design of Experiments.* New York: McGraw Hill, 1964.

Holusha, J. "Improving Quality the Japanese Way." *The New York Times,* July 20th, 1988, D7.

John, P. *Statistical Analysis and Design of Experiments.* New York: Macmillan, 1971

Lipson, C. and Sheth, N. *Statistical Design and Analysis of Engineering Experiments.* New York: McGraw-Hill, 1973.

Ross, P. *Taguchi Techniques For Quality Engineering.* New York: McGraw Hill, 1987.

Roy, R. *A Primer on the Taguchi Method.* New York: Van Nostrand Reinhold, 1990.

Phadke, M. *Quality Engineering Using Robust Design.* Engelwood Cliffs, NJ: Prentice-Hall, 1989.

Shina, S. "Reducing Solder Wave Defects Using the Taguchi Method." In *American Supplier Institute, Sixth Symposium,* Dearborn, Michigan, October 1988.

Shina, S. and Capulli, K. "Alternatives for Cleaning Hybrid Integrated Circuits Using Taguchi Methods." In *Nepcon East Conference,* Boston, MA, June 1990.

Shina, S. and Wu, J. "Optimization the new HOLLIS wave solder machine." In *American Supplier Institute Seventh Symposium,* Phoenix, Arizona, October 1989.

Taguchi, G. *Introduction to Quality Engineering.* Tokyo: Asian Productivity Institute. 1986.

Taguchi, G. El Sayed, E., and Hsiang, T. *Quality Engineering in Production Systems.* New York: McGraw-Hill, 1988.

Taguchi, G. *System of Experimental Design.* White Plains, NY?: NIPUB—Kraus International Publications, 1976.

Young, H. *Statistical Treatment of Experimental Data.* New York: McGraw-Hill, 1962.

# Six Sigma and Its Use in Analysis of Design and Manufacturing for Current and New Products and Processes

The strategy for the implementation of six sigma quality in current product manufacturing is quite different form the strategy of setting and achieving six sigma quality goals for new products. Half of the six sigma ratio, the product specifications, is usually fixed for current products, since the cost of altering them and retesting the product designs would be too prohibitive. Reaching six sigma for current products that were designed without a formal quality program is difficult, especially since the other half of the six sigma ratio, that of reducing variability, is the only option available. For new products, the opportunity to influence both sides of the six sigma ratio is much greater; hence, achieving six sigma is easier. This chapter will focus on achieving six sigma quality for both current and new products. The topics discussed in this chapter are:

1. Current product six sigma strategy. In Section 8.1, the strategy to attain six sigma for current products is developed by gradually using different tools as the manufacturing quality improves with time. Examples will be given in several manufacturing areas to demonstrate the evolution of quality in manufacturing.
2. Transitioning new product development to six sigma. In Section 8.2, the efforts required by the manufacturing and design teams to

quantify the steps necessary to attain six sigma in new products are shown. Some of these efforts include process capability studies, whereas others are more qualitative and involve design guidelines for reducing defects in new designs

3. Determining six sigma quality in different design disciplines. Six sigma is a tool for both design and manufacturing. Previous chapters have shown how to determine six sigma from the product specifications and manufacturing variability, as well from manufacturing reject rates. In Section 8.3, designs from different disciplines will be analyzed for six sigma quality and their capability calculated with detailed examples.

4. Using six sigma quality for new product introduction. In Section 8.4, the use of six sigma to determine overall new product introduction strategy and the use of quality tools to help achieve six sigma quality and defect removal goals will be shown at the product and system levels.

## 8.1    Current Product Six Sigma Strategy

The quality of current products and manufacturing processes is dependent on their history and original design parameters. In many cases, the products and the manufacturing operations used to produce them were not created with six sigma quality in mind. It is very difficult to achieve six sigma when that was not one of the goals at the very start of the product development process.

The road to higher quality begins with understanding current quality levels, then working with a plan to incrementally increase quality until the goal is achieved. A hierarchy of tools could be used at different stages of quality. Figure 8.1 is a good example of successive quality improvements that can be used as a roadmap for improving quality in current operations. It was used to improve soldering process quality from unacceptable defect rates to six sigma quality. The progression was accomplished through several phases:

1. The TQM (total quality management) phase. This phase is shown on the left of Figure 8.1 and should be used in situations where it is obvious that the manufacturing quality is out of control. This may be due to a large influx of new production operators or a ramp-up of production volume. The goals of this phase should be to stabilize the quality of production by investing in operator training and the operational aspect of production. New support staff should be recruited, process documentation inspected and improved, and training of operators and line management increased. The quality methodology

PPM

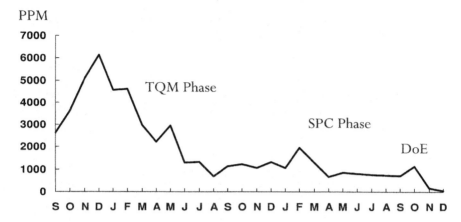

**Figure 8.1**  Progression of quality tools for existing products.

to be used in this phase is the process improvement tools discussed in Chapter 3. The improvements in this phase are in operational issues, and hence tend to be gradual. They can reach a plateau if no changes are made to the material, equipment, or processing parameters of production. At the end of this phase, it is expected that a quality plateau to be reached will be around the three sigma quality or Cpk = 1. This would result in defect rates of 300–3000 PPM. To ensure outgoing quality in this phase, a large inspection and test staff is used to remove defects generated by production. It is estimated that up to 40% of the direct labor expended in this phase is consumed by inspection, rework, and testing.

2. The SPC (statistical process control) phase. In this phase, the manufacturing process is stabilized and control methods discussed in Chapter 3 are used to ensure that the process remains in control. Tools such as control charts to monitor production quality and sampling methods for incoming inspection are used to ensure that defects in material or lapses in processing methods are caught early in the manufacturing cycle and corrected promptly. The management goals in this phase are to increase the communications between the different production operations, the supply chain, and the customer. This will allow for quick reaction to quality problems throughout the organization, and reveal long-term trends. The TQM efforts will continue in this phase, improving the quality and reducing defect rates incrementally. The quality levels and defect rates will continue at the same rate if no additional investments are made in materials and equipment.

3. The DoE (design of experiments) phase. In this phase, the management sets more aggressive quality improvement goals. There are greater investments in several areas, including more statistical and complex quality training tools such as DoE, DFM (design for manufacture), and QFD (quality function deployment). More technical support staff, such as process engineers, are hired and trained to use these tools. Mandated quality improvement projects are prescribed, such as one DoE experiment for every team at least once per year, or performing process DFMA on every production operation. Purchasing of new equipment or materials is encouraged when economically justified. Communications loops are tightened, and reactions to quality problems are expected to be instantaneous. Examples would be the use of red lights in production to summon engineers and managers in case of a problem; production line stopping authority given to certain operators when they detect problems; quality alerts to the field and customers, and instant or 24 hour supply chain communications to share information on quality problems and design changes. The typical goals set for this phase are at four sigma quality or Cpk = 1.3. That results in a defect rate in the range of 20–200 PPM. Once this level is achieved, focused quality projects should be initiated to target specific defect problems and bring the quality closer to six sigma, as explained in the next section.

### 8.1.1  Process improvement in current products

In current products, most of the process and product improvements should be concentrated on specific high-defects problems. A Pareto chart should be made of the top ten problems, and projects such as DoE or process DFMA initiated to rectify these problems. In many cases, good results can be quickly achieved using these tools, especially if they focus on a particular problem that requires more specialized operating conditions than the rest of production.

An example of a focused problem that can be resolved by a quality improvement project is a PCB assembly produced with special requirements. Such a case is outlined in Table 8.1 and Figures 8.2 and 8.3. This case study involves PCBs that are double-sided with mixed technology of through-hole and SMT components. The PCBs were wave soldered, resulting in poor quality. A cause-and-effect analysis shown in Figure 8.2 was performed on the problem and it was concluded that SOT-23 SMT bottom side components were the most likely reason for the defects because they resulted in a shadowing effect on the rest of the PCB components. It was decided to perform a DoE on the solder operation for this particular PCB to see if it required a

**Table 8.1**  Design and analysis of DoE for mixed technology PCBs

Factors selected: | | Levels of each factor:
---|---|---

| | | | | | |
|---|---|---|---|---|---|
| A = Preheat temperature | | 400 | 425 | 450 | Degrees |
| B = Belt speed | | 2.5 | 3.0 | 3.5 | FPM |
| C = Wave height | | 4 | 5 | 6 | Setting |
| D = Solder pot temperature | | 470 | 480 | 490 | Degrees |

| Exp. # | A | B | C | D | A | B | C | D | Defects/PCB |
|---|---|---|---|---|---|---|---|---|---|
| | | L9 (3 × 4) | | | Orthogonal array saturated design | | | | |
| 1 | 1 | 1 | 1 | 1 | 400 | 2.5 | 4 | 470 | 7.6 |
| 2 | 1 | 2 | 2 | 2 | 400 | 3.0 | 5 | 480 | 11.8 |
| 3 | 1 | 3 | 3 | 3 | 400 | 3.5 | 6 | 490 | 2.6 |
| 4 | 2 | 1 | 2 | 3 | 425 | 2.5 | 5 | 490 | 3.8 |
| 5 | 2 | 2 | 3 | 1 | 425 | 3.0 | 6 | 470 | 4.4 |
| 6 | 2 | 3 | 1 | 2 | 425 | 3.5 | 4 | 480 | 15.2 |
| 7 | 3 | 1 | 3 | 2 | 450 | 2.5 | 6 | 480 | 0.6 |
| 8 | 3 | 2 | 1 | 3 | 450 | 3.0 | 4 | 490 | 6.0 |
| 9 | 3 | 3 | 2 | 1 | 450 | 3.5 | 5 | 470 | 12.6 |

Averaging all experiments with the same factor levels:

| Factor | A | B | C | D |
|---|---|---|---|---|
| Level 1 | 7.3 | 4.0 | 9.6 | 8.2 |
| Level 2 | 7.8 | 7.4 | 9.4 | 9.2 |
| Level 3 | 6.4 | 10.13 | 2.53 | 4.13 |
| Average | 7.2 | 7.2 | 7.2 | 7.2 |

ANOVA analysis

| Source | DOF | Sum SQ | Mean SQ | F value | SS' | p% |
|---|---|---|---|---|---|---|
| A (error) | 2 | 3.04 | 1.52 | — | 12.2 | 6 |
| B (speed) | 2 | 56.6 | 28.3 | 18.5 | 53.6 | 27 |
| C (height) | 2 | 97.12 | 48.56 | 31.9 | 94.1 | 47 |
| D (pot T) | 2 | 43.2 | 21.6 | 14.2 | 40.2 | 20 |
| Total | 8 | 200.0 | 25.0 | | 200.0 | 100.0 |

Level selection for lowest defects = Preheat A3 (450°F), speed B1 (3.0 RPM), height C3 (6), and pot temperature D3 (490°F).
Average from all experiments = 7.18 defects.
$EV$ = experiments average – (B1 + C3 + D3) contribution.
$EV = 7.18 - (7.18 - 4.0) - (7.18 - 2.53) - (7.18 - 4.13) = -3.7.$

different operational setup of the solder wave machine than the rest of the PCB population.

Four factors were selected (preheat, belt speed, wave height, and pot temperature), and an orthogonal array L9 with three levels was chosen for the DoE. The quality characteristic selected was smaller is better defects. Five production PCBs with SOT-23 were used for each

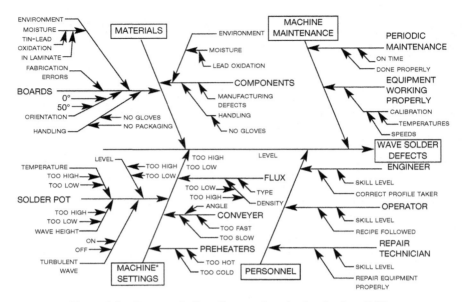

**Figure 8.2**   Cause and effect diagram for mixed technology PCBs.

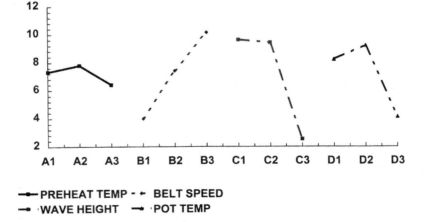

**Figure 8.3**   Graphical analysis of DoE for mixed technology soldering of PCBs.

experiment in order to generate enough defect opportunities to allow for statistical analysis of the defects. The design of the experiments and the analysis of data are shown in Table 8.1.

The factors selected for this DoE proved very easy to manipulate. The second level for each factor was the current soldering process operational settings. Preheat temperature could be set automatically using the machine setting. Special wax temperature indicators that would melt at the specified temperature were placed on the PCBs to indicate the proper preheat levels just before reaching the soldering wave. The belt speed in feet per minute was adjusted by using a potentiometer setting in the machine. The solder pot recirculating pump was adjusted with a potentiometer setting of 4, 5, or 6 to control the solder wave height. The solder pot temperature was varied in increments of 25°F. Because of the thermal mass of the solder pot, this operation took a long time, and the experiment lines sharing the same solder pot temperature were run in sequence. For example, when the solder pot temperature was set at 400°F, experiments 1, 2, and 3 were run sequentially, although DoE practitioners recommend a random order when running the experiments. In addition, the choice of levels for this experiment has to be within the operating range of the process. If the solder pot temperature is too high and the conveyor speed is set too slow, the components could sustain thermal damage.

The graphical analysis in Figure 8.3 shows the relative importance of the factors and levels that were selected. The ANOVA analysis at the bottom of Table 8.1 shows the distribution of factor effects, with factor $A$, the preheat temperature as being the least significant, and therefore used as the error source for the $F$ ratio analysis. A more in-depth statistical analysis could collect the errors according to each PCB, and then calculate the error variance based on the repetitions of the experiment.

The expected value ($EV$), which is −3.7; is much lower than the lowest defect average obtained in experiment line 7, which was 0.6 per PCB. During the conduct of the experiment, it was very difficult to convince the production operators not to forgo the mathematical analysis and immediately switch to the levels used in experiment line 7. As can be seen from the recommended levels, none of them matched the current process. The negative value of the $EV$ is obviously within the confidence limits, since there is no such concept as negative defects. The $\sigma_{error}$ can be quickly calculated from the square root of the variance error or mean square of the error. This is not available in this experiment as it should be derived from the replication errors, not from the assuming that the factor with the smallest $SS_F$ is the error.

It is obvious that this experiment could be successful in achieving

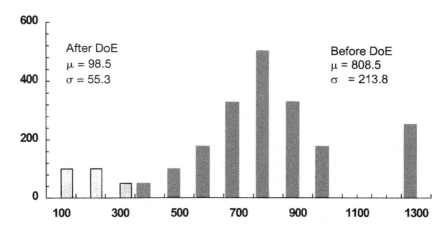

■ BEFORE EXPERIMENT □ AFTER EXPERIMENT

**Figure 8.4**  Histogram of solder defects distribution 6 months before and after DoE.

zero defects by using the graphical analysis only. The production process was changed for this one PCB to the levels recommended, and it resulted in near zero defects in the short term. For the medium term (up to 6 months after the process change) a histogram should be kept of the process before and after the DoE. In the process outlined in Figure 8.1, the histogram of the solder process defects for 6 months before and after the implementation of the DoE is shown in Figure 8.4. It can readily be seen that the average and standard deviation of the defect distribution has shifted dramatically to left, with much lower defect rates. The zero defects obtained from the DoE were not sustained over time because of the variability of the materials and new operators. The end average defect rate was 100 PPM (four $\sigma$ quality), with a maximum rate of 300 PPM.

## 8.2  Transitioning New Product Development to Six Sigma

The implementation of a six sigma program in an organization necessitates several major activities: understanding the design quality of new products as measured by six sigma, knowing the capability of current manufacturing processes, as well as being ready to adopt more capable processes for new products. In this section, each issue will be explained in detail with examples and case studies. Special examples will be given in discipline-specific designs in the next section.

### 8.2.1 Design analysis for six sigma

When a six sigma program is agreed to in the development of new products, the design team has to consider developing quality measures for all new designs. These include the design quality level for each element to be designed, as well as the quality level for module units and systems. The quality level could be expressed by any of the measures that were introduced in previous chapters, including units of sigma designs, Cpk, DPU (PPM), or FTY. It is important to note that the design quality measure is due only to the design as expressed in terms of component specifications, and not to the physical implementation of the design in manufacturing such as PCBs. The design defects are due only to improper designs, not to any variation in production. These will be calculated separately and combined in an overall new product quality, including design and manufacturing, as shown in Figure 8.5.

The application of six sigma in design is different than in manufacturing, since it will be based on the design components' specifications and the proper use of components in the design. In order to obtain a good estimate of the quality of the design, the component specifications must be known, and the design has to be modeled to obtain a distribution of the performance based on the component tolerance distribution.

The six sigma design estimate can be made of typical components as the design nominal and the components worst-case conditions as

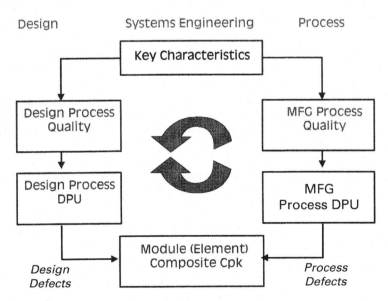

**Figure 8.5**  Overall new product quality, including design and manufacturing.

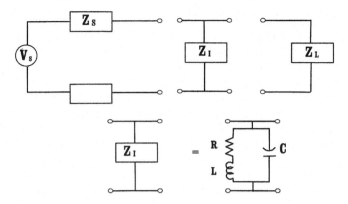

**Figure 8.6**   Design six sigma example—bandpass filter.

the specification limits. Components could be modeled as linear or normal distributions of values between the specification limits for one- or two-sided tolerances. Modeling results of Monte Carlo simulation using random selection of uniform data could be used to show a distribution of results of the design versus its specifications. An example of this process is given for a simple bandpass filter (Figure 8.6) whose specifications are described in Table 8.2.

Using the method outlined above, the results of the six sigma quality study are shown in Table 8.3, expressed as Cpk. These results are based on simulation of the design and a Monte Carlo distribution for each component, as shown in Table 8.3. The simulation results are recorded in terms of average and standard deviation for each of the bandpass parameters. Based on the specification and results of the simulation, the Cpk can be obtained for each of the specification parameters, as well as the defects per unit (DPU) and the expected first-time yield (FTY). The FTY is based on the design and component selection, and does not contain the defects due to the manufacturing process variability.

The total expected quality of the bandpass filter is determined by either adding the defects (DPUs) or multiplying the yields for all of

**Table 8.2**   Specification for bandpass filter example

| | |
|---|---|
| Target frequency $f_0$ | $= 110$ MHz |
| Output ratio $[V_L/V_S]$ | $< 0.15$ dB @ $f_0 \pm 200$ KHz |
| Insertion loss $(I_L)$ | $> 6$ dB @ $f_0 = 90$ and $130$ MHz |
| | $< 2$ dB @ $f_0 = 110$ MHz |
| Conditions | $Z_S = Z_L = 50$ ohm |

**Table 8.3**  Simulation results for Cpk analysis of a bandpass filter

| Component | Value and tolerance | Distribution for simulation |
|---|---|---|
| $R$ Nominal | 0.1 ohms | Linear distribution |
| $L$ Nominal | 15 nH ±10% | Normal with $\sigma$ = 5.00E-10 |
| $C$ Nominal | 140 pf ±10% | Normal with $\sigma$ = 4.67E-12 |

| | Insertion loss | | | $V_L/V_S$ deviation, dB |
|---|---|---|---|---|
| | 90 MHz | 110 MHz | 130 MHz | |
| Average ($\mu$) | 7.18 | 0.83 | 6.31 | 0.119 |
| Sigma ($\sigma$) | 0.69 | 0.55 | 0.74 | 0.0184 |
| Specification | > 6 | < 2 | > 6 | < .15 |
| Cpk | 0.57 | 0.71 | 0.14 | 0.56 |
| $Z$ (Cpk · 3) | 1.71 | 2.13 | 0.42 | 1.68 |
| DPU [$f(-z)$] | 0.0436 | 0.0166 | 0.3372 | 0.0465 |
| FTY (FTY = $e^{-DPU}$) | 95.73% | 98.35% | 71.38% | 95.46% |
| | | | | |
| TFTY | 64.15% | | | |
| TDPU | 0.444 | | | |
| Composite design Cpk | 0.26 | | | |

the design parameters. A composite design Cpk for the bandpass filter is calculated to be 0.26. Obviously, this does not meet the six sigma requirements, and selection of the tolerance of the components has to be tightened considerably.

### 8.2.2   Measuring the capability of current manufacturing processes

Up-to-date capability data for the manufacturing processes to build new products have to be available to the new product design team. The data can be used to calculate the design and manufacturing quality of the new product. The data should contain the process average and standard deviation, as well as design guidelines for design for manufacture (DFM) and early supplier involvement (ESI). In addition, the data has to be updated regularly, typically every quarter, so that the design team is working on the latest capability of the manufacturing processes. These processes can be divided into two categories:

1. Processes that are used to build current products similar to the new one, with adequate process capability. These processes should have long-established capability of meeting six sigma (or specific Cpk) requirements. They should also include guidelines for DFM and ESI.

2. Processes building current products that are not capable for all operations. In this case, the manufacturing process engineers should collect a list of alternative manufacturing processes available that can make products with varying quality depending on the specified parameters.

An example of quality data collected for PCB assembly is shown in Table 8.4. This example is for a mix of surface mount technology (SMT) and through-hole (TH) design. Several options are available to the design team for specifying certain manufacturing process parameters. For example, specifying laser stencil or paralene conformal coatings will result in greater quality than etch stencils or acrylic spray coating. The design team has to select process and material parameters based on the quality and cost goals of the new product.

Once these process quality parameters have been identified, a measure of typical defect rates for PCBs can be generated. Any new PCBs to be designed can be analyzed for quality, given the component count. The defect rate is normalized by the number of opportunities based on terminations of leads or solder joints per component, as well as the DPMO method, discussed in Chapter 4. A quality analysis for a new PCB is shown in Table 8.5, with typical quality levels for the various PCB assembly operations. The PCB is two-sided, with many components of various technologies, including automatic insertion of through-hole (TH) and placement of surface mount technology (SMT).

**Table 8.4**  Quality data for PCB assembly manufacturing processes

| Operation | Process parameters | Attributes | Cpk |
|---|---|---|---|
| SMT forming | Standoff | Height = 0.005″ | 0.96 |
| | | Height = 0.002″ | 1.48 |
| | Lead length | | 1.72 |
| | Toe–toe | | 1.64 |
| | Coplanarity | 50M | 2.55 |
| | | 25M | 1.80 |
| SMT | Placements | | 1.25 |
| | Reflow solder shorts | Etch stencil | 1.05 |
| | | Laser stencil | 1.20 |
| Through-hole | Autoinsertion | | 1.32 |
| | Lead length | | 1.25 |
| Solder shorts | Solder shorts | | 1.30 |
| | Hand solder | | 0.97 |
| Miscellaneous ASY | | | 0.97 |
| Wash | Cleanliness | Ionograph | 1.31 |
| Coating | | Paralene | 1.4 |
| | | Acrylic | 0.6 |

**Table 8.5**  Quality analysis of a two-sided PCB with TH, SMT, and mechanical assembly and multiple components and leads

| Assembly operation | Opportunities | Cpk | PPM | DPU | FTY $(e^{-DPU})$% |
|---|---|---|---|---|---|
| Autoinsertion/lead | 620 | 1.32 | 74 | 0.05 | 95.34% |
| SMT place/lead | 4400 | 1.25 | 176 | 0.77 | 46.30 |
| Solder/wash/lead | 5020 | 1.30 | 96 | 0.48 | 61.88 |
| Hand solder/joint | 40 | 0.97 | 3620 | 0.14 | 86.94 |
| Mechanical ASY/part | 20 | 0.97 | 3620 | 0.07 | 93.24 |
| PCB total | 10,100 | 1.26† | 150* | 1.51 | 20.80% |

*The 155 PPM for the total PCB was obtained from dividing the defects by the opportunities = 1.51 · 1,000,000/10,100.
†The Cpk for the PCB was calculated by:
Defect rate (one-sided) = $f(-z) = f[150/(2 \cdot 1,000,000)] = 0.0000755$
$z = 3.79$
$Cpk = z/3 = 1.26$

In addition, the PCB has 40 components leads to be hand soldered and 20 mechanical parts to be assembled. The PCB will also have to soldered and washed. The component counts have been translated to defect opportunities depending on assembly operations such as the number of component leads, solder joints or mechanical assembly. The resultant quality level is 1.51 defects/PCB or 22% FTY. This is very poor quality and will necessitate extensive testing. An exercise such as this example might prove to be very positive for increasing the design team focus on the quality drivers for PCBs discussed next.

For the cases where the quality of current operations are not adequate, a list of drivers should be generated to alert the design team to the critical attributes of the design that will influence the quality of manufacturing. The design team can thus focus on modifying the design to allow manufacturing to build the new product to the specified level of quality (six sigma or Cpk target). An example of such a list for PCBs is shown in Table 8.6. In many of the items in that table, the geometric properties of the components of PCB layout or the PCB warp specifications are shown to be important in increasing the quality of the PCB assembly. Unfortunately, it is difficult to generate a precise amount of quality improvements associated with items on this list.

### 8.2.3  Investigating more capable processes for new products

When some of the current and alternative manufacturing processes are not capable, additional manufacturing options in materials and processes have to be explored. A common solution is to invest in new plants and equipment, or to select new suppliers that can offer

**Table 8.6**   Quality drivers for printed circuit board (PCB) assembly

| Operation | Process parameters | Quality drivers |
|---|---|---|
| SMT forming | Standoff, lead length Toe–toe, Coplanarity | Incoming components specifications Handling and packaging methods |
| SMT place | Solder paste height Placement accuracy | PCB warp, handling specifications Component footprints, size specifications |
| Through-hole assembly | Autoinsertion | Component mix |
| Wave solder | Lead length, shorts | PCB warp specifications Solder mask specifications Lead and PCB hole specifications |
| Miscellaneous assembly | | Component footprint specifications |
| Coating | Thickness | Masking specifications |
| Cleanliness | Ionograph measurement | Solder mask specifications Lead and PCB hole specifications |

greater quality in manufacturing. In many of these cases, the benefit cost analysis of these higher-capability processes is not known. A DoE could be used to determine the relative importance of quality improvements using these new processes. The DoEs used in this case are more general and should optimize a universal manufacturing process, in contrast to the focused DoEs for improving current processes, such as the one discussed in Section 8.1. The DoEs tend to be survey-related, scanning the current spectrum of materials and how to process them in order to quantify the quality improvements. They are combination designs, or successive investigations for narrowing down the material or process alternatives, then optimizing the final few selections with a more in-depth DoE.

### 8.2.4   Case studies of process capability investigations for manufacturing: Stencil technology DoE

Process surveys to investigate recent advances in materials and processing techniques should be undertaken regularly by process and manufacturing engineers to make current processes more capable. The capability of current processes should be the ultimate arbitrator on deciding what processes to investigate first. The aim of these investigations is to reduce the variability of the current processes by investigating new materials, equipment, and processing parameters. The investigation should be universal in nature, affecting as many existing and new products as possible.

An example is an investigation into the technology of solder deposition using stencils for SMT PCBs. The DoE should examine alternative technologies from different suppliers, and rate the soldering quality produced by the stencil types. The following is a discussion of the issues encountered and decisions to resolve them in the DOE. These issues could be useful when conducting similar survey DoE's:

- The quality characteristic was the height of the solder "bricks" formed by the solder deposition operation through the stencil. Minimum variability of the solder brick height was shown to result in good soldering with reduced defects. The height proved to be difficult to measure because the solder brick top surface was not uniform, and individual readings of the solder brick height varied according to the presence of solder spheres in the paste. The volume of the bricks proved easier to measure by a laser detection system, and it was decided to measure the volume of the solder bricks as well. A combination of the two, the area of solder, was chosen as the quality characteristic; it is equal to the volume/height. The measurement of the variability of the solder areas was repeated several times and transformed as a S/N value. The statistical analysis was performed on the single number representing the variance of solder areas for each experiment line.

- The factors chosen were the stencil technologies available. They differed in the creation of the holes for depositing the solder paste on the surface of the PCB's. The technologies included band, chemical etching, laser drilling, and electroforming. Several suppliers for each technology were included for a total of eight levels of stencils. Other factors included the use of paste with or without aqueous cleaning (C or NC) after soldering, snap-off distance (5 or 0 mils), squeegee pressure (35 or 30 lbs), lead orientation angle of the components (90 or 0 degrees from the squeegee travel), and lift-off pressure (75 or 60 lbs).

- A specially made test PCB used was used, with 208 components, 19.37 mils lead pitch on each PCB.

- Other known factors affecting solder deposition were fixed for this DoE. They included the stencil thickness of 0.006" (6 mils), stencils with aperture sizes of 10 × 55 mils, and using the same stainless steel squeegee for all experiments. One squeegee pass at the same speed was used for all experiments. These factors were determined to be not significant in earlier experiments.

- The stencils were used to deposit solder paste on bare copper substrate for all experiments.

- An L16 orthogonal array was used. The factor assignments are shown in Table 8.7. The selected factors were assigned to specific

**Table 8.7**  DoE stencil technology experiment factor and level selection

| Primary factors | Levels | L16 col. no. |
|---|---|---|
| 8 stencil types | 4 technologies, 8 suppliers | 2, 4, 8 |
| Solder pastes | Aqueous clean/no clean | 1 |
| Snap-off | 5, 0 mils | 12 |
| Orientation | 90, 0 degrees | 6 |
| Squeegee pressure | 35, 30 lb | 10 |
| Lift-off pressure | 75, 60 lb | 15 |

| Interaction columns | | |
|---|---|---|
| Paste × stencil (3 columns) | 3 (1 × 2), 5 (1 × 4), 9 (1 × 8) | |
| Stencil × lift-off (3 columns) | 7 (8 × 15), 11 (4 × 15), 13 (2 × 15) | |
| Paste × lift-off (1 column) | 14 (1 × 15) | |

| Confounding of primary factors | | |
|---|---|---|
| Orientation | vs. | interaction of stencil types (6 vs. 2 × 4) |
| Snap-off | vs. | interaction of stencil types (12 vs. 4 × 8) |
| Squeegee pressure | vs. | interaction of stencil types (10 vs. 2 × 8) |

columns so that the interactions of interest were isolated. The eight levels of stencil technology were used in a multilevel combination column, consisting of the columns 2, 4, and 8. The interaction of the columns forming the stencil levels were confounded with some of the other factors, as shown in Table 8.7

- A total of 480 points were measured for the experiments, consisting of 10 replications of the 16 experiments in the L16 design. Each replication consisted of taking 10 points measured three times on the deposited solder pattern on bare copper substance.

The L16 design is presented in Table 8.8, with the interaction columns not shown. The ANOVA analysis is shown in Table 8.9, with the percentage contributions only. The total degrees of freedom (DOF) is equal to number of experiment lines minus 1. The DOF of the stencil factor is equal to three, since three columns with two levels each were used. The interaction of the paste and stencil was also equal to three, since three stencil columns were used.

The analysis clearly indicates the importance of the stencil technology in the quality of the solder deposition. Of the factors selected, snap-off, lead orientation, and squeegee pressure were significant, whereas paste selection and lift-off pressure were not significant. Only one interaction, paste versus lift-off pressure, proved significant, even if paste was not considered significant.

The cost of the different stencil technologies is variable and more analysis is needed to quantify the trade-off of increased quality for

**Table 8.8**   Stencil technology DoE L16 design

| | | | Column levels | | | | | | Factor assignments | | | | | |
|---|---|---|---|---|---|---|---|---|---|---|---|---|---|---|
| | | | | | | | | | 2,4,8 | 1 | 6 | 10 | 12 | 15 |
| # | 1 | 2 | 4 | 6 | 8 | 10 | 12 | 15 | Stencil | Paste | Ortn | SQPr | Snap | Lift |
| 1 | 1 | 1 | 1 | 1 | 1 | 1 | 1 | 1 | 1 Band 1 | C | 90 | 35 | 5 | 75 |
| 2 | 1 | 1 | 1 | 1 | 2 | 2 | 2 | 2 | 2 Band 2 | C | 90 | 30 | 0 | 60 |
| 3 | 1 | 1 | 2 | 2 | 1 | 1 | 2 | 2 | 3 Chem 1 | C | 0 | 35 | 0 | 60 |
| 4 | 1 | 1 | 2 | 2 | 2 | 2 | 1 | 1 | 4 Chem 2 | C | 0 | 30 | 5 | 75 |
| 5 | 1 | 2 | 1 | 2 | 1 | 2 | 1 | 2 | 5 Laser 1 | C | 0 | 30 | 5 | 60 |
| 6 | 1 | 2 | 1 | 2 | 2 | 1 | 2 | 1 | 6 Laser 2 | C | 0 | 35 | 0 | 75 |
| 7 | 1 | 2 | 2 | 1 | 1 | 2 | 2 | 1 | 7 Electro | C | 90 | 30 | 0 | 75 |
| 8 | 1 | 2 | 2 | 1 | 2 | 1 | 1 | 2 | 8 Chem 3 | C | 90 | 35 | 5 | 60 |
| 9 | 2 | 1 | 1 | 1 | 1 | 1 | 1 | 2 | 1 Band 1 | NC | 90 | 35 | 5 | 60 |
| 10 | 2 | 1 | 1 | 1 | 2 | 2 | 2 | 1 | 2 Band 2 | NC | 90 | 30 | 0 | 75 |
| 11 | 2 | 1 | 2 | 2 | 1 | 1 | 2 | 1 | 3 Chem 1 | NC | 0 | 35 | 0 | 75 |
| 12 | 2 | 1 | 2 | 2 | 2 | 2 | 1 | 2 | 4 Chem 2 | NC | 0 | 30 | 5 | 60 |
| 13 | 2 | 2 | 1 | 2 | 1 | 2 | 1 | 1 | 5 Laser 1 | NC | 0 | 30 | 5 | 75 |
| 14 | 2 | 2 | 1 | 2 | 2 | 1 | 2 | 2 | 6 Laser 2 | NC | 0 | 35 | 0 | 60 |
| 15 | 2 | 2 | 2 | 1 | 2 | 2 | 2 | 2 | 7 Electro | NC | 90 | 30 | 0 | 60 |
| 16 | 2 | 2 | 2 | 1 | 1 | 1 | 1 | 1 | 8 Chem 3 | NC | 90 | 35 | 5 | 75 |

certain stencil technologies versus the additional cost of the technology. Using the quality loss function (QLF), discussed in the last chapter, might improve the stencil technology selection. The formula for the loss function (QLF) for this case study is given in Table 8.10. Two costs of quality are derived. One is the traditional quality loss when a solder short results from a solder brick area 50% larger than the target, which is stencil aperture. The other cost is associated with stencil

**Table 8.9**   Stencil technology percent contribution analysis of average solder deposition area

| Factor | DOF | Percent contribution |
|---|---|---|
| 1 Stencil | 3 | 33 |
| 2 Paste type | 1/0 | pooled |
| 3 Snap-off pressure | 1 | 17 |
| 4 Squeegee pressure | 1 | 14 |
| 5 Stencil × Paste | 3/0 | pooled |
| 6 Lead orientation | 1 | 13 |
| 7 Paste × lift-off | 1 | 17 |
| 8 Stencil × lift-off | 3/0 | pooled |
| 9 Lift-off pressure | 1/0 | pooled |
| Pooled Error | 8 | 6 |
| | | |
| Total | 15 | 100% |

**Table 8.10**    Stencil technology quality loss function (QLF) formula

$$L = [A_1/\Delta^2] \cdot [\sigma^2 + (\mu - t)^2] + A_2/C$$

| | |
|---|---|
| $A_1$ | Loss due to variability (solder short) |
| $A_2$ | Loss due to stencil clogging (area larger than 50% above aperture area) |
| $\sigma$ | Standard deviation solder brick area |
| $\mu$ | Average of solder brick area |
| $t$ | Target solder brick area (stencil aperture) |
| $\Delta$ | Deviation at maximum loss |
| $C$ | Number of PCBs printed that would cause stencil to clog |

cleaning when the stencil is clogged with solder paste. This methodology can help make clear decisions as to what stencil technology to use for new products, and the monetary impact of the decision. Actual results of the experiments are not shown because of the continual evolution of the stencil technology and the varied claims made by competing suppliers.

## 8.3    Determining Six Sigma Quality in Different Design Disciplines

The design six sigma is a measure of the design quality: how the design meets it intended specifications, regardless of the manufacturing steps necessary to produce the product or system. It is determined by the variability of the components specified in the design versus the overall design performance to its specifications.

The application of the six sigma design is based on the selection of components for the design. In order to obtain a proper estimate of the design quality, the components' specifications must be known, and the design has to be modeled to obtain a distribution of the performance based on its components' tolerance distribution.

In many cases, the distribution of the components' characteristic values is not known, though the worst-case conditions are readily available. This has led to worst-case analysis, in which the design performance is evaluated when the components' characteristics are assumed at their specification limits. When using six sigma design, the span of the components' specifications to the nominal value is considered to be 6 $\sigma$, and the nominal becomes the component average value $\mu$.

### 8.3.1    Mechanical product design process

The product design process starts with the concept models, proceeds to prototype models, and then on to production. The concept and prototype models are primarily made in the companies' model shop or outside machine shops where most of the individual components are fabricated by one or two toolmakers. The emphasis is to prove the concepts. The toolmakers work with the designer and are given some latitude in order to make the parts fit together.

The product is still being designed at this stage, so changes are frequent and the parts are altered to fit. Because of time constraints, the changes are drawn by freehand sketches and given to the toolmakers. When the models are completed and assembled, they go through extensive testing. More changes are made and incorporated. After testing is completed, drawings are updated to reflect the changes. Suppliers are selected and orders are issued to produce parts.

When parts are received for the first time and are assembled for production, it is frequently discovered that they do not fit. At this time, it can also be discovered that detailed tolerance analysis was not performed due to schedule pressures. Drawings made and released based on the concept and prototype models did not account for the manufacturing process variability.

### 8.3.2  Mechanical design and tolerance analysis

No manufacturing process can a make a part to exact dimensions. Hence maximum and minimum limits of dimensions (or tolerances) are specified with two goals in mind:

1. The limits must be set close enough to allow functioning of the assembled parts (including interchangeable parts).
2. The limits must be set as wide as functionally possible, because tighter limits call for expensive processes or process sequences.

Once the limits (or tolerances) are set by the designer, all parts or components are manufactured to those specified limits. Assembly of the parts causes tolerances to accumulate, which can adversely affect the functioning of the final product. In addition, tolerance accumulation can also occur, based on the method by which the parts are dimensioned. Tolerance accumulation that occurs in the assembly of parts is sometimes referred to as "tolerance stackup."

To make sure that parts successfully mate at subassembly or final assembly, and the products function per the design intent, an analysis is performed to uncover the existence of any interference. It is referred to as "tolerance analysis." The following is a brief review of tolerance issues.

*Tolerance* (per ANSI Y14.5M) is the total amount by which a specific dimension is allowed to vary. Geometric tolerance is a general term applied to the category of tolerances used to control form, profile, orientation, location, runout, etc. Tolerances are primarily of two types: tolerance of size and tolerance of form.

*Tolerance of size* is stated in two different ways:

1. Plus-or-minus tolerancing, which is further subdivided into bilateral and unilateral tolerancing. Bilateral tolerance is applied to both sides of a basic or nominal dimension. Examples are:

   0.375 ± 0.010

   0.375 + 0.005/–0.002

2. Limit dimensioning is a variation of the plus-or-minus system. It states actual size boundaries for a specific dimension. It will eliminate any calculation on the part of the manufacturer. Limit dimensioning is practiced in two ways: Linear or one next to another, and dimensions placed one above the other. Examples are:

   0.625 – 0.635

   0.635

   0.625

   When one dimension is placed above the other, the normal rule is to place the larger dimension above the smaller.

   There are no rigid guidelines regarding tolerancing techniques. The choice depends on the style of the designer and very often both types of tolerancing methods (the plus-or-minus and limit dimensioning) are used in the same drawing.

*Tolerance of form* includes location of geometric features and geometric properties such as concentricity, runout, straightness, etc.

### 8.3.3   Types of tolerance analysis

It is important to note that the parts used in a product are divided into standard or off-the-shelf parts and nonstandard or designed parts. Examples of standard parts are bearings, shafts, pulleys, belts, screws, nuts, snap rings, etc. These parts come with the manufacturer's specified tolerances. The designer does not have any latitude in changing these limits. Nonstandard or designed parts are custom made for the product. Hence, the designer can specify wider or narrower limits based on the functionality requirements. There are two types of tolerance analysis: extreme case tolerance analysis and statistical analysis.

*Extreme case analyses* are further subdivided into two categories: best-case analysis and worst-case analysis.

- Best-case analysis describes a situation in which the parts fall on that side of the tolerance (positive or negative) in which there is no chance of interference in the assembly of these parts.
- Worst-case analysis is the study of a situation wherein the parts produced are assembled as per the worst case. The probability of interference is certain or unity.

The extreme case analysis method is the most widely used method for tolerance analysis. Most designs are analyzed using this concept and have worked successfully. The method is simple to apply and consists of designing the parts to nominal dimensions and then assigning tolerances in such a way that if tolerances accumulate in one direction or the other, the assembly continues to meet the functional requirements of the design. This method, though ensuring that all parts will always be able to be assembled correctly, has a built-in waste mechanism. Designs can be overly conservative, leading to high product costs by assigning tighter tolerance zones. By using statistical analysis based on six sigma, a more reasonable understanding of the design specifications and how parts will be assembled will be demonstrated in the next section.

### 8.3.4 Statistical tolerance analysis for mechanical design

Statistical analysis involves the application of statistical probability distributions to analyzing tolerances for assemblies. It will prevent overly conservative designs, which can increase the cost of the product without adding to quality. With statistical analysis, tolerances can be widened, readily achieving six sigma.

Statistical tolerance analysis is based on the assumption that most mechanical parts are made to normal probability distributions within their specified tolerance limits. The distributions of individual parts can be combined into a normal distribution, representing the variability of parts from their nominal dimensions. In six sigma quality, the nominal dimension of a part is set to its average, and the specified tolerance limits of that part at $\pm 6\ \sigma$.

### 8.3.5 Tolerance analysis example

An example is given in this section to demonstrate some of the concepts of tolerance analysis. An assembly consisting of three parts or rectangular blocks is to be assembled together into a box cover (see Figure 8.7). Their critical dimensions (mating surfaces) and their specified tolerances are also shown in Figure 8.7. If the box cover for

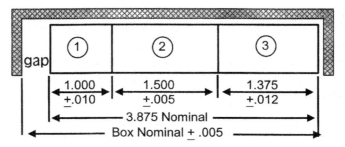

**Figure 8.7**   Tolerance analysis example, three square parts (all dimensions in inches).

these three parts is be designed, what specifications should be assigned to the box for these three parts to fit? The problem will be solved using worst-case analysis and then by statistical analysis.

For the worst case analysis in Table 8.11, Case 1, the cumulative dimension of the three parts is at a maximum of 3.902 inches. It is comprised of the individual maximum dimensions of the blocks. The minimum dimension of the box should be set at 3.903 inches, ensuring a minimum clearance gap of 0.001. Assigning a box tolerance of ±0.005 inches, the nominal dimension for the box is 3.903 + 0.005 = 3.908 inches, and the maximum box dimension is 3.913 inches. This shows that there could be a maximum gap of 3.913 − 3.848 = 0.065 inches, and average gap of 3.908 − 3.875 = 0.033 inches. Having such a wide variation (0.055 to 0.001 inches) may not be acceptable as functional requirement for the assembly of the box and three blocks. If this assembly were part of a front panel, having a gap average of 0.033 inches might not be aesthetically pleasant and could convey the impression of poor quality.

**Table 8.11**   Tolerance analysis for three-parts example, worst-case analysis

|  | Case 1. Normal tolerance | | | Case 2. Tight tolerance (± 0.002) | | |
| --- | --- | --- | --- | --- | --- | --- |
| Part | Dimension | High | Low | Dimension | High | Low |
| P1 | 1.000 | 1.010 | 0.990 |  | 1.002 | 0.998 |
| P2 | 1.500 | 1.505 | 1.495 |  | 1.502 | 1.498 |
| P3 | 1.375 | 1.387 | 1.363 |  | 1.377 | 1.373 |
| Total | 3.875 | 3.902 | 3.848 |  | 3.881 | 3.869 |
| Box | 3.908 | 3.903 | 3.913 | 3.884 | 3.882 | 3.886 |
| Gap, box to parts |  | 0.001 | 0.065 |  | 0.001 | 0.017 |
| Average gap, box to parts | | 0.033 | | | 0.009 | |

To reduce the variation, a logical approach might be to tighten the tolerances. Table 8.11, Case 2, gives the result of having all the parts made to closer tolerances of ±0.002 inches. In this case, the minimum dimension of the box is 3.882 inches. If this dimension is assigned a ±0.002 tolerance, then the nominal dimension becomes 3.884 inches, with a maximum of 3.886 inches. The gap maximum is 3.886 − 3.869 = 0.017 inches, and the gap minimum is 3.882 − 3.881 = 0.001 inches (by design), with an average of 3.884 − 3.875 = 0.009 inches. This is more acceptable than the case with normal tolerances above, but it comes at a higher cost. As a result of narrowing the tolerance band, the normal manufacturing process is not used. Narrower tolerance requires extra time in equipment setup, increased inspection, and increased defect rate due to parts made out of tolerances. This is an example of an overly conservative design.

### 8.3.6  Statistical analysis of the mechanical design example

Using statistical analysis, all three three parts are assumed to be made to a normal probability distribution within their specified tolerance limits. For six sigma, the parts are assumed to be made to ±6 $\sigma$, so that the tolerance of ±0.010 of Part 1 results in a standard deviation $\sigma$ of 0.00167. The remaining calculations are shown in Table 8.12, using the RSS of the $\sigma$ calculations:

$$\sigma_{\text{system}} = \sqrt{(\sigma_1^2 + \sigma_2^2 + \sigma_3^2 + \ldots)} \tag{8.1}$$

It can be seen from Table 8.12 that for six sigma design, the box nominal is six sigma away from the average of the three blocks P1, P2, and P3. For six sigma design, the box dimensions are 3.892 ± 0.005 inches, somewhere in the middle of the first two worst-case de-

**Table 8.12**  Tolerance analysis for three-part example, six sigma statistical analysis. Case 3: statistical tolerance

| Part | Nominal | Tolerance | Six sigma | One sigma |
|------|---------|-----------|-----------|-----------|
| P1 | 1.000 | ±0.010 | 0.010 | 0.00167 |
| P2 | 1.500 | ±0.005 | 0.005 | 0.000833 |
| P3 | 1.375 | ±0.012 | 0.012 | 0.002 |
| Box | ? | ±0.005 | 0.005 | 0.000833 |

$\pm 1\ \sigma_{\text{gap}} = \sqrt{0.00167^2 + 0.000833^2 + 0.002^2 + 0.000833^2} = \pm 0.002859$

Nominal box = nominal of 3 blocks + 6 $\sigma_{\text{gap}}$ = 3.875 + 6 · 0.002859 = 3.875 + 0.01716
  = 3.892″

Gap Nominal = 0.017″

signs discussed earlier. This design will produce defects on the order of 3.4 PPM.

There are two important items of interest in statistical design for mechanical parts. First, the defect rate is only one-sided around the normal distribution, since only one-half of the interference will occur when the box is too small. If the box is too large, then there are no defects. The second item is that the RSS analysis will produce the same results given any assumption of the number of sigmas for the design. The results will be the same for the box nominal if the parts were assumed to be made with three sigma instead of six sigma.

### 8.3.7    Tolerance analysis and CAD

There is increasing use of computer aided design (CAD) systems for mechanical design. They make it convenient to present drawings in 3D to the designer, and hence offer a better view of parts mating together at assembly. Parts are drawn in CAD to nominal dimensions and can be checked for interference, using extreme case tolerances.

Not all CAD systems can be used for tolerance analysis; some of the mechanical parts can be created using different CAD system formats, and cannot be mated together as an assembly on master CAD screens without the use of expensive translators. Tolerance analysis involves the understanding of the functionality of the product and knowledge of the processes that are used in making the parts. Even if the CAD system has been used for making the prototypes, and parts were mated successfully at that stage, a separate tolerance analysis study should be done to ensure high quality in production.

### 8.3.8    Tolerance analysis and manufacturing processes

A product is made of many parts that have been made from different materials. Many electronic products use parts made from plastics, sheet metal, machined parts, rubber, castings, etc. Parts made from these processes have unique properties, and manufacturing dictates the tolerances that can be specified with these parts. Sheet metal parts require a much wider tolerance band compared to machined parts. Plastic molded parts, made from mature molds and processes, have low variation in dimensions from batch to batch. Machined parts will vary from batch to batch, and many operators tend to make parts on the high side of the tolerance, so they can be able to reduce some dimensions in the future if the parts fail to pass inspection tests.

Knowledge of manufacturing processes, and proper use of six sigma design for mechanical parts will reduce the need for conservative designs, thereby decreasing the costs of the product as well as providing for high-quality parts.

### 8.3.9  Mechanical design case study

In mechanical design, statistical design analysis can be substituted for worst-case tolerance analysis. A case study in mechanical design tolerance analysis is that of a typical vibrating probe that is used for angioplasty medical operations. As shown in Figure 8.8, it consists of a vibrating element of wires wound around a magnetic barrel, and a cover to enclose the assembly. The vibrating barrel has an outside diameter 0.0075 ± 0.0002 inches, and the winding coil around the barrel has an outside diameter equal to 0.0027 ± 0.0002 inches. The wires and vibrating barrel were purchased from outside suppliers, and therefore had fixed tolerances. The designer is faced with a dilemma: If the cover specifications are too loose, then the mechanical assembly gap between the cover, barrel, and the wires is too large, causing the assembly to come apart. If the cover specifications are too tight, then the mechanical assembly design has interference. The statistical analysis allows for the best design to meet this contrasting set of conditions.

Using statistical design analysis, based on the RSS values of $\sigma$, the design quality prior to manufacturing can be calculated as follows (Table 8.13):

$$\sigma_{system} = \sqrt{(\sigma_1^2 + \sigma_2^2 + \sigma_3^2 + \ldots)}$$

Cover nominal = barrel nominal + 2 wire nominal + gap (6 $\sigma_{system}$)

Using this RSS technique, it can be seen from Table 8.13 that the gap should be set to 0.0004 inches, regardless of whether one assumes

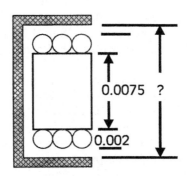

**Figure 8.8**  Mechanical design of a typical vibrating angioplasty probe.

**Table 8.13**  Statistical design analysis of angioplasty probe

---

$\sigma_{GAP} = \sqrt{\sigma_{Barrel}^2 + \sigma_{Cover}^2 + 2 \cdot \sigma_{Wire}^2}$

$\sigma_{GAP}$ (6σ parts) $= \sqrt{(0.0002/6)^2 + (0.0002/6)^2 + 2 \cdot (0.0002/6)^2} = 0.000067$

$\sigma_{GAP}$ (3σ parts) $= \sqrt{(0.0002/3)^2 + (0.0002/3)^2 + 2 \cdot (0.0002/3)^2} = 0.000133$

Average (μ) of gap between wired barrel and cover (6σ parts) = 6 · 0.00067 = 0.0004

Average (μ) of gap between wired barrel and cover (3σ parts) = 3 · 0.000133 = 0.0004

Nominal of cover = nominal of barrel + nominal of 2 wires + gap = 0.0075 + 2 · 0.0027
  + 0.0004 = 0.0133

---

three or six sigma incoming parts. The nominal of the cover will be equal to the nominal of the components of the assembly plus the gap, or 0.0133 inches. If the cover is given a similar tolerance of ±0.0002 inches as the purchased parts of barrel and wires, then the minimum of the cover (0.0131 inches) is in interference with the maximum of the assembly by 0.0004 inches (maximum barrel + maximum 2 wires = 0.0077 + 0.0029 · 2 = 0.0135 inches). The expected defect rate is half the normal, since defects only occur on one side of the gap distribution.

### 8.3.10  Thermal design six sigma assessment example

Many electronic products require thermal cooling systems to remove heat generated in the electronic boxes and to maintain proper operating temperatures in the transistor junctions within the PCBs. Several techniques have been developed to achieve thermal cooling for electronics. They include adding fans to the electronic box or using PCBs with thermally conducting cores. Several core materials are used, including aluminum and composite materials with high thermal conduction properties.

Thermal cooling systems can be overdesigned using these techniques. Cost reduction in thermal designs can be achieved by using six sigma quality principles. Statistical data such as best- and worst-case design conditions have to be gathered from different sources including current thermal designs and thermal modeling of new designs.

A summary of an example thermal design assessment is given in Table 8.14. In this case, the thermal cooling system for an electronic box is specified to maintain the transistor junctions (Tj) on the PCBs to less than 105°C when the inlet air temperature into the electronic box is given at 55°C. The current design meets these specifications with two cooling fans and composite core PCBs. An analysis of the variability of current boxes indicates that σ = 1.2. Table 8.14 lists the total temperature rise starting from the inlet air, through the differ-

**Table 8.14**    Thermal design six sigma assessment

| | Current design, 2 fans, composite core | | Modified design | |
| --- | --- | --- | --- | --- |
| | Best case | Worst | Al core, worst | 1 fan worst |
| Electronic box | | | | |
| Inlet air temperature (start point) | 55°C | 55°C | 55°C | 55°C |
| Box air conduction temperature rise | 6.5 | 10 | 10 | 14 |
| Convection rise | 8 | 12 | 12 | 16 |
| Chassis conduction | 2.5 | 3 | 3 | 3 |
| PCB Interface | 3.5 | 5 | 5 | 5 |
| PCB level | | | | |
| Board edge temperature level | 75.5°C | 85°C | 85°C | 93°C |
| Core conduction | 5.5 | 7 | 11 | 7 |
| Conduction through PCB | 2 | 3 | 3 | 3 |
| Component attachment | 1 | 3 | 3 | 3 |
| Component case to transistor junction | 3 | 7 | 7 | 7 |
| Junction temperature $T_j$ (endpoint) | 87 | 105 | 109 | 113 |
| Average $T_j$ = (worst case − best case)/2 | | 96 | 98 | 100 |
| Design quality ($\sigma = 1.2$, $T_j$ spec. $\leq 105$) = (spec. − average $T_j$)/$\sigma$ | | 7.5 $\sigma$ | 6 $\sigma$ | 4 $\sigma$ |

ent heat transfer mechanisms in the box and PCBs, to the transistor junctions (Tjs). Best- and worst-case conditions are given for the current design, resulting in an average $T_j = 96$°C and a 7.5 $\sigma$ design quality based on the maximum specification of 105°C.

Two assumptions are made in this assessment: the best-case condition and the variability of the current design remain the same for all scenarios. Several modified design scenarios are explored in Table 8.14. One is to substitute aluminum for the composite core, while the other is to use only one cooling fan. The first option increases the thermal core conduction through the PCB from 7 to 11, while the second increases the conduction and convection air temperature rise by 4°C each. The effect of the modified designs is to decrease the heat transfer of the design. This raises the average transistor junction temperature, hence reducing the distance fromm the average to the 105°C $T_j$ specification. It can be seen from Table 8.14 that substituting the composite core with the aluminum core and the current two fans will achieve six sigma quality, whereas removing one fan from the original design will result in a four sigma design.

In this example, six sigma can be used as a tool by the design team to explore alternatives to reduce the cost of overdesigned systems while maintaining high quality and low defect rates.

## 8.3.11   Six sigma for electronic circuits with multiple specifications

The design quality calculation for a single electrical circuit can be expanded to include several design targets for the circuit. To obtain a design for quality assessment of the circuit, it has to be simulated using any of the available software packages such as SPICE for analog or OMNISYS for microwave circuits. The resulting performance characteristics of the circuit simulation have to be matched to the design specifications. The design quality, expressed as Cpk for each of the design key characteristics, has to be accumulated into a composite Cpk according to the methodology presented in Chapter 5.

A detailed example of an electronic analog circuit design is given in Table 8.15 for a design analysis of a RF amplifier. The data developed for this example are based on a Monte Carlo simulation on the individual components that make up the RF amplifier. The phase specification exhibits a low Cpk value, which in fact is the major contributor to the composite Cpk value. In addition, some of the characteristics have two-sided specifications, whereas others are one-sided. The composite Cpk is calculated on the basis of two-sided specifications. This could possibly lead to the condition that the composite Cpk implies a better quality indicator than those of the individual characteristic Cpks.

In the example of Table 8.15, both the worst-case Cpk and the composite Cpk are shown. The worst-case Cpk could help in focusing the design team on where to improve the design. The composite Cpk can be reported back up into the total product design, if this circuit is part of a module to be combined with other modules to complete the design of a product or subsystem. Alternately, the composite Cpk could be reported if the circuit represented a complete unit or subsystem function.

**Table 8.15**   Composite Cpk design analysis of an RF amplifier

|  | Characteristics | | | Specifications | | | Design analysis | | |
|---|---|---|---|---|---|---|---|---|---|
| Key | Minimum | Maximum | Units | Mean | $\sigma$ | $z$ | Cpk | FTY | DPU |
| Gain | None | 2 | dB | 0 | 0.6 | 3.33 | 1.11 | 0.99957 | 0.00043 |
| Noise | None | 30 | dB | 24 | 2.0 | 3 | 1.0 | 0.99865 | 0.00135 |
| Delay | 0 | +10 | ns | 4 | 1.0 | 4, 6 | 1.33 | 0.999936 | 0.000064 |
| Phase | −5 | +5 | degrees | 1 | 2.0 | 2, 3 | 0.67 | 0.97613 | 0.02416 |
|  |  |  |  | Composite CCpk* | | | 0.74 | 0.9743 | 0.026 |
|  |  |  |  | Worst-case Cpk | | | 0.67 | 0.9761 | 0.024 |

*Note that the composite CCpk assumes two-sided normal distribution with no shift of the average. In this case it shows better quality than the worse-case Cpk

### 8.3.12 Special considerations in Cpk for design of electronic products

In many cases, it is not possible to obtain Cpk analysis of the electronic design because of the functionality of the circuit or module, or the need to have complete certainty in the output of the design. Some of these cases are as follows:

- Designs that perform emergency actions such as shutdowns, switching to alternate power, or sensing alarm conditions. These designs could have a desired very high Cpk value or have a sequential control scheme where one function cannot proceed until another has been positively completed.

- Synchronized digital electronic designs, where electronic signals are propagating in the circuit according to clocked conditions. Normally, the variability in the circuit performance is due to the turn on or off times of electronic gates. If not properly designed, the circuit could exhibit "race" conditions, where spurious signals are being generated in the circuit. However, the designer can use a variety of techniques to eliminate this condition, such as the use of a very fast clock to enable gate transitions or changing the phase of the signal to ensure that other derived signals in the circuit do not interfere with the original signal.

- Software designs or modules that perform specific functions. Since the software is translated into hardware-based machine instructions, and is normally duplicated every time it is run, it is difficult to quantify any variation of design. Software defects, which are measured in defects per lines of code, result from coding errors, not from the variability of the software compilers or hardware instructions.

- Mechanical or electrical designs in which the functional continuity is interrupted with adjustments or limit stops. In these cases, the tolerance analysis or stackup is not allowed to accumulate. In mechanical designs, this is referred to as breaking the tolerance loop. Although these designs remove the necessity for tight tolerances, they are much costlier to produce because of operator adjustments and additional test equipment. The policy of using adjustments in design should be addressed in the DFM or ESI phase of the design.

In these above conditions, the design six sigma quality analysis should be performed at the higher level of the design, such as the module or systems design and architecture. The interface schemes between these design elements and the product design can be evaluated through the design quality techniques mentioned in this chapter.

### 8.3.13  The use of design quality analysis in systems architecture

In many of these cases, although the individual function or design cannot be assessed by the design six sigma quality methodology, the system architecture can be evaluated using six sigma techniques. For example, a combined software and hardware system, such as a communication system for downloading data to many remote locations, can have specifications such as total new information download time to all receivers or subscribers in less than a specified time.

In a typical communication system, variables could be software-based, such as the data transfer rate, the maximum message size, the actual data packet per frame, and the size of overhead and control bits. Hardware-based issues could also be varied such as the number of control units to channel the messages, the cabling scheme to distribute the messages, the interface converters for each receiver, and the number of receivers per interface.

An example six sigma analysis could be based on the overall specification to download all information from the central node to all receivers in the system to less than one minute. The design of the system architecture, such as different frame sizes, content overhead, timing allocation through the system, and cabling schemes can be examined, and trade-offs made to achieve maximum design quality or six sigma. The design quality characteristic to be used in six sigma calculations is based on download time versus the number of receivers in the system.

## 8.4  Applying Six Sigma Quality for New Product Introduction

Currently, many electronic products are designed concurrently with new manufacturing processes to produce them. The overall quality of the design and manufacturing processes have to be determined, and an overall quality plan has to be in place in preparation for new product introduction. This quality plan should include the design review and selection of the most cost-effective product and constituent parts and assembly specifications, using tools such as QFD, discussed in Chapter 1. The design quality analysis of major circuits, subassemblies, and modules discussed in this chapter has to be performed to determine design-related defects. These design analyses tend to be discipline-specific, and the final product analysis could include trade-offs in the design quality of these elements. In addition, current and new production lines should be optimized for least variability using DoE to ensure the attainment of the six sigma goals in design and manufacturing.

Six sigma strategy for new product introduction includes making sure that new manufacturing processes are optimized for meeting design goals and producing the least amount of variability, as well as examining the total defects generated by new product design and manufacturing in the prototype stages of design. These defects could be reduced by redesigning the lowest-rated designs, or by optimizing the final product through trade-offs in the different design disciplines.

The test strategy for the quality plan includes where and how the defects will be removed and using the most economic methods of removal. Many issues in the test strategy for PCBs and products were discussed in Chapter 4.

### 8.4.1  Optimizing new manufacturing processes

The process for implementing new manufacturing lines is almost the reverse of the one for current manufacturing lines shown in Figure 8.1. TQM and SPC tools and charts should be in place for new six sigma products. In the event that six sigma has already been achieved, and control charting is not being used because of low defect rates, other tools of TQM should be implemented, such as run charts and Pareto diagrams for monitoring and reviewing DPU (PPM) and DPMO levels of defects in manufacturing.

New production lines should be optimized for selecting the most favorable equipment, material, and processes for achieving six sigma. As discussed in the previous chapter, design of experiments (DoE) tools could be used effectively to originally design as well as survey the marketplace for the optimum choices of material, equipment, and processes. Several approaches discussed earlier could be used:

- Large experiments to evaluate different materials and the processes needed to produce them concurrently. This is the most comprehensive approach but could be difficult to achieve because of the time pressures involved in new product introduction

- Successive smaller experiments leading in the direction of steepest ascent. This could be applicable if the effort to perform the experiments and the measurement of the quality characteristic is relatively easy and quick.

- A screening experiment to quickly determine the most likely significant factor alternatives, and then a more in-depth experiment to select the best levels or processing parameters of significant factors. This is the preferred method since it offers the most efficient approach to process and product optimization.

### 8.4.2    New process optimization example: Target value manipulations and variability reduction DoE

A good example of new process optimization is the introduction of fine pitch SMT into the manufacturing process. Fine pitch SMT requires a smaller solder paste deposition of solder bricks with a target height of 0.005″, and the quality is enhanced by the variability reduction of the process. The fine pitch SMT project is a succession of small DoEs that leads to achieving the new product quality target. It can be summarized as follows, with the data information listed in Table 8.16 for the average and Figure 8.9 for variability of the SMT processing parameters:

1. The quality characteristics were defined as achieving a solder paste height in the solder deposition process with a target of 0.005″, with minimum variability.
2. The quality characteristics were measured on a test PCB containing many of the fine pitch components used.
3. Solder paste thickness was the average of four measurements in each PCB, measured at the corners of specific components. The corner represents the most difficult location in which to achieve uniformity.
4. The measurements were repeated on two PCBs, for determining variability. They were expressed as S/N for the smaller-is-better case, which is the same as −10 log variance. This S/N level was used instead of the S/N nominal formula since there were two separate analyses, one for average and the other for variability.
5. A full factorial L8 orthogonal array DoE was initially used to select the material supply for the process. Factors included the selection of the paste, stencil thickness, and the squeegee hardness.
6. For the processing methods selection, an L9 orthogonal array was used in saturation design, with four factors at three levels, including squeegee speed, pressure, down-stop, and snap off distance. The same experiment was used to analyze average and variability data.
7. The stencil was wiped off between successive prints on the PCBs. An automatic height laser machine recorded the measurements.
8. Average and variability analyses were calculated for the experiments as shown in Table 8.16 and Figure 8.9. Some of the data indications are:
   • The S/N for variability reflect mostly negative numbers due to the −10 log formula for variability conversion. The desired outcome for each factor is the level with the most positive value in all cases of variability analysis.

**Table 8.16**   Fine pitch SMT processing parameters DoE

| Factors selected: | | | | Levels of each factor: | | | |
|---|---|---|---|---|---|---|---|
| A = Squeegee speed | | | | 0.5 | 1.5 | 2.5 | ips |
| B = Squeegee downstop | | | | 0.030 | 0.060 | 0.080 | Inches |
| C = Snap-off distance | | | | 0.010 | 0.020 | 0.030 | Inches |
| D = Squeegee pressure | | | | 30 | 45 | 60 | lbs |

| | L9 (3 × 4) | | | | Orthogonal array saturated design | | | | | |
|---|---|---|---|---|---|---|---|---|---|---|
| Exp. # | A | B | C | D | A | B | C | D | Solder height | | S/N |
| 1 | 1 | 1 | 1 | 1 | 0.5 | 0.03 | 0.01 | 30 | 7.0 | 7.4 | −17.15 |
| 2 | 1 | 2 | 2 | 2 | 0.5 | 0.06 | 0.02 | 45 | 5.5 | 5.7 | −14.97 |
| 3 | 1 | 3 | 3 | 3 | 0.5 | 0.08 | 0.03 | 60 | 6.2 | 4.2 | −14.48 |
| 4 | 2 | 1 | 2 | 3 | 1.5 | 0.03 | 0.02 | 60 | 5.2 | 5.8 | −14.82 |
| 5 | 2 | 2 | 3 | 1 | 1.5 | 0.06 | 0.03 | 30 | 5.5 | 5.7 | −14.97 |
| 6 | 2 | 3 | 1 | 2 | 1.5 | 0.08 | 0.01 | 45 | 5.8 | 5.6 | −15.12 |
| 7 | 3 | 1 | 3 | 2 | 2.5 | 0.03 | 0.03 | 45 | 6.8 | 6.6 | −16.52 |
| 8 | 3 | 2 | 1 | 3 | 2.5 | 0.06 | 0.01 | 60 | 5.2 | 5.2 | −14.32 |
| 9 | 3 | 3 | 2 | 1 | 2.5 | 0.08 | 0.02 | 30 | 5.6 | 5.4 | −14.81 |

Averaging all experiments with the same factor levels:

| Factor | A | B | C | D |
|---|---|---|---|---|
| Level 1 | 6.00 | 6.47 | 6.03 | 6.10 |
| Level 2 | 5.60 | 5.47 | 5.53 | 6.00 |
| Level 3 | 5.80 | 5.47 | 5.83 | 5.30 |
| Average | 5.80 | 5.80 | 5.80 | 5.80 |

Set parameters to levels yielding closest to 0.005, and with minimum variation = speed 2 (1.5 ips), down stop 2 (0.060″), snap-off 3 (0.03″), and pressure 3 (30 lbs).

Contribution is additive yielding expected value ($EV$):
$EV$ = experiments average − ($B2 + C3 + D3$) contributions from significant factors
$EV = 5.8 − (5.8 − 5.47) − (5.8 − 5.83) − (5.8 − 5.3) = 5.0$ (target)

| Source | DOF | Sum SQ | Mean SQ | F value | SS' | p% |
|---|---|---|---|---|---|---|
| A (speed) | 2 | 0.48 | 0.24 | pooled | | |
| B (stop) | 2 | 4.00 | 2.00 | 7.75 | 3.48 | 35.26 |
| C (snap) | 2 | 0.76 | 0.38 | 1.47 | 0.24 | 2.47 |
| D (press) | 2 | 2.28 | 1.14 | 4.42 | 1.76 | 17.85 |
| Repetition error | 9 | 2.36 | 0.26 | | | |
| Total error | 11 | 2.84 | 0.26 | | 4.39 | 44.43 |
| Total | 17 | 9.88 | 0.58 | | 9.88 | 100 |

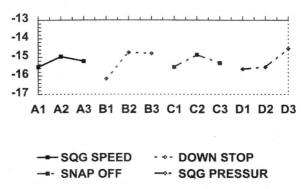

**Figure 8.9**   S/N analysis for fine pitch SMT processing variability.

- Levels 2 and 3 of Factor B (squeegee down-stop) have the same average but different variability effects.
- Experiment line 8 had the least variability, but not at the target value (0.005).
- One additional row in the ANOVA analysis, not shown in the previous Chapter 7, is the error due to replication. It is calculated from the subtraction of the total $SS_F$ of the factors from the $SS$ of the total.

$$SS_{\text{Error}} = SS_{\text{Total}} - SS_A - SS_B - SS_C - SS_D$$

- Factor A (squeegee speed) was not significant and was pooled into the error.
- The S/N graphical analysis data in Figure 8.9 was obtained from analyzing the experiment S/N data versus each experimental line, similar to the average analysis.
- Levels selected to reach the target value of 0.005 include those levels with the lowest variability. For example, $B2$ was selected instead of $B3$ because it was more positive in the S/N calculations, even if both scored the same value for the average analysis.
- Subsequent DoEs are needed to continue to account for the 44% error of the experiment. Interactions of the factors should be studied as well as more repetitions of the experiments and more factors and levels.

It can be seen that the manufacturing process was manipulated to produce an average of 0.005, equivalent to the process target. The factor and selections were made from the average table (Table 8.16), and tempered by using data from the variability graph of Figure 8.9 for those levels with lowest variability.

### 8.4.3 Trade-offs in new product design disciplines

Six sigma design quality analyses for new products include the modeling and simulation of major circuit components and assemblies during the early phase of the design. The state of the design analysis tools allows for good determination of worst-case as well as statistical analysis of designs in individual disciplines, but not for multidisciplinary analysis in complex products. The integration of the various elements and disciplines of product design is mostly performed in the prototype phase of the design cycle. The potential defect rate for the overall product manufacturing can be surmised from the prototype data. Defect reduction for the production phase might involve interdisciplinary analysis of trade-offs of design elements. The DoE tools offer good resolution of some of these problems, as shown in the case study presented in the next section.

### 8.4.4 New product design trade-off example— Screening DoE followed by in-depth DoE for defect elimination in thermal printer design

The design of new thermal printer for foil printing involved a team of many disciplines: mechanical, electrical, and software engineering. Each team member contributed to module design in his or her own discipline. When the team completed the design phase and began the verification phase prior to product introduction, it was discovered that the print quality defect level was too high. The team considered several alternatives to improve design quality. They decided to use DoE techniques in order to quickly resolve the design quality problems and deliver the new product to the customer on time and with high quality. The techniques they selected consisted of using a screening DoE to narrow down the list of possible design quality improvement factors, followed by a more in-depth DoE to optimize the remaining factors. The plan to achieve the design quality goals was as follows:

1. Identify the quality characteristics. Foil printing defects were classified and defined by the team according to three major categories: voids, which are defined as no printing when it is required; fills, which are defined as printing when none is required; and adhesion problems. Each classification was in turn divided into smaller categories. The number of defect opportunities was defined as the maximum allowed per foil card printed. It was decided to print 100 foil

**Table 8.17**  Defect classifications for printer DoE

| | |
|---|---|
| 1. Voids (lack of printed materials) | |
|    a. Transitions lines (parallel to print line) | One per card |
|    b. Nontransition lines (parallel to print motion) | One per card |
|    c. Perpendicular line missing | One per card |
|    d. Edge voids (mostly leading edge) | One per card |
|    e. Fine detail missing (lines and dots) | Circle, cross, lines, dots |
|    f. Voids (other) | One per frame and corner each |
| 2. Fills (excess printed materials) | |
|    a. Bleeding fills (excess material next to lines or shapes) | |
|    b. Bridging fills (excess material between lines or shapes) | |
| 3. Adhesion (pigment does not stick to substrate) | |
|    a. Number of dots removed per Scotch tape pull | One per card |

cards per experiment line. Only team members and not production workers inspected the cards for defects and classified them. The quality defect classification is shown in Table 8.17.

2. The team agreed on a pattern to print on the foil cards. The pattern was very difficult to print and it was designed in such a manner as to generate as many defects as possible. The pattern consisted of a completely filled square in the corner, small dots and empty squares at the lowest print resolution possible, as well as adjacent slanted lines and a cross inside a circle as close as possible to the print resolution. The pattern is shown in Figure 8.10.

3. The team decided on a screening DoE with an L8 orthogonal array with saturated design of seven factors. Other factors that the team considered not significant were kept constant through the screening DoE. Many of the levels selected were exploratory and deter-

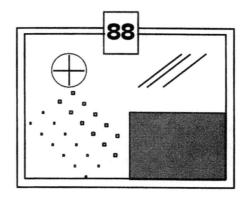

**Figure 8.10**  Printer quality DoE test pattern.

mined by varying the current design up or down by a small percentage to test its effect on the print quality. A summary of the factors, levels, and the screening experiment layout are given in Table 8.18.

4. The screening experiment was run with 100 cards for each experiment line for a total of 800 cards. The team was pleased with the amount of defects generated, as they could be analyzed for better quality. The distribution of defects was different than anticipated, and therefore some of the defects that were classified earlier were combined in the seven categories shown in Table 8.19. The average number of defects per 100 cards was 73.

5. The defect data were analyzed for each type of defects, as shown in Table 8.20. Each set of three rows is a defect-type analysis, with the preferred level for low defects shown in the top line, the actual preferred level value in the middle line, and the percent contribution in the third line. For nonsignificant factors, the rows were left blank. It was decided from the data to narrow down the number of factors to four, and fix the other three factors to the level recom-

**Table 8.18**  Printer quality screening DoE L8 design

| Factor | Symbol | Type | Level 1 | Level 2 |
|--------|--------|------|---------|---------|
| A | HA | Head Alignment | 0 | −.020 |
| B | PE | Print energy (% of normal $N$) | L(0.97%$N$) | H(1.03%$N$) |
| C | RO | Roller hardness | 45 | 60 cps |
| D | FT | Foil tension | Normal ($N$) | Reduced (0) |
| E | DC | Dot compression software | Off | On |
| F | HT | Head temperature | 35°C | 45°C |
| G | HF | Head force | $N$ (normal) | $N - 4$ lb |

Confounding (three-way) HF versus interaction of HA × PE × FT
(two-way) roller and interaction (HA × PE), DC and interaction (HA × FT), HT and interaction (PE × FT)

Fixed factors: foil pretravel (1/8″), foil material (Parker, gold), peel angle (guided), temperature/humidity (ambient, recorded), print speed (1″/sec)

| Exp. # | HA | PE | RO | FT | DC | HT | HF | Results |
|--------|------|---------|----|----|-----|------|------|----------|
| 1 | 0 | 0.97$N$ | 45 | N | Off | 35°C | $N$ | |
| 2 | 0 | 0.97$N$ | 45 | O | On | 45°C | $N$-4 | |
| 3 | 0 | 0.103$N$ | 60 | N | Off | 45°C | $N - 4$ | |
| 4 | 0 | 0.103$N$ | 60 | O | On | 35°C | $N$ | |
| 5 | 0.020 | 0.97$N$ | 60 | N | On | 35°C | $N - 4$ | 2nd best |
| 6 | 0.020 | 0.97$N$ | 60 | O | Off | 45°C | $N$ | |
| 7 | 0.020 | 0.103$N$ | 45 | N | On | 45°C | $N$ | Best |
| 8 | 0.020 | 0.103$N$ | 45 | O | Off | 35°C | $N$-4 | |

**Table 8.19**  Printer quality screening DoE defect results

| Experiment # | 1 | 2 | 3 | 4 | 5 | 6 | 7 | 8 |
|---|---|---|---|---|---|---|---|---|
| Line defects | 33 | 33 | 33 | 21 | 0 | 7 | 1 | 2 |
| Fine details | 44 | 44 | 44 | 41 | 13 | 28 | 5 | 34 |
| Voids | 22 | 22 | 22 | 22 | 1 | 4 | 0 | 9 |
| Squares filled | 11 | 11 | 11 | 10 | 3 | 11 | 8 | 11 |
| Bleeding | 0 | 0 | 0 | 3 | 7 | 4 | 6 | 0 |
| Total defects | 110 | 110 | 110 | 97 | 24 | 54 | 20 | 56 |

Average for all experiments = 73 defects

Line defects = vertical + horizontal (transitional and nontransitional)

Voids added = edge + interior voids

Bridging across small gap considered outside of capability of printer
Small squares were considered outside of the capability of printer

mended by the experiment. The three factors fixed by the experiment were:

   i. Roller hardness was set to level 1 (45 cps) since it was significant only in two defect types. Level 1 had the highest significance in bleeding defects.

**Table 8.20**  Printer quality screening DoE results analysis

| Defects | Head align | Print energy | Roller hardness | Foil tension | Dot compression | Head temperature | Head force |
|---|---|---|---|---|---|---|---|
| Lines | L2 | x | x | x | L2 | L1 | x |
|  | 020 | x | x | x | On | 35 | x |
|  | 91% | x | x | x | 3% | 2% | x |
| Fine details | L2 | x | x | L1 | L2 | x | x |
|  | 020 | x | x | N | On | x | x |
|  | 66% | x | x | 13% | 17% | x | x |
| Voids | L2 | x | x | L1 | L2 | x | x |
|  | 020 | x | x | N | On | x | x |
|  | 93% | x | x | 2% | 2% | x | x |
| Squares filled | L2 | x | L2 | L1 | L2 | L1 | x |
|  | 020 | x | 60 | N | On | 35 | x |
|  | 19% | x | 5% | 19% | 28% | 5% | x |
| Bleeding | L1 | x | L1 | L2 | L1 | x | L2 |
|  | 0 | x | 45 | 0 | Off | x | $N-4$ |
|  | 40% | x | 12% | 7% | 30% | x | 7% |
| Recommended settings | L2 020 | 100% | L1 45 | L1 N | L2 On | L1 35 | L2 $N-4$ |
| Factor carried to next DoE | Yes | Yes | No | No | No | Yes | Yes |

ii. Foil tension was set to level 1 (normal) since it was significant on most defect types except for bleeding defects.

iii. Dot compression was set to level 2 (software on) since it was significant on most defect types except for bleeding defects.

Print energy was carried on to the next DoE even if it was not significant to any defect type, because the design team wanted to explore wider variations in the print energy than the 3% used in the screening experiment

6. The in-depth DoE was performed for the remaining four factors at three levels. An L9 orthogonal array was used in saturated design. Additional levels were used to further explore the design space—two factors within the two levels of the screening experiment (head force and head alignment), and two other factors (print energy and head temperature) explored wider alternatives to the ones used in the screening experiment. The in-depth DoE is shown in Table 8.21. Each experiment line was repeated by printing 100 foil cards for a total of 900 cards.

7. The results of the in-depth DoE are shown in Table 8.22. It can be readily seen that zero defects can be obtained for certain defect types at different printer settings. Obviously, a compromise setting will have to be made for the printer and zero printing defects will be difficult to achieve.

**Table 8.21**  Printer quality, second DoE design

Set following factors: roller hardness = 45, Foil tension = $N$, Dot compression = On

| Factors Selected | | Levels of each factor | | |
|---|---|---|---|---|
| $A$ = Head alignment | 6 | 16 | 26 | Deg |
| $B$ = Print energy | 165 | 175 | 185 | mw |
| $C$ = Head temperature | 35 | 45 | 55 | °C |
| $D$ = Head force | 4 | 6 | 8 | lb |

| | L9 (3 × 4) orthogonal array | | | | Saturated design | | | | |
|---|---|---|---|---|---|---|---|---|---|
| Exp. # | $A$ | $B$ | $C$ | $D$ | $A$ | $B$ | $C$ | $D$ | Results |
| 1 | 1 | 1 | 1 | 1 | 6 | 165 | 35 | 4 | 2nd best |
| 2 | 1 | 2 | 2 | 2 | 6 | 175 | 45 | 6 | |
| 3 | 1 | 3 | 3 | 3 | 6 | 185 | 55 | 8 | |
| 4 | 2 | 1 | 2 | 3 | 16 | 165 | 45 | 8 | |
| 5 | 2 | 2 | 3 | 1 | 16 | 175 | 55 | 4 | Best |
| 6 | 2 | 3 | 1 | 2 | 16 | 185 | 35 | 6 | |
| 7 | 3 | 1 | 3 | 2 | 26 | 165 | 55 | 6 | |
| 8 | 3 | 2 | 1 | 3 | 26 | 175 | 35 | 8 | |
| 9 | 3 | 3 | 2 | 1 | 26 | 185 | 45 | 4 | |

**Table 8.22**   Printer quality screening DoE defect results

| Experiment # | 1 | 2 | 3 | 4 | 5 | 6 | 7 | 8 | 9 |
|---|---|---|---|---|---|---|---|---|---|
| Lines | 0 | 0 | 0 | 10 | 3 | 6 | 1 | 0 | 8 |
| Fine details | 1 | 40 | 39 | 8 | 0 | 9 | 0 | 9 | 0 |
| Voids | 9 | 20 | 20 | 0 | 4 | 0 | 0 | 0 | 0 |
| Squares filled | 3 | 0 | 0 | 10 | 2 | 10 | 10 | 10 | 9 |
| Bleeding | 0 | 0 | 0 | 2 | 1 | 1 | 7 | 1 | 8 |
| Total defects | 13 | 60 | 59 | 30 | 10 | 26 | 18 | 20 | 25 |

8. The in-depth DoE defect analysis and final recommendations are given in Table 8.23. Results concurred with the screening experiment findings, and included additional information when more levels were selected within the factor design space. The recommended levels of print energy, head temperature, and head force were the same as for the screening experiment. The recommended level of

**Table 8.23**   Printer in-depth DoE analysis and final recommnedations

| Defects | Head align | Print energy | Roller hardness | Foil tension | Dot compression | Head temperature | Head force |
|---|---|---|---|---|---|---|---|
| Lines | L1 | L2 | | | | L3 | x |
| | 006 | 175 | | | | 55 | x |
| | 47% | 15% | | | | 29% | x |
| Fine | L3 | L1 | | | | x | L1 |
| details | 026 | 165 | | | | x | 4 |
| | 41% | 10% | | | | x | 21% |
| Voids | L2 | L1 | | | | L1 | x |
| | 016 | 165 | | | | 35 | x |
| | 75% | 12% | | | | 7% | x |
| Squares | L1 | L2 | | | | L3 | x |
| filled | 006 | 175 | | | | 55 | x |
| | 66% | 7% | | | | 6% | x |
| Bleeding | L1 | L2 | | | | L1 | L1 |
| | 006 | 165 | | | | 35 | 4 |
| | 52% | 5% | | | | 5% | 50% |
| Final | L2 | L1 | L1 | L1 | L2 | L1 | L1 |
| recommended | 016 | 165 mw | 45 cps | $N$ | On | 35°C | 4 lb |
| setting | | | | | | | |

Confirming test with final recommended settings: 100 cards printed with recommended settings, 7 defects total (all large squares filled only), reduced from 73 in the first experiment average.

head alignment was in the middle of the two levels explored in the screening experiment.

9. The final recommendations of factor levels were run with 100 foil cards. The cards had only one failure type. Seven large squares were filled. The large squares were not really large but were printed with 3 × 3 elements of the smallest resolution possible (small squares were 2 × 2 the smallest resolution). This is quite a quality improvement from the original design settings in the screening experiment, with an average defect of 73, or a ten times quality improvements. The design team decided that this defect type is one that will rarely be used by the customer. No further experiments were planned and the product was ready for delivery to the customer with low defects and no engineering changes. Material selection and geometrical part adjustments could accomplish the changes suggested by the DoE experiments.

### 8.4.5  New product test strategy

Six sigma quality analyses of new product elements will produce the total expected defects from design and manufacturing. These defects will have to be removed by test systems at various locations in the manufacturing cycle. In most cases, there will be a postfinal assembly test, including tests for burn-in and other environmental conditions such as humidity and vibrations, to further remove latent defects and improve the reliability of the new product. The product test strategy should build on the PCB test strategy outlined in Chapter 4, and present an overall product defect removal analysis.

### 8.4.6  New product test strategy example

For electronic products, most of the defects will be generated in the PCBs, as discussed in Chapter 4. The final product might consist of many other mechanical and electrical parts, including sheet metal, plastics, and connections to other electronic boxes, signal inputs, and display units. For new product introduction, the total defects expected from all of the product components could be tallied using the design and manufacturing quality analysis, and then a plan for removing them could be implemented. A test strategy could be developed to have the proper balance between investing in improving the PCB assembly process capability or performing additional tests and troubleshooting to remove defects generated. Cost modifiers such as equipment investment and volume adjustment would certainly affect this balance.

An overall example of a product test strategy is shown in Figure

10 PCB's - 16 defects

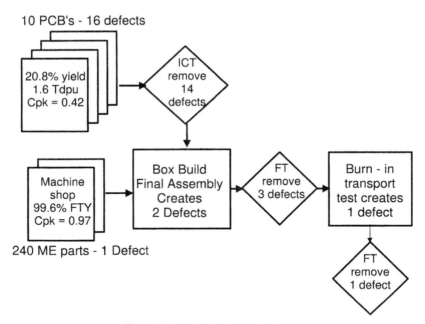

240 ME parts - 1 Defect

Figure 8.11    Product test strategy.

8.11, based on the sample new PCBs outlined in Table 8.5. The choice of removing these defects could be decided by the test strategy: which PCBs will undergo in-circuit tests, and what type of functional or system tests should be performed. The example system is made up of 10 PCBs and 240 mechanical parts and assemblies. Figure 8.11 shown an optimized defect removal scheme based on six sigma quality analysis of defects generated by design and manufacturing.

## 8.5  Conclusions

The application of quality and cost improvement techniques to the design process requires an assessment of design quality as well manufacturing process capability of the product creation life cycle. Six sigma design quality analysis can be performed at all levels including systems, modules, and printed circuit board designs as well as part selection and specifications. Examples of using quality-based analysis at each level of mechanical and electrical products and systems were shown. This statistically based analysis contrasts with the traditional worst-case analysis of design, and is shown to be compatible with six sigma design for quality techniques. In addition, special considerations such as synchronized designs, emergency shutoffs, and software modules can be examined at the system level where a six sigma analysis could be performed using the system architecture. Finally, the use of six sigma tools such as DoEs can be performed to improve multidisciplinary trade-offs in the design and analysis for high quality and low defects in new products and systems.

# Six Sigma and the
# New Product Life Cycle

The revolution in the high-technology industries has shrunk design and use product life cycles to a period of weeks and months through concurrent engineering. At the same time, traditional design and manufacturing cycles in electronics circuits, tooling, and packaging have had to be modified or outsourced to keep up with the pace of new and lower-cost product introductions. The design team has been extended through the ubiquitous Internet to include collaborative activities within the company, its customers, and suppliers. This chapter will investigate current trends in design, manufacturing acceleration, and achieving world class quality in order to establish best practices for the high-technology industries and to avoid the pitfalls of early adopters of these methodologies.

The major premises of concurrent engineering have mostly been achieved, in terms of faster time to market, colocation of the various product creation team members to increase communications and feedback, and the use of design and quality metrics to monitor and improve the design process. The challenge is how to maintain and improve these gains by leveraging the trends in the globalization of design and manufacturing resources, and the wide use of the Internet as a communication tool.

This chapter is divided into three sections:

1. Background: concurrent engineering successes and new trends. Section 9.1 is a review of the recent trends in new product creation,

including the impact of using the Internet for communications among global resources in design and manufacturing.

2. Supply chain development. The advent of the supply chain provides for new emphasis on the need to make sure that six sigma goals are achieved in a decentralized environment. The supply chain development, communications, qualifications, and management are discussed in Section 9.2 relative to achieving the overall quality goals of new products and potential problems of using the supply chain, including the issues of trading competency versus dependency.

3. Product life cycle and six sigma design quality issues. The total product life cycle stages are discussed in Section 9.3 in terms of six sigma and communications within the enterprise and expectations and goals for each stage.

### 9.1    Background: Concurrent Engineering Successes and New Trends

Concurrent engineering principles came to the fore as a strategic set of four goals for new product development: high quality, low cost, reduced engineering change orders (ECOs) and time to market, and customer satisfaction, as presented in Figure 9.1. These goals were sup-

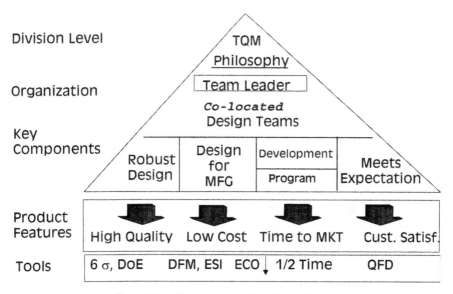

**Figure 9.1**  Concurrent engineering culture.

ported by a set of methodologies and tools such as empowered colocated cross-functional teams, integrated project management, total quality management, six sigma, design for manufacture (DFM), and quality function deployment (QFD) among many others. In addition, enabling technologies such as CAE/CAD and enterprise resource planning (ERP) allowed for large improvement in performance for design and manufacturing. Typical recorded results of concurrent engineering include:

1. Faster new product development time and a corresponding reduced design effort by at least 50%. This is the most visible outcome of concurrent engineering—allowing companies to emulate the earlier Japanese model of fast product introductions and many focused products for greater customer satisfaction.

2. Increasing quality to a level of factory defects in parts per million and a corresponding improvement in reliability, with the gradual adoption of six sigma quality and its derivatives by large corporations such as Motorola, Xerox, and GE, as well as the auto industry and many other companies.

3. Decreasing manufacturing cycles and inventory level by the application of zero inventory techniques and Kanaban (just in time) systems. The reduction of durable goods inventory ratios was from 16.3% of annual shipments in 1988 to 12% in 2000, producing a capital opportunity of $115 billion per year.

4. Although most major companies have achieved these benefits and more, their emphasis on core competency and recent trends in globalization have led to significant changes in the way business is conducted and how new products are developed and managed in most companies. The reorganization of the engineering function into distributed virtual teams, and the emphasis on keeping these teams "lean and mean," has resulted in a decline in the need for traditional discipline experts or "gurus." Nonproject design engineering positions such as "consulting engineers" or "engineering fellows" are being reduced, while more emphasis is being placed on either accessing the expertise of these individuals from one of the company's locations through the Internet or purchasing needed skills from consulting individuals or companies. Concurrently, engineering analysis tools have improved greatly, allowing for design analysis and validation in a host of different electrical and mechanical disciplines, including analog and digital circuit, mechanical strength, thermal, flow, and vibrations analysis. Initial analysis can be performed by the engineering team members, whereas in-depth analysis using advanced software packages can be de-

ferred to experts either in-house or from outside the company, since these analyses do not occur frequently enough for all engineers to master them easily.

The increase in outsourcing and supply chain growth has resulted in having many companies discard their manufacturing capability, hence becoming dependent on outside suppliers for manufacturing resources. At the same time, the cost of acquiring expensive modern manufacturing equipment has become prohibitive, and the pace of new manufacturing technology has quickened, making discreet product companies or original equipment manufacturers (OEMs) reluctant to invest in their manufacturing plants, lest they become obsolete in a short time. In addition, the advent of global competition for quality and cost has increased the need for new product design teams to incorporate design and manufacturing feedback through early supplier involvement (ESI) as well as design for manufacture (DFM) into the design of new products.

This trend toward outsourcing selected portions of design and manufacturing competency has been happening at different rates, depending on the industry sector and the maturity of the product offerings in that sector. Table 9.1 is a summary of data collected from 30 companies and 50 interviews conducted by the author. It shows that manufacturing and design outsourcing correlates strongly with the time to market pressures in the particular industry. Military program development tends to be long-range and dependent on the use of proven technology. This is contrasted with increased outsourcing in the communications and electronics (C/E) industries, which are under greater pressure to reduce time to market and are early adopters of new technology. The C/E industries are also heavy users of value added manufacturing outsourcing, in which a primary manufacturing service

**Table 9.1**  Status of companies outsourcing hardware design and manufacturing capabilities

| Sector | Number of companies Interviewed | Design outsourcing | | Manufacturing outsourcing | |
|---|---|---|---|---|---|
| | | Design | Analysis | Cost driven | Value added |
| Communications | 2 | 35% | 20% | 0% | 100% |
| Computer | 4 | 25% | 10% | 40% | 40% |
| Consumer | 2 | 15% | 25% | 50% | 0% |
| Industrial | 5 | 10% | 10% | 25% | 0% |
| Military/medical | 4 | 0% | 5% | 10% | 0% |
| Design services | 3 | N/A | N/A | 50% | 50% |
| MFG services | 10 | 10% | 0% | N/A | N/A |

provider delivers supply chain management for the total product material procurement, assembly, and test cycle. The consumer and industrial sectors rely on their manufacturing competency in their own plants, which is difficult to outsource, since they are specific to each product, and not typically electronic boxes.

Manufacturing service providers are moving up and down the supply chain to provide more value to their customers. Although most suppliers are cost-driven, focusing on one type of manufacturing competency, many are adding more value for their customers either by providing design services or managing other suppliers in the supply chain. A plastic supplier indicated: "We can turn a 25 cent plastic part into a 5 dollar assembly with the additions of electronics and cabling."

Contract design companies are also leveraging specialized core competencies in order to offer design and manufacturing services to the their customers, such as engineering analysis and access to tooling and manufacturing outsourcing in low-cost countries. In many cases, they can provide complete design and manufacturing resources for specialized subcomponents such as printers, motors, and electronic box packaging.

These rapid changes have combined with the growth of global economy to create globally competitive companies with design and manufacturing sites in many countries, simultaneously launching worldwide products over a wide spectrum of countries and customers. These companies are partnering with global suppliers to achieve the best strategy for worldwide design and manufacturing optimization of their operations. An example of consolidation of many suppliers into a few global ones is found in the auto industry, where the parts industry will shrink from 1000 first-tier companies to as few as 25 well-financed global suppliers in the future.

The principles of six sigma have also become increasingly important as a communication tool between engineering and manufacturing, as well as companies and their supply chain. Six sigma quality levels are being specified as part of the contractual agreements with the supply chain, just as are cost and delivery information.

### 9.1.1 Changes to the product realization process

The product realization process has undergone several changes with the advent of concurrent engineering. The change from a serial process of product development to a more parallel process has resulted in the need for new paradigms. Clearly, the impact of these new products is very critical, as indicated by vintage charts at different companies. In many high-technology companies, 70% of the total revenues of the company come from products introduced during the last few years. In the communications industry, it is widely recognized that the first company to market a new product captures 70% of market share. This is the result of shorter life cycles for products, as shown in Figure 9.2. It can also be seen in the figure that the R&D investment as a percentage of total sales also increases with the decrease in life cycles.

Traditional product development required top-down control of the various activities of product creation. Very formal organizational structures were developed and managed with a phase review process. Plans and milestones had to be completed at the end of each phase of product development, and were subject to several levels of management reviews. After each review, the project was allowed to proceed and be funded until the next review.

The pressure toward shorter project time frames, global teams, quality, and design and manufacturing outsourcing have resulted in significant changes in the relationship between the company personnel and their suppliers, with more frequent communications occurring earlier in the product development cycle. These suppliers and their own subsuppliers are called the supply chain. The changes can be summed as follows:

1. Less frequent formal milestones in the development process, but many more smaller informal meetings, most of which are one-on-

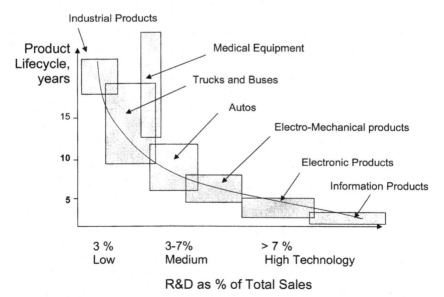

**Figure 9.2** Life cycle models for different products.

one engineering interactions discussing the merits of design functionality, engineering analysis, design validation, quality goals, and manufacturability feedback. There is a need to quickly have these meetings available to all interested engineers without having to hop on the next plane and meet in person. The Internet provides a good environment for communications between members of a virtual design team. As one design manager explained, "OK, I have a problem . . . let's put an agenda together, and let's get together (on the Internet)."

2. The trend toward faster time to market has necessitated earlier than usual release of hard tooling. Although investing in rapid prototyping and soft tooling methodologies can minimize risks, production tooling is being launched much earlier than before. One manager of a plastics supplier reports, "Steel is being cut for the tooling before design reviews are completed."

3. The selection of the potential supplier(s) has to occur very early in the design stage, without the benefits of being able to select suppliers from a bidding process on a complete set of product documentation. Therefore, the selection process will depend on intangible issues such as the history and financial position of the suppliers, their communication methodologies with their customers, and their demonstrated quality levels and cost models.

4. Though supplier exchange networks have sprung up in many industries, their focus will be on commodity items, not on outsourced designs. The product companies (OEMs) are using various methods to obtain manufacturability feedback prior to awarding contracts, such as inviting a selected list of potential suppliers to provide ESI information before contracts are awarded. These additional ESI costs will have to be negotiated between the suppliers and the OEMs or become part of the operating overhead structure of the suppliers.

5. The trend toward lower manufacturing costs, greater quality in six sigma, and design for manufacture (DFM) is changing new designs into fewer but more complex major parts. In addition, the pressure toward faster product development is also increasing the amount of information available on engineering drawings. Engineers prefer a smaller number of drawings with the maximum amount of information attached to them, such as showing as many parts as possible in one drawing, as well as adding assembly information and special instructions.

6. As a result of the above items, the need for increased communications between distributed project teams, design analysis and support experts, and manufacturing resources in the company and

## Traditional Communications:

*Well defined transfer milestones - Formal communications*

## Concurrent Engineering Communications

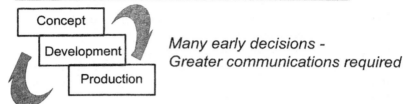

**Figure 9.3** Traditional versus concurrent engineering project communications.

their supply chain is expanding. There have been many technological developments in communications between the various stakeholders in the design process, which will be discussed later in this chapter.

A new model of the product realization process is summarized in Figure 9.3, showing the transition from the traditional serial development process to concurrent engineering product realization, with less formal reviews and much greater communications required to speed up product development time.

### 9.2 Supply Chain Development

The use of the supply chain is changing as the manufacturing services (or the supply chain) sector continues to grow. Chip foundries that are providing the baseline silicon for major OEMs such as Motorola and Texas Instruments are expected to increase their share of world semiconductor production from a fraction today to 35% by 2010. In addition, contract electronic manufacturers (CEMs) are expected to increase their share of electronic products assembly in similar fashion.

The major supply companies have mimicked the OEMs reach by distributing their manufacturing centers globally, to be near their customers' sites. In this manner, supply companies can service global OEMs. The issues of the global supply chain can be summed as follows:

1. The suppliers are focused on their customers' issues. Time to market is one of the most important goals for the communications and electronics industries. The supplier companies have set up plants around the world to react quickly to their customers' demands. One OEM CEO declared, "In all of our facilities around the world, . . . with the exception of burned in products . . . , nothing takes longer than a day to build."

2. The location of the supply companies has also to do with their ability to provide increased service during the prototype phase of product development. One manager of a CEM indicated that, "Locating my plant near my customer can save two days of FedEx time for prototypes."

3. The OEMs are forcing their suppliers to conform to their design specifications. For example, most OEMs will specify that the design and manufacturing documentation from suppliers must conform exactly to their in-house CAE/CAD systems, including the system type and model number. Since most OEMs collectively use at least half a dozen design systems, this is forcing their suppliers to maintain several CAE/CAD systems with operators knowledgeable in more than one system.

4. The OEMs are also asking their suppliers for final testing, including troubleshooting of their products and systems. In these cases, suppliers are absorbing the cost of the training programs for test technicians. To make matters worse, these technicians are being wooed by competitive suppliers, and even by the OEMs themselves. This is resulting in wage competition, raising labor costs for skilled supplier personnel.

5. Increased dependence on supplier quality and lower cost goals have resulted in eliminating incoming inspection for parts, making companies vulnerable to spurious quality problems in the supply chain.

6. The trend toward increasing the links in the supply chain by further subcontracting to achieve even lower-cost manufacturing has resulted in low-technology suppliers getting into the manufacturing cycle for high-technology products. These suppliers do not have the sophisticated technology or the controls in place to make sure that all necessary specifications are inspected and variances in quality are promptly reported up the supply chain. For an OEM, a poorly managed supply chain is vulnerable to quality problems if changes are made in the subcontractor chain without their approval or notification. It is recommended that the supply chain not extend beyond three levels down from the final assembly.

### 9.2.1  Outsourcing issues

As companies rush to outsource design and manufacturing, they are concerned about several issues. Among them, does outsourcing really save money? What should be outsourced and when. What should be done with the remaining competencies in-house. How should some of the dependency problems of outsourcing be avoided?

One of the engineers for a major OEM that explained, "I do not see how design outsourcing makes any sense. We normally have two or three MEs [mechanical engineers] designing the "box," . . . now it takes two or three MEs just to manage the design contractor!" Another manufacturing engineer said, "My company has decided to contract manufacturing outside; we want to send out our oldest products . . . and no contractor wants to do it at a reasonable savings to us."

Both these quotes highlight a common problem with outsourcing. Upon further investigation, both cases are the same: the tendency of OEMs to begin outsourcing either older products or designs that are just about completed, leaving new designs and products to remain within the company. The difficulty then is the transfer of large amount of knowledge about these older products, together with their nonstandard methodologies and operations. It is best to begin outsourcing at the beginning of a new product cycle, so that the competencies of both the OEMs and their suppliers are maximized.

Outsourcing should be implemented in various steps, according to the company's needs. Table 9.2 is a summary of the common issues in outsourcing, ranked by importance, both in terms of what to out-

**Table 9.2**  Common issues in selecting outsourced products and competencies

| What to outsource | Minimum competencies needed to manage supply chains |
| --- | --- |
| Commodity product/assembly with standard interfaces, manufacturing, cost, and quality requirements | Identify/select qualified suppliers Write appropriate system specifications |
| Product/assembly with well-defined interfaces | Evaluate incoming bids |
| Product/assembly requiring multidisciplines | Validate deliverables and meet specification |
| Product assembly containing new technology | Questionable in-house competencies |
| Product/assembly with associated system integration | Improve submitted bids |
| Basic core competency product or assembly | Help supplier technically in design/manufacturing |
| Basic new product with many characteristics needing company evaluation, justification, and ROI | Help supplier operationally; training Improve deliverables after receipt |

source, and what are the necessary remaining competencies in the company for managing the supply chain.

Outsourcing for design or manufacturing occurs for two basic reasons: capacity or competency. If the design of the product or system is well partitioned, then outsourcing for capacity is easily accomplished with minimum effects on the company retaining its competency. Outsourcing for competency should be carefully selected, so that the company retains its desired or core competencies.

### 9.2.2 Dependency versus competency

The advantages of outsourcing are many: the enterprise does not have to keep abreast of noncore manufacturing or development technologies, saving engineering resources, capital equipment, and large payrolls. The enterprise can easily expand and contract in response to business conditions and product introductions without the burdens of hiring or firing, and is free to seek the lowest-cost contractor, especially those that can leverage their size into low-cost material procurements. As one communications OEM CEO remarked, "We can grow to $1 billion and never have to spend a cent to expand a plant or upgrade a computer."

Outsourcing can take many different forms, depending on the company's willingness to increase its dependency on its suppliers, starting with manufacturing then moving on to development and design. As each outsourcing scenario is embraced by the enterprise, the concern to maintain competency and reduce dependency has resulted in maintaining some unnecessary competencies in-house.

In the initial stages of manufacturing outsourcing, Most OEMs are concerned that unused competency will disappear. While direct labor is contracted away, more skilled resources, such as manufacturing and process engineers, are kept on to manage in-house prototype shops or outside supply chains. Lamented one manufacturing manager in a telecommunication company, "We used to design our manufacturing process for our products to last for 40 years, . . . the lifecycle of new products has shrunk now to months instead of years."

In some cases, as more manufacturing is outsourced, there is no comparable reduction in the manufacturing staffing and expertise within the company, reducing the benefits of outsourcing. The author interviewed many companies who are successfully outsourcing without the benefit of internal competency in manufacturing processes, relying on early supplier involvement for DFM feedback and rapid prototype services.

A similar effect is also occurring in design outsourcing. As an example, OEM design engineers will perform all of the design specifications, connectivity, testing, and analysis required for the outsourced product component or assembly. They will specify the design space of the outsourced component, model and analyze the internals, and study the mechanical interference and performance specifications in relation to the overall product. They design the mechanical interface and electrical connection to the rest of the product, then perform validation for the design. More efficiency could be achieved by allowing the suppliers to increase their contribution to the design. An ME manager from a consumer company who contracted the design of a battery system said, "We defined the envelope; the design house can decide on the geometry inside the envelope, what size of pins, connectors, etc. . . . They send us back an FEA based on their design. We approve the design. Then they can go ahead and tool-up and produce qualifications and validations reports."

### 9.2.3  Outsourcing strategy

Business concerns should be paramount in developing a good outsourcing strategy. There should be synergy between the business model of the company and its outsourcing efforts. Some of the issues of proper outsourcing strategy and the selection of a supply chain are:

1. The types of customers in the industry that the company operates in. Issues such as customer expectation in cost, quality, reliability, life cycle, support, deployment, and speed of delivery are paramount in selecting the proper supplier that is focused on these issues, and may be supplying other companies in the same market.

2. The type of market environment, including other competitors supplying similar products. This would include typical financial metrics prevalent in the particular OEM industry, such as profit margins, material overhead and efficiencies from contracting, cost of goods sold, product turnover timelines, development and technology investments, workforce skill level required, and timely cost improvement efforts.

3. The expectations of product operations, such as six sigma design, reliability, cycle time, system assembly and configurations, revision control and upgrade policies, as well as any design, manufacturing, or quality standards required.

4. Business concerns have to be addressed in terms of forecasting the impact of outsourcing on the business indicators for the company. Such indicators include absorption ratios of overhead, materials, and direct labor; cost of goods sold ratios; material and labor efficiencies from supplier material procurement and equipment lever-

age; as well as adequate planning and reporting of cost improvement projects.

5. Legal issues should be considered carefully to eliminate potential liability. These include the instructions to bidders and subcontractor documentation such as their quality plan and certificates of compliance, warranty for delivered parts and products, subcontractor management, delivery schedules, liability for late delivery, and conditions of forecast demand and how to manage changes to the forecast.

A plan to formulate the outsourcing strategy can be divided into three parts: an outsourcing competency matrix, a selection process, and a communications plan.

A competency matrix summary is presented in Figure 9.4, which shows the organizational competencies needed for manufacturing the product. The content and resources for each competency should be examined as outsourcing decisions are made, including whether to outsource a particular competency partially or fully, the impact of outsourcing on staffing levels for each competency, and the need to support and manage outsourced competencies.

The selection of the supply chain should occur at the concept stage of the product. The selection and bidding processes should be based on historical relationships. Suppliers could be initially qualified and then bid their cost model for the design and manufacture of the product.

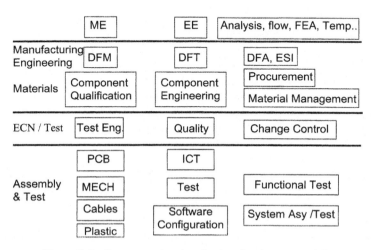

**Figure 9.4**  Core competencies chart and outsource matrix.

The supply chain qualification process is dependent on industry standards of cost, quality, and timeliness. Cost should be well quantified through a cost model based on activity-based costing. Quality should be quantified by six sigma, which outlines comprehensive methodologies for controlling, maintaining, and implementing quality programs. Timeliness should be measured in terms of turnaround time of orders and historical delivery performance of the supplier, and should include any incentives for early delivery. The supply chain should also provide the company with their technology and equipment acquisition plans, as well as plans for reducing cost over time through more efficient process operations.

The interface between the supply chain and the company's remaining core elements should be the same as for in-house capabilities. The supply chain should provide for ESI and design guidelines to their manufacturing capabilities and constraints very early in the product realization process. The communication links between the supply chains and the enterprise should be as easy to implement as the in-house ones, including regularly scheduled meetings such as design reviews, as well as on demand meetings to discuss problems and their resolution. The communications should be instantaneous as well as simultaneous; supplier company engineers should be equipped with personal communications devices for 24/7 access. As one design house manager put it, it should be possible to "meet your consulting engineer anytime you desire."

### 9.2.4 Supply chain communications and information control

Company/supplier communication process have developed over time to provide human facilitation as much as possible. Suppliers have placed their plants strategically near their customers, and have encouraged the temporary placement of employees at each other's plants. They have scheduled frequent meetings to address cost, quality, and delivery issues. Table 9.3 describes of these communication process changes with use of the Internet for the supply chain, a process referred to as e-supply.

One result is the increase of virtual meetings using the Internet, involving smaller groups or engineer-to-engineer interactions, and a decrease in the bigger meetings involving larger groups discussing scheduled topics such as quality, performance, or design reviews. By making data available on-line, suppliers can quickly discuss pertinent engineering issues and update their performance indicators for remote access by their customers.

Tooling suppliers can communicate the latest information about

**Table 9.3**  Supply chain communications

| Traditional | e-Supply |
|---|---|
| Suppliers placing personnel in OEM factories | Suppliers communicate through the Web |
| Suppliers/OEM meet face-to-face on a regular basis | They meet as needed via the Web |
| Small number of formal design reviews | Many minireviews, with smaller groups |
| Suppliers/OEM meet regularly for quality and cost data | Data supplied through remote access |
| Large companies do not engage small design houses | Design houses use the Web to leverage partners |
| ECN changes signed off and distributed in days | ECN changes dissemination in hours |
| Operational data flows link-to-link through supply chain | Operational data shared immediately |
| Engineering data (DFM/ESI) provided serially by phone/fax/e-mail | Decisions and data Web shared |
| Problems in the supply chain resolved serially | Immediate problem resolution process |

hard tooling and suggested improvement to designs. One electronic company was able to reduce monthly trips to their plastic and tooling supplier in Asia from once per month to once per quarter, using Internet-based, engineering data collaborative systems. Given that the project manager and two engineers traveled each month, the savings could be substantial over the lifetime of the development project, which is estimated at 18 months.

Using the same communications technology, small design service companies can enhance their services by augmenting their competency with connections to their own supply chain to provide design, analysis, tooling, and production capabilities. In this manner, a design service company with less than a dozen engineers can deliver global design and manufacturing resources to Fortune 500 companies.

With the advent of Internet-based engineering communications, ECO processing could change from days to hours, given that all parties in the supply chain can communicate effectively and in real time using advanced engineering-based communication tools. This will help companies in the supply chain improve their services by reducing scrap as well as purchasing the proper materials on time.

Supply chain management involves the outsourcing of all of the control and communication issues for a large portion of the design and/or manufacturing of a new product to one or a small number of major suppliers. This supplier in turn can outsource some of the subcomponents of the system to a subcontractor with a special competency in a particular component or subsystem. If not managed properly, suppliers are free to achieve the lowest cost by subcontracting some of their design and manufacturing to the lowest bidder worldwide, with possible negative consequences.

The management of the supply chain in many companies is through the commodity management model. The companies retain control of the supply chain by identifying commodity engineers and managers (usually former production staff from manufacturing operations that were shut down) to manage the supply chain by commodity or discipline. This model is very inefficient, as the information to the supply chain has to be distributed, then funneled through these individuals. This causes delays and bottlenecks, especially when there are engineering changes and quality issues. The commodity managers tend to stay focused on their own commodity disciplines and not have a broad overview of problems and their possible impact on other areas. The sequential supply chain model, first practiced by the auto industry, involved a hierarchy of suppliers called tiers. The OEM company manages the tier-one suppliers, who in turn manage several tier-two suppliers and so on until the third tier. These relationships are shown in Figure 9.5. The communication system to manage all of the information for the total supply chain is very important, so that all elements of the supply chain can instantly react to quality problems and engineering changes to rectify them.

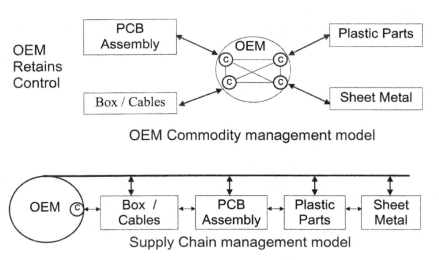

**Figure 9.5** Supplier management models.

It is readily apparent that the complexity of a high-technology products increases as the assembly level increases. At the same time, it is normal to expect that a subsupplier to another supplier would possess little or no electronic or high-technology comprehension or knowledge. In fact, most of the subsuppliers would not recognize or fully appreciate what their products might be used for. This lower level of competence is contrasted with the need for much more quality at the assembly level, and is of most concern for the product system test, where in-depth knowledge of system requirements such as cabling and interfacing with other electronic products is required. A well-managed supply chain is required to ensure conformance and quality within all of the supply chain links, and to limit the depth of the link to a maximum of three levels of tiered suppliers.

### 9.2.5 Quality and supply chain management

There are two guiding principles of quality practices for high-technology supply chains: (1) do not generate any defects within your span of control, and (2) do not pass on any defects to the next link in the chain. Quality is controlled by the use of adequate tests at different stages throughout the supply chain. In addition, it is generally assumed in the electronics and high-technology industries that the lowest total cost of quality is afforded by testing at the lowest level of assembly possible. Defects are more expensive to find and remove at higher levels of product assembly, so they should be found and removed at the lowest level. Unfortunately, that level has the lowest competency of the chain.

The supply chain management system is critical to overseeing and controlling the competency and operational data for the chain. The ability to handle and distribute technical as well as operational data instantly throughout the chain is very critical. When that communication is not properly enabled, or it breaks down, results could be very catastrophic in terms of unusable product.

In one particular case that the author is familiar with, a supply chain with four companies contributing to the build-up of the final electronic product was faced with a serious manufacturing quality problem. The four links represented the manufacturing and testing of printed circuit board lamination, fabrication, assembly, and final product assembly. The bottom link of the chain, the laminator, subcontracted some of the work to another company, which did not perform to specifications. Due to the lack of instant engineering communications between the links, this nonconformance problem was not detected until the product was in the customer's hands. This rendered the total inventory in the supply chain defective. In addition, communication problems made the eventual discovery of the cause more difficult, and hence valuable sales and recovery time were lost. The disposition of millions of dollars of unused product had to be determined through legal deliberations.

The supplier management program is a cornerstone of quality assurance in a supply chain. It consists of many steps including:

1. Qualification. The qualification process begins with a list of potential suppliers, which are then audited either by reviewing their documentation (both financial, manufacturing, and quality procedures) or visiting their manufacturing facilities. During the visit, the team can review the supplier procedures and conformity to international standards of manufacturing and quality, such as six sigma, ISO 9000, good manufacturing practices (GMP), and the various IPC standards.

2. Ratings. After the initial approval, the supplier is usually placed in an approved status, and purchase orders can be placed and goods received with good incoming inspection and testing procedures.

3. After a period of time, and with increased communication and confidence that the supplier has demonstrated their capabilities and quality in a consistent and continuous manner, a supplier might be placed in a preferred status. The company would be motivated to place orders with this supplier, knowing that less testing and inspection of incoming products would be required.

4. The next step in supplier status is full partnership or sometimes a preferred ranking such as "A" preferred. In this case, there might be close ties with the supplier/partner; purchase orders might be placed without bids and supplier parts might go directly into the company's stock as "ship to stock" or "ship to dock."

5. Audits. In the process of qualifying or rating suppliers, quality audits are performed to ensure conformance with industry and company standards. The quality audits could take the form of visits, actual or through the Internet, to the suppliers to check on performance, material testing, and inspection above and beyond normal test and inspection procedures. Dimensional data can be directly transferred from the company's CAD system to the inspection equipment.

The level of incoming inspection is dependent on the status of the suppliers. For a nonqualified or new supplier, there should be an extended incoming inspection and testing program. On the other hand, a preferred supplier should be in a position to ship to stock (if agreed), where their materials are received directly into the stockroom with no incoming tests or inspections. In addition, a supplier whose materials are to be altered, such as fabricated PCBs, should have a greater level of inspection and testing, whereas a supplier of materials that will stay intact in the product, such as PCB assemblies, might be subject to a reduced incoming inspection program. Another method of substituting for the incoming inspection function is to ask the supplier to provide certificates of compliance and/or testing to specifications. The requirements or test certificates should be mutually agreed upon to include the specifications that are relevant to the product being manufactured.

### 9.2.6  Supply chain selection process

The qualification process outlined earlier is useful for selecting bidders for built-to-specifications or custom designs in an e-supply marketplace. In these cases, two issues are immediately apparent: first, how to bid on product specifications when the detailed design is not available, and second, how to factor in the DFM and ESI issues. Several methods can be used for both cases: the bidders can quote against an older but similar in functionality design, and the bidders can input their ESI feedback on the new design specifications in a collaborative session with no audit trail, as mentioned earlier. In the latter case, the cost of ESI input is included in the overhead burden of the bidders.

In most cases, the differences in bids can be very small, especially in manufacturing outsourcing, and the selection process could be based on other intangibles such as the willingness of the company to absorb some of its overhead in order to ensure winning the bid.

A selection process consisting of two steps should be applied: qualifying potential suppliers who meet a minimum set of financial, operational, and technology requirements, and then comparing the qualified suppliers through a supplier matrix, shown in Table 9.4. The matrix is based on a criterion rating system of comparing alternatives. The maximum score is the value that can be attained by a supplier for a particular criterion, if they have met all expectation. For each supplier, the score is multiplied by the weight of each criterion, under the supplier column. Each criterion is composed of many subcriteria in order to render a complete analysis of the decision. An example of a subcriteria matrix for quality is given in Table 9.5

An example of a supplier selection for the assembly of PCBs for the communications industry is given in Table 9.6. A sample PCB was provided to the bidders, who were asked to break down their costs to include additional information such as the material costs, the NRE (nonrecurring expenses of tooling), their corporate materials leverage,

**Table 9.4** Weighted criteria for supplier selection matrix

|  |  |  | Alternative suppliers | | | |
|---|---|---|---|---|---|---|
|  | Weight | Maximum score | A | B | C | D |
| Quality | 30% | 96 | 91/27.3 | 90/27 | 88/26.4 | 48/14.4 |
| Process | 15% | 120 | 110/16.5 | 110/16.5 | 93/14 | 68/10.2 |
| Service | 10% | 58 | 44/4.4 | 47/4.7 | 53/5.3 | 39/3.9 |
| Delivery | 20% | 63 | 49/9.8 | 50/10 | 44/8.8 | 29/5.8 |
| Cost | 25% | 57 | 43/10.8 | 39/9.8 | 45/11.3 | 35/8.8 |
| Total | 100% | 78.5 | 68.8 | 68.0 | 65.8 | 43.1 |
| Percentage of possible maximum score | | | 87.6% | 86.6% | 83.8% | 54.9% |

cost reduction schedule over time, and warranty period. It was assumed that DFM and ESI will be accomplished through having the new design follow established industry standards, and that special components such as proprietary ICs are to be supplied by the company or its own supply chain.

It is apparent from Table 9.6 that the two lowest bidders are very close in their submissions, the difference being lower than half a percent. In this case, other intangible factors could leverage the bid selection process, or the company could go back to the two finalists and ask for an additional bidding cycle.

When transferring these procedures to the Internet using e-supply, it is expected that large savings will be realized. They will occur through quicker decision making in the selection and negotiations processes. This is accomplished by using only approved suppliers with design and manufacturing procedures and systems that are compati-

**Table 9.5** Weighted quality criteria for supplier selection matrix

|  |  | Alternative suppliers | | | |
|---|---|---|---|---|---|
| Criteria | Maximum score | A | B | C | D |
| ISO 9000 certified | 10 | 10 | 10 | 10 | 3 |
| Manufacturing standards | 10 | 10 | 10 | 8 | 8 |
| Statistical process control | 10 | 8 | 9 | 10 | 1 |
| Quality diagrams available | 10 | 10 | 10 | 10 | 5 |
| Test failure reporting | 10 | 8 | 7 | 10 | 5 |
| Incoming inspection | 10 | 10 | 10 | 10 | 4 |
| Defect analysis | 8 | 8 | 8 | 5 | 8 |
| Obsolete material | 8 | 7 | 6 | 8 | 4 |
| Design for manufacture | 10 | 10 | 10 | 7 | 5 |
| Continuous quality improvement | 10 | 10 | 10 | 10 | 5 |
| Total | 96 | 91 | 90 | 88 | 48 |

**Table 9.6** Comparison of PCB assembly costs

| Metrics | Supplier 1 | Supplier 2 | Supplier 3 | Supplier 4 |
|---|---|---|---|---|
| Cost of sample board | $700.00 | $728.52 | $703.00 | $738.91 |
| Materials/Sample board cost | N/A | $531.60 | $521.06 | $518.67 |
| Materials divisor | N/A | 0.7297 | 0.7412 | 0.7200 |
| NRE for sample board | N/A | $15,924.00 | $6,325.00 | $19,289.00 |
| Material vendor warehousing | Yes | Yes | Majority | Minority |
| Purchasing leverage | $225M | $500M | $15M | $120M |
| Frequency of cost reductions | Quarterly | Quarterly | Quarterly | Quarterly |
| Typical reductions achieved | 4–6% | 4–5% | N/A | 10% |
| Workmanship warranty | 12 months | 30 days | 12 months | Negotiable |

ble with the company's. In addition, the communication loops will be made much shorter through use of data-rich engineering collaboration systems, resulting in quick decisions and lower-cost designs through reduced engineering changes.

It is also expected that the engineering effort for e-supply will be reduced through the use of formal procedures for specifying designs and awarding contacts. Nonengineers such as procurement personnel can replace engineers in selecting design and manufacturing suppliers, thus reducing the cost of new products. In the long term, subcontracting design resources from the company specifications will be a more efficient process, because the DFM/ESI feedback will be wholly within the supplier domain.

Shifting design engineering resources to a supplier might result in a competitor engaging that supplier and benefiting form the expertise gained from the company's design practices. This concern is lessened in mature technology products, such as the auto industry, where the emphasis is on design and manufacturing standards as well as lowering costs. In the fast changing technological markets such as electronics and telecommunications, competing companies can leverage each other's competencies through the use of common suppliers, leaving them to concentrate on core competencies and new technologies.

## 9.3 Product Life Cycle and Six Sigma Design Quality Issues

The need to develop new products at an accelerated rate and the shortened life cycle of many electronics products have led to increased need for good quality design evaluation through the six sigma methods discussed in this book. This shorter life cycle is caused by the speed of technology improvements and competitive factors. New products are replacing existing products with more capability and performance at a lower cost and higher quality, while expanding the market by satisfying more customers.

An example of this technology impact has been quite evident in the personal computer industry and is shown on Figure 9.6. It can be seen that new products are released to the marketplace in an accelerated fashion due to improving technology. Customer expectations are raised for the new technology, and once the new product is announced, the sales for the older technology disappear. Therefore, the new product has to ramp-up to mature volumes very quickly.

There is little time for quality defects or the resulting engineering changes. Quality problems can do great damage to a company's reputation as they force a delay in introducing the new product while demand for older products evaporates. In addition, as mentioned earlier, in some technology-based industry segments, the company with the first product to market gets 70% of market share. A major engineering change order (ECO) for some of these companies could result in a loss of acquiring the current technology, and the company might have to delay the product introduction until the next technology cycle is available. This is because the end of each product generation life cycle is fixed, determined by technology improvement, market forces, and competitive factors. This is very costly, as a one month slip in product introduction is one less month of sales, as well as loss of customer satisfaction.

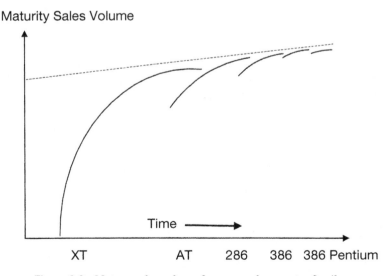

**Figure 9.6**   Mature sales volume for personal computer family.

Six sigma quality analysis methods discussed in this book can reduce defects in design and manufacturing and protect the company at the critical transition time between products in a product family life cycle plan.

### 9.3.1    Changes in electronic product design

During the last decade, advances in high-technology industries have accelerated. The price performance ratios continue to follow the industry idioms of more performance for lower price. Intel's Gordon Moore first proposed the law that bears his name in the late 1960s: chip complexity (as defined by the number of active elements on a single semiconductor chip) will double about every device generation, usually taken as about 18 calendar months. This law has now been valid for more than three decades, and it appears likely to be valid for several more device generations. The capacity of today's hard drives is doubling every nine months; and the average price per megabit have declined from $11.54 in 1988 to an estimated $0.02 in 1999.

Similar improvement has been occurring in the field of communication, both in the speed and the availability of the Internet. It is estimated that global access to the Internet has increased from 171 million people in March 1999 to 304 million in March 2000, an increase of 78%.

At the same time, the requirements for developing new products in high-technology industries have followed these improvements, with faster product development and shorter product life cycles. Many of the leading technology companies have created a "virtual enterprise," aligning themselves with design and manufacturing outsourcing partners to carry out services that can be performed more efficiently outside the boundaries of the organization. These partnerships enable a company to focus on its core competencies, its own product brand, its customers, and its particular competency in design or manufacturing.

These newly formed outsourcing companies are providing for cost-effective and timely services. In manufacturing, they provide multidisciplinary manufacturing, testing, and support services, including printed circuit board (PCB) assembly and testing, packaging technology such as sheet metal and plastic injection molding, and software configuration and support services such as repair depot and warranty exchanges. They also offer lower cost, higher flexibility, and excellent quality, eliminating the need to spend money on capital equipment for internal capacity. This new outsourcing model allows all links in the supply chain to focus on their own core competencies while still reducing overall cycle times.

In design outsourcing, the supply chain offers the flexibility of single or multiple competencies, including specialized engineering analysis and design validation, testing, and conformance to design standards for multiple countries or codes. In addition, suppliers can offer their own supply chain of strategic alliances in tooling and manufacturing services worldwide. Most of these outsourcing companies offer design feedback in terms of design for manufacture (DFM) through early supplier involvement (ESI). These design service providers have reduced the need for high-technology companies to purchase or maintain expensive engineering and design competencies, some of which are used infrequently in project design cycles.

### 9.3.2  Changing traditional design communications and supplier involvement

The advent of Internet communications and the supply chain have provided an opportunity to increase design efficiency while maintaining the gains achieved from concurrent engineering. Distributed resources can be easily accessed to collaborate with and augment the design effort. These include country-specific requirements for global products, outside design and analysis expertise, and design and manufacturing service providers' ESI feedback.

It is important that the distributed design team, with associates in other locations, as well as the design and manufacturing service providers, collaborate *real time and at the same time.* Collaboration should be real time and not asynchronous, able to reach anyone, anywhere in the world efficiently. A problem should be quickly resolved by bringing the team virtually together before it becomes a "show stopper." As one design house CEO proclaimed: "one shared mind, one shared moment . . . minds being there when you need them."

The design process phases in most industry sectors are similar. With increased collaboration, each phase can be optimized. A summary of attributes and metrics of success for each design phase is given in Table 9.7.

**Product concept phase.** This is the period of market research, product functionality definition, development process methodology, performance measures, supplier selection, and developing initial cost, sales, and profit targets for new products. During this phase, only a small team of engineers and marketers is working on the product idea and they keep improving on it. This period is completed when product feasibility is demonstrated and approval is received from management for commitment to the final concept(s) and proceeding to development. At that time, resources are identified and committed in terms of personnel and equipment, a return on investment (ROI) analysis is

**Table 9.7**  Attributes and metrics of success for each design phase

| Phase | Necessary attributes for success | Metrics of success |
|---|---|---|
| Concept | Fast access to technology resources<br>Communication with customers/marketplace<br>Feedback evaluating design alternatives<br>Select and negotiate with suppliers<br>Reduce concept ideas to 1 or 2 ideas | High success switching to development<br>Number of inputs: manufacturing, design, sales, and service<br>Maximize number of design alternatives<br>Leverage of new technology in product<br>Concept phase duration |
| Output early development | Prototype functionality demonstration<br>Design analysis and simulation<br>Standardized design methodologies<br>Less formal design reviews<br>Rapid prototyping/soft tooling<br>Minimum design changes<br>Early supplier involvement/DFM<br>Component qualifications | On-time design reviews<br>Metrics (DFM, DFX, cost models)<br>Frequent informal design reviews<br>Turnaround time for prototypes<br>Fast implementation of changes<br>Communications with suppliers<br>Component engineering process |
| Output late development | Final concept validation<br>Detail design models<br>Quick resolution of design/MFG problems<br>Hard tooling commitments<br>Environmental/life testing plan and analysis<br>Material procurements | Interfacing models to other systems<br>Communication process for resolution<br>Communications with suppliers<br>Communication with test houses<br>Communications with global suppliers |
| Output | Product launch plans: Material, manufacturing, and distribution | |

completed, a design schedule is agreed upon, and deliverables outlined for the design completion.

An important change to this phase is the concept selection process. Many ideas are floated to achieve the product in look and feel as well as functionality. These ideas should be solicited from everyone in the company, including overseas sites, as well as customers and the supply chain. As many product ideas as possible are encouraged, to explore the maximum range of thought and techniques. Quipped one design manager, "no idea is a bad idea at this stage." These ideas are distilled down to one or two that are then developed.

Free flow of communications between the core concept team and the other stakeholders of the design process—marketing, various engineering expertise, manufacturing, and the supply chain—is important at this phase. Although there are no detailed technical drawings of the product, rough CAD-generated outlines or wire mesh frames can be used to model the look and system layout of the product. These outlines can be the subject of ESI and DFM discussions with suppliers and manufacturing, and local and country-specific inputs can be solicited in order to provide for global rollout of single products. In addition, the concepts that did not make it to development should be carefully documented so that they can be evaluated in later-generation products.

**Product development phase.** This phase is usually divided into two parts:

1. Early development phase. During this phase, activities are initiated to complete the detailed design model on CAD, so that engineering analysis of various disciplines can be performed for mechanical strength, vibration, thermal characteristics, and drop performance of the design. Prototypes are made with rapid prototyping or soft tooling in order to evaluate the fit and performance of the product and validate the design concept. Depending on the desired level of testing and tolerances, machined parts or stereo-lithography prototypes are made to simulate final plastic, cast, or forged parts. Details of the final design can be omitted or substituted in this phase to accelerate the testing. Examples of this could be to substitute screw holding or gluing for snap fit parts in the prototypes. Most OEMs and suppliers do not use the drawings to actually make the parts, relying mostly on 3D models and electronic data transfers to manage the manufacturing of parts. Paper drawings are mostly used for annotating features for incoming inspection. Communications at this stage are accomplished with detailed initial CAD models of the product being transferred back and forth from the design team to the design stakeholders. Design experts analyze the design and making suggestions for securing the subassemblies or cables better so that the assembly could pass the drop and transportation tests. Suppliers provide for ESI and DFM feedback such as asking the design team to move some of the features around so that plastic molds would be easier to tool.

   At the same time, the initial bill of materials (BOM) is being loaded into the procurement system, and purchasing is trying to find suppliers for these materials and perform part qualification tests for new parts. Some part suppliers have augmented their services with the Internet. These e-suppliers have greatly improved their service, adding engineering staff with ready expertise in parts attributes such as life cycle stages, applications, suggested lower-cost replacements, and the likelihood of obsolescence.

   The use of engineering communications and fast access to analysis and supplier expertise has reduced the need to monitor and measure the engineering change orders (ECOs) that were very common in concurrent engineering as a techniques to reduce development time. This is coupled with the increased capability of product data management (PDM) systems, which allow for warehousing of documentation and historical progression of product design. Since most of the changes are to the engineering model, with no hardware being built, there is less delay resulting from ECOs.

2. Late development stage. During this stage, sometimes called the pilot stage, hard tooling is committed and parts are made in quantities that are significant enough to test the manufacturing process capability and readiness. There might be several different pilot runs, with products being made for life testing, local and global regulatory agencies, and preferred customers for field trials. Hard tooling may be started before design reviews are completed to meet tight schedules. Product launch and roll-out plans are made for global products at different plants in different countries simultaneously.

Communications in this phase are very critical and involve complete product CAD models between the different manufacturing suppliers such as toolmakers, production shops, and the design team. This is helpful to lessen the risks of early hard tooling commitment. When a problem is detected at this phase, very fast secure and instant communications are needed to solve problems, because of materials and production schedule commitments. These might involve a second round of expert analysis and redesign.

Table 9.8 is a summary of the changes from the traditional design process to the new methodologies mentioned above.

### 9.3.3  Design process communications needs

Table 9.9 is a communications summary for the different phases of design. There are various technologies for communications through the phone system and the Internet. These include teleconferencing, videoconferencing, e-mail, web-based meetings, electronic bulletin boards and notebooks, white boards, and collaborative geometric model modifications. These could be combined in various ways to allow the stakeholders in the design process to collaborate together.

From earlier discussions, it is apparent that collaborative and geometric model modification is the primary tool suited for all engineering design stages, including the concept phase, interfacing with the virtual design team, experts, and the supply chain. It is the most effective means for quick engineering problem resolution after the virtual meeting. Participants can change the model on-line, evaluate the

**Table 9.8**   Changes from traditional engineering to new methodologies

| Phase | Traditional engineering | New methodology |
|---|---|---|
| Concept | Design for country-specific market | Design for global market |
| | Limit the number of concepts to explore | Explore maximum number of concepts |
| | Concepts made from paper sketches | Concepts from wire mesh CAD |
| | Inputs from in-house and customers | Input from global company resources |
| | Suppliers bid on detailed design | ESI on concepts before bid is awarded |
| | Suppliers selected after final concept | Suppliers bid on cost of model/similar part |
| | Unused concepts discarded | Unused concepts documented for future |
| Early development | Design analysis and simulation | In-depth and system analysis available |
| | Drawings are used to make prototypes | Prototypes are made from 3D models |
| | Limit and control the number of engineering changes | Engineering changes on CAD model |
| | Purchasing selects part suppliers | e-Suppliers provide more information |
| | Models routed and discussed serially with experts, suppliers, other teams | Models routed synchronously, discussed on-line |
| Late development | Hard tooling commitment after review | Tooling started earlier with communication |
| | Quick problem resolution process | Secure communications for remote resolution |
| | Material acquisition and production plans | Global acquisition and manufacturing plans |
| | Regulatory, environmental/life testing plans | Testing for global requirements |
| | Material procurements | Communications with global suppliers |

results, reach a satisfactory compromise, and then go back to their individual systems to permanently record the results.

## 9.4   Conclusions

The impact of the Internet and the global reach of the supply chain in design and manufacturing is changing the product development process. The implementation of six sigma has to be considered in light of the decentralized organizations that are making design decisions anywhere and building anywhere. The successful implementation of six sigma in this fast-changing environment requires an appreciation of the dangers and trade-offs of global design and manufacturing.

**Table 9.9**  Communications summary for design phases

| Communication | Application/drawbacks | Best-use method |
|---|---|---|
| Teleconference | Telephone line dialing to a predetermined number<br>Lack of prompting to speak<br>Lack of ability to specify an area in a drawing | Adjunct to other Internet systems |
| Video conference | Telephone or Internet dialing<br>Can see the other parties<br>Details of drawings sketchy (even if projected)<br>Show physical part or production area | Supplier negotiations and first introductions |
| e-Mail and derivatives,<br>Web-based groups,<br>Web bulletin boards | Universal in reach<br>Provides a written record<br>Limited interactive capability<br>Asynchronous (serial) in nature<br>Large text files, graphics, and CAD models can be attached | Noncritical communications<br>Prompt for net meetings |
| Whiteboards and electronic notebooks | Allow for sketches and notes to be recorded<br>Allow for multimedia and archiving<br>Capture meetings action items and resolution<br>Asynchronous (serial) in nature | Concept design capture<br>Decision capture |
| Web conferencing | Host and manage on-line conferencing<br>Security concerns are addressed<br>One user is the host of the session<br>Different access levels<br>Graphics/data and CAD model viewed/manipulated<br>Ideas recorded/exchanged | Scheduling and negotiations |
| Collaborative and geometric model modification | Users allowed to share and record data/graphics<br>Interact full CAD models from different systems<br>Ability to identify particular geometry locations<br>Security concerns are addressed<br>Different levels of permissions allowed<br>Data can be text, spreadsheet, and CAD models<br>Interfaces to enterprise systems PDM/ERP<br>Neophyte (CAD) users can manipulate data<br>Session recorded and archived | All design stages<br>Engineering problem resolution<br>Manufacturing quality and inspection |

## 9.5    References and Bibliography

American Society for Quality Control (ASQC). ANSI/ASQC Q90. *Quality Management and Quality Assurance Standards—Guidelines for Selection and Use."* Milwaukee, WI: ASQC. 1987, p. 6.

Bylinsky, G. "Heroes of U.S. Manufacturing." *Fortune,* March 20, 2000.

Fliman, H. "You Order, They Will Make It." *Business Week,* May 29, 2000.

Galuszka, P. et al. "Big Dents in Auto Parts." *Business Week,* April 12, 1999.

Lipis, L. et al. "Putting Markets into Place: An eMarketplace Definition and Forcast." *Bulletin 22501.* IDC, Framingham, MA, June, 2000.

Meieran, E. "21st Century Semiconductor Manufacturing Capabilities." *Intel Technology Journal,* 4th quarter, 1998.

Moore, J. and Burrows, P. "The Info Tech 100/Sevices Headlines. TSMC: Silicon Central and Job Shops take Center Stage." *Business Week,* June 21, 1999.

Price, L. *Digital Economy 2000.* U.S. Department of Commerce, June, 2000.

Research Report, "Success Strategies for High Technology Companies." Boston, MA: *American Electronics Association,* 1997.

Shah, J. "For Auto Makers, e-Chain is About More Than Lower Prices." *Electronic Buyers' News,* April 21, 2000.

Shankland, S. "High-Tech Manufacturers Add Brains to Brawn." *CNET News.com,* August 18, 2000.

Taylor, A. "Detroit Goes Digital." *Fortune,* April 17, 2000.

*Bear Stearns Companies.* "Third Annual Electronics Manufacturing and Supply Chain Survey," June 29, 2000.

Toigo, J. "Avoiding a Data Crunch." *Scientific American,* May, 2000.

Wheelwright, S. and Clark, K. *Revolutionizing Product Development.* New York: The Free Press, 1992.

# New Product and Systems Project Management Using Six Sigma Quality

This chapter outlines a methodology for new system and product design and development using six sigma quality-based project management. The method consists of determining the quality and capability of each level of the system design and using this information to guide allocation of specifications to individual subsystems and modules. In addition, this methodology can drive trade-off decisions in system architecture, component selection, and manufacturing and testing operations. Several tools such as composite Cpk and design quality matrix are discussed to aid system, design, and manufacturing engineers in achieving a quality-based new system and product design process. The chapter is divided into two main sections:

1. The quality system review and quality-based project management. In Section 10.1, the traditional view of enterprise quality management and new product development project management are discussed, as are tools and techniques used to ensure successful product introductions, including methods for project tracking and control, using formal milestones as well as informal status meetings.

2. Technical design information flow and six sigma system design. In Section 10.2, a methodology for six sigma based system design and project management is outlined, using tools such as composite Cpk and Cpk tree to manage system and project quality. Key character-

istics are selected to track and focus on the quality of module and system design.

## 10.1   The Quality System Review and Quality-Based Project Management Methodologies

Currently, the method for adopting quality advocacy at major U.S. corporations is the quality systems review. This procedure is used to assure that the corporation's quality system is effective in achieving total quality and customer satisfaction. The historical focus on corporate regulatory, product quality, and reliability issues is augmented by quality advocacy at each functional entity. At the company's highest management levels, there is an emphasis on facilitating an organization-wide adoption of quality methods such as total quality management (TQM), with a process rather than a product focus. The role of the corporate quality function is a consulting one assigned to assist other entities in integrating quality methods into their day-to-day operation.

The quality systems review is an assessment vehicle to evaluate the status of quality in each function and department. The review defines the quality vision of how business should be conducted, sets a common goal of quality, and provides an awareness of quality requirements across the organization. The quality system review process should be used as a measure of the progress toward quality, to provide opportunities for exchanging ideas and to refocus each part of the organization on the basic issues of quality.

### 10.1.1   The quality-based system design process

The quality system review can be used to drive quality into the new product design process. In most instances, quality goals for new products and systems are given as six sigma or Cpk values for individual parts and processes, with design engineering providing the product specifications while the manufacturing operations are calculating the process averages and variability in order to meet the six sigma goal.

This six sigma quality assessment methodology works well at the micro level, with individual product part or component specifications and their manufacturing steps. It is the purpose of this chapter to outline a procedure for using this methodology at the macro level, with multiple specifications and designs leading to the system performance requirements. The six sigma quality-based design methodology could also be used for systems architecture and partitioning of hardware and software, design trade-offs, manufacturing and test plans, and monitoring the systems design performance relative to its requirements and specifications.

The principles, tools, and examples used in this quality-based system design for six sigma methodology were developed for high-technology companies that are pioneering the use of design for quality methods to augment their traditional design process. They have been able to successfully develop cost-effective new products and systems using this methodology for allocation of module specifications and the trade-off of defects generated through partially conforming designs versus the cost of removing these defects in manufacturing.

### 10.1.2  Six sigma quality-based system design process benefits

The use of six sigma quality-based system and product design processes can be beneficial to the general design process by making decisions that are based on sound quantitative analysis and not solely by the individual designers' experience or their "gut feeling." Six sigma quality analysis can quantify the design's ability to meet customer requirements by performing six sigma analysis at each level of system and module design. It can be used to analyze design alternatives and to focus the design team on what elements of the design need to be enhanced to meet the overall system or product specifications. It can also be used to establish a common language among design functions and engineering disciplines. It can also provide a common set of background data for resource allocation, and add more information to other functions of project management such as cost trade-offs and risk assessment.

Six sigma quality-based product design can provide intangible benefits in project management. It requires and promotes teamwork, inherent in the two parts of the six sigma equation—product specifications and process variability—making systems, design, and manufacturing engineers work together. It can provide an objective basis for negotiations between customers or marketing and the design team, and between design and manufacturing engineers. It helps in focusing the entire organization on the common goal of six sigma, and it encourages sharing of information, decisions, and case studies of six sigma successes across organizational and discipline boundaries.

It is important to note that six sigma quality in design is not a method to achieve "zero defects" at any cost. It should focus on price–performance relationships. It is not a substitute or compensation for poor engineering or design. Nor should it be used to assign blame or point to poor performance. It should be used as a positive problem solving tool for achieving high-quality, low-cost products in design and manufacturing.

### 10.1.3 Historical perspective of project management

New products are developed in companies based on long-range strategic and business plans for the market segment in which the company wants to operate. The new product strategy plan is very dependent on the product/market life cycle phase: whether the product is in the start-up, growth, mature, or commodity phase will greatly influence the capability and performance versus price range and timing of the product introduction.

To formulate a new product strategy that is cohesive with the rest of the enterprise, it is important to begin with the company goals. They should outline which business segment is targeted and the boundaries of that segment. In addition, they should define the competitive advantage of the company; e.g., innovation and technology, cost and manufacturing technology, quality effort, customer satisfaction, and organizational flexibility.

The components of the product strategy should include a a hierarchy of elements that define the strategy over time:

- Mission statement—A broad statement for the next 10–20 years.
- Intermediate-range plan—A more detailed plan outlining goals and actions to be taken for the next 3–10 years.
- Product plan—Products (performance and price) to be introduced over the next 1–5 years.
- Tactical plan—Action plan for the short range of 12–18 months to accomplish objectives.

The new product development process varies from company to company depending on market requirements, competitive pressure, and the internal company strategy and methodology. Most companies have "go/no-go" decision points at various points of the concept through the development stages, with checkpoint meetings and targets to be met and revisited.

The concept stage begins with the identification or the creation of a specific project or product team. The product idea can develop from several sources: competitive evaluations, marketing and customer surveys, and, most importantly, from the product champion. The product champion, whether a high-ranking executive or a staff engineer on the bench, plays a very important part in the introduction of new products. He or she has the vision, the focus, and the determination to carry through his or her new product ideas to fruition. His or her risks are high but the rewards can be great. If the product champion's ideas are not implemented, he or she might leave the company to start up a new venture.

Once the product is identified, an iterative process begins to take shape in order to further refine, clarify, and specify the product definition. Market research, surveys, and economic justifications enhance this phase. The development and the performance measures of the product are identified in terms of product specifications, potential revenues, costs, product life cycle, and impact on current company products.

The results of this iterative process are built into a business plan. The plan is the blueprint for developing, manufacturing, and selling a new product in the marketplace. It is imperative to develop business plans for all new products. Elements of the business plan are:

- The market analysis for the market segment targeted by the product, in terms of market development stage, competitive analysis, and potential volume.
- The marketing strategy for penetrating the target market—whether to compete on price, features, performance, or quality.
- The development plans in terms of the chosen technology and architecture, people and equipment, tooling, and material requirements.
- The manufacturing plan on how and what is required to produce the product, the supply chain strategy, the fabrication and assembly processes and equipment, and the test and quality plans.
- The product support plan in terms of field support and training, product repairs and warranty strategy, and impact on support for existing products.
- The financial analysis and projected return on investment for the new product: the product development costs, the expected manufacturing and support costs, the warranty and service levels, as well as the economic impact on existing products.

In addition to the business plan, the development of a prototype or mock-up for the product is important to demonstrate the idea or product feasibility to management and give potential customers a chance to comment on the utility of the product. There are various rapid prototype techniques using physical mock-ups, software for screen generators, and command and transaction modeling. The advent of advanced three-dimensional mechanical computer aided design (CAD) stations with rendering capabilities can produce a three dimensional image of the product on a computer screen.

There are certain criteria for "go/no go" decision points to proceed into the development stage, as shown on Table 10.1. These criteria were discussed previously. Although there are no rules as to the cor-

**Table 10.1**   Total product development process concept-to-development criteria

- Market requirements identified
- New product definition and its release schedule meet market needs
- Chosen technology and architecture are acceptable
- Technical feasibility demonstrated through a working model or prototype
- Planned levels of price, performance, and reliability are acceptable
- Adequate project return on investment (ROI)

rect method for ensuring that a product is accepted into development, many factors predominate: the current financial status of the company, whether the management is willing to take a chance at this particular time, the credibility and previous track record of the project team and its leader, and the current competitive situation in the industry.

It is well understood that a certain percentage of products do not go into development at this point, even if the product idea is sound, because of the company's current financial condition or competitive situation, and sometimes because of poor preparation by the project team. The company managers are looking for a particular return on investment (ROI) which is in line with the financial conditions in the industry, but would tolerate a lower rate in the hope of landing a stellar performing product in the future.

Figure 10.1 is a simplified flow diagram of a new electronic product development process, showing the phases of a new product development from concept to production. It divides the process into two major

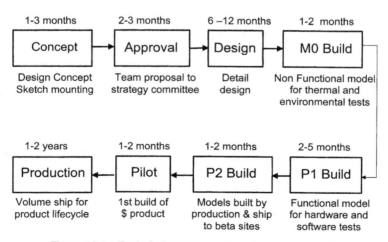

**Figure 10.1**   Typical electronic product development cycle.

phases: concept and design/development. There is only one approval "go/no-go" decision point, which occurs when deciding to go from the concept to development phase. Other phases build successive models for the product for differing reasons:

- M0 build is for a few nonfunctioning units, to ensure that the design is verified for thermal and environmental testing, such as for elevated temperature, humidity, RFI, vibrations, and transportation simulations.
- P1 build, which could be done concurrently with the M0 build, is for a few functioning units consisting of modules and PCBs for hardware and software integration and verification of product performance to design specifications.
- P2 build, sometimes called the beta phase, consists of units to be shipped to selected customers for verification that the product meets the intended customer needs. Feedback from customers is solicited and evaluated for possible incorporation into the design through revisions to the product.
- Pilot run. The volume of this run is dependent on the production volume. For high-volume consumer products, a run of 100–300 could be made. Pilot units are made in order to test the production tooling and methods. They are eventually sold to paying customers.
- Production volume is initiated after pilot run is completed successfully, and could be made anywhere based on the tooling that was tested in the pilot phase.

### 10.1.4  Project management of the product development process

Project management for electronic product development is organized with well-developed tools and procedures. The first part of planning a successful new product is to lay a good foundation of knowledge about the project, including:

- Good identification of customer needs. Customers can be internal or external, drawn from the installed base or targeted for a new product.
- Customer needs should be converted to new product requirements and specifications through interviews, focus groups, and structured methods such as quality function deployment (QFD), discussed in Chapter 1.
- A competitive position analysis and how it meshes with the company strategy should be performed. Issues such as the position of the

new product on the price–performance curve should also be considered, as well as technology improvement cycles.

- Regulatory and industry standards issues are also important in clarifying the tasks to be accomplished. In addition, industry standards and environmental and ergonomic factors should also be considered in the design of the new product.
- A risk assessment for the project should be undertaken, in terms of technological obsolescence, competitive factors, and alternate or disruptive technologies on the horizon that might have an impact on the marketplace.
- Trade-offs should be quantified, including a good understanding of the issues of cost, quality, schedule, and performance.
- Constraints that could trigger adverse consequences, including functional and resource constraints in design, manufacturing, sales, support, distribution, and service.

The project planning methodologies consists of several iterations of the schedule to allow for input from the various parts of the company, and to reach consensus on a schedule that is agreeable to both management and engineers. These iterations will take the following form:

1. Task selection and definition for the project. In this phase, product design activities are broken down into small well-definable tasks, each with a start point, resource requirements, and an endpoint. The tasks should be small enough to be assigned to one team member, with an identified deliverable to be produced within a short time. Task duration for each step is estimated, depending on the quality of the personnel assigned to accomplish the tasks. Historical data should be used, depending on project type, technology used, and personnel skills available. The estimate should be tempered with a probability level for assured delivery (90% probability of completion on time) and aggressive schedules (50%), the latter resulting from successfully using new innovative methods and design ideas. The estimate should allow for turnover, training, and nonproject variables, especially if design engineers are supporting current products or field problems.
2. Bottoms-up collation of task definitions and time for each task. All tasks are collected in a task list and then plotted in a critical path for the overall estimated duration of the project. The accumulation of tasks' duration determines the endpoint of the project (manufacturing release). Obviously, this time will be on the conservative side.
3. Top-down scheduling. Management determines the endpoint (manufacturing release and shipment to customers), then attempts to

work backwards to see how the project could be accomplished to meet the marketing timetable. This estimate is usually at odds with the bottom-up estimate.

4. Synchronization. This is when the two estimates of bottom-up and top-down scheduling are combined to produce a consensus on the final project schedule by negotiations between the management and the design team. Trade-offs are considered in scope, performance, and resources. Additional opportunities are considered, including prediction of possible new process improvements with tools and technology. The result of this process is an agreed upon final schedule between the management and the design team.

5. Risk analysis and contingency planning. After the final schedule has been decided, the project team and the management should consider the adverse consequences of the schedule decisions. By considering "what if" scenarios, possible conditions for project delay should be outlined as well as critical project parameters that should be continuously monitored to ensure that the project in on track. In addition, plans should be in place for quick reaction when the project is at risk.

Once the project is launched, several tools are used to maintain control and timing. They involve tracking tools and charts, as well as milestones meetings and specialized reporting such as:

• PERT charts, which document the relationship of different tasks in order of start and finish times. The charts will also show the latest start and earliest finish time for each task, and the critical path that will keep the project on time, with no slippage in the delivery date.

• GANTT charts document the schedule, events, activities, and responsibilities necessary to complete the development project.

• Milestones or project phase completion meetings are predetermined endpoints of product development phases. They are integration points to ensure synchronization of all functions in the project. They serve as management checkpoints to meet with the team to review progress and funding before proceeding to the next phase, by showing in tabular form all of the task start and end dates and persons responsible.

• The timing of the milestone meetings should be variable, with milestones more frequent as the project nears completion. Historically, engineers have complained that management does not really pay attention to the project until the release date is fast approaching. Figure 10.2 shows a timeline for project phases and milestones, with general goals and driver outlines for the phases.

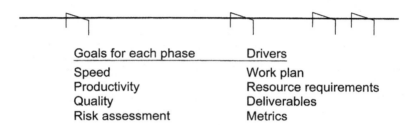

| Goals for each phase | Drivers |
| --- | --- |
| Speed | Work plan |
| Productivity | Resource requirements |
| Quality | Deliverables |
| Risk assessment | Metrics |

**Figure 10.2**  Development project time line: phases and milestones.

- Communication within the project teams is also important. These communications should be encouraged to be informal as well as independent of the milestone meetings, which are major meetings and presentations to management requiring extensive preparations. The informal meting should focus on the status of the project, and could be presented by project discipline groups to report progress toward achieving interim goals, accomplishments worthy of note to others, decisions made, and new concerns raised. Any slip in the interim goals should be clearly outlined and its possible impact on the overall project schedule indicated. Any previous concerns should remain on the agenda until adequate resolution. An example of a monthly project communications meeting is given in Figure 10.3

The design phase is usually broken down into small steps, with the completion of each step recorded either in a formal checkpoint meeting or at the completion of a particular task or milestone. It is important to have each checkpoint or milestone be of some significant and measurable progress in the project, to add to the project team's and management's confidence in the progress made toward achieving the product goals.

Historically, design project tasks have been divided along engineering disciplines: electrical, mechanical, software, and manufacturing. These disciplines were formed into distinct project groups, each with engineering responsibility over the discipline. This has sometimes led to interdisciplinary friction and factionalism. Another technique is the grouping of the engineers along project tasks or subparts, with interdisciplinary teams to encourage communication. Sometimes this is called matrix management.

The organization of the project team is very much dependent on the company's management philosophy, and some companies have found

| Project_____ Manager_____Date_____ | |
| --- | --- |
| **Previous month goals:** | **Current month goals:** |
| Complete layout Rev #2   ✓ | Release prototype |
| Complete material list   ✓ | Purchase pre-production material |
| Schedule design review   → | Schedule resign review |
| **Accomplishments** | **Decisions** |
| MFG efficiency score>50% | Build prototype outside |
| **Previous concerns** | **New concerns** |
|  | Memory IC's obsolete 05 |

**Figure 10.3**   Project communications monthly meeting example.

that both methods of organizing a project, along engineering disciplines or using interdisciplinary teams, have been equally effective. It is important to manage these different activities positively by specifying the project interface among the different groups, and increasing the communication links by formal and informal meetings, updates, and demonstration of phase and milestone completions. These milestones and status meetings for the project teams are important tools that can be used to update all project team members and project management on the team's progress.

## 10.2   Technical Design Information Flow and Six Sigma System Design

In a typical project management scheme, there are myriad sets of communication topics necessary to successfully implement the project. They include product and process requirements, design specifications, manufacturing process capabilities, supply chain resources, and equipment purchase plans. System, design, and manufacturing engineers may not communicate effectively or use uniform language. Six sigma offers an opportunity to develop a common set of communication tools and standards for successful project planning and execution.

Historically, systems engineers or discipline managers flowed project requirements to design engineers. Systems engineers may not have been sure how conservative or risky the requirements for each task were, and in most cases, design engineers felt that the requirements were overly difficult. Once the design engineers completed the preliminary design, they sent the design performance back to the system engineers. Typically, their estimates were worst-case designs and overly conservative. There was little acknowledgement of either exceeding or easily meeting the specifications. Opportunities for reallocation of the design performance over the different product modules were lost. In addition, system and design engineers did not have the necessary information from manufacturing as to their future plans for improving process capabilities.

At the same time, manufacturing engineers were interested in designs that were manufacturable. They were willing to launch projects for quality improvements, but did not have guidance from the system engineers on which processes needed improvements in order to meet certain module design specifications. Given these conditions, many product or system designs were unable to perform within the specified requirements.

### 10.2.1  Opportunities in six sigma for system or product design improvements

Many of today's electronic products and systems are technology driven, trying to garner as much advantage as possible from leading-edge technology. Project managers deal with a complex array of design trade-offs in new product performance, project cost, product cost, and delivering the product to the marketplace on time. The most important ingredient in project success is the skill and experience of the design team. These engineers have traditionally been conservative, relying on worst-case design analysis, providing adequate design margins and good allocation of product specifications among the design modules. As it is difficult to staff all projects with good experienced engineers, six sigma based design analysis can quantify the experience base. Six sigma can be the cornerstone of design analysis, replacing the experience base with an unbiased metric for analysis.

The six sigma design analysis process can proceed as follows:

- Each discipline participates in a six sigma design process that optimizes quality, cost, and the performance of their individual designs.
- All parts, modules, subassemblies, and systems should be analyzed for design and manufacturing quality using six sigma techniques for yield and defects.

- Six sigma is used as a guide in setting systems requirements based on cost and performance.
- Six sigma is used to focus on low yields or marginal designs.
- Six sigma is used to negotiate trade-offs in systems, design, and manufacturing.
- Six sigma helps focus on a limited set of a few important parameters.
- Six sigma analysis is performed at all stages in the design, and is reviewed regularly at design checkpoint meetings.
- Six sigma is used in negotiations with the supply chain and technical customers.

### 10.2.2   The system design process

Traditionally, system design launch begins after the system or product overall specifications have been agreed upon. These may have been developed through negotiations with the marketing department and after consideration of customer expectations or competitive factors. Several tools such as QFD, discussed in Chapter 1, have been widely used to implement a quality-based specification process.

The system designers then partition the system into subsystems or modules according to the customer requirements. A system architecture is also developed through which particular customer requirements are achieved by defining how modules should be designed and integrated with each other. The modules could be discipline-specific, such as electrical, mechanical, or software, and the architecture could define the method of connecting them together to achieve system performance.

Normally, the system designers depend on their previous experience and knowledge of system and module design to properly partition and distribute the total system performance across the module specifications. This partition involves a negotiation process between the systems engineers and the module designers on individual module specifications that will have to roll up into the systems requirements. Module designs that are difficult to achieve are given wider specifications and vice versa. A Six sigma or Cpk based negotiation process could be useful in formally achieving good system partitions and overall performance.

### 10.2.3   The System design steps

The quality-based system design methodology is an overall system analysis based on modeling results from the system module design process and the process capability for manufacturing these modules. It is an interactive process by which the system engineers allocate the design margin to the design engineers through requirements based on experience and prior system design knowledge.

The design engineers then analyze their design performance and the test strategy, and feed back the results to the system engineers. Working together with the manufacturing engineers, they develop the manufacturing process flows, select the manufacturing processes and equipment, determine which manufacturing steps are critical, estimate the quality and cost of these processes, then input the results back to the systems engineers for final architecture and module specifications allocation plans.

Three parts are required to make this system design methodology work effectively in the system design:

1. The use of a composite Cpk metric to measure the design quality and manufacturing capability.
2. The narrowing down of the system specifications to between 3 and 10 key characteristics to perform the systems Cpk analysis.
3. The use of standardized procedures in design and manufacturing to determine the composite Cpk of each design and its manufacturing capability.

It is important that these key characteristics be independent of each other to eliminate the potential problems of multivariate quality control. This independence of characteristics will have to be determined empirically by the system design engineers, since their selection is made at the concept stage of system design and cannot be established statistically.

### 10.2.4  Composite Cpk

The composite Cpk (CCpk) is a back-calculated number obtained from the total defect rate (TDPU) or from total rolled yield $Y_R$. The TDPU is calculated by the summation of the individual design characteristics or manufacturing steps DPUs, and the total module or system yield (TFTY) is calculated by multiplying the individual FTYs. A composite Cpk is back-calculated from the TDPU, usually from a standard normal curve, assuming that the process is centered and specifications are available at both the high and low limits.

The CCpk is a representation of the total quality of design or manufacturing, but cannot be related directly to any individual step. The CCpk is calculated separately for each module design and manufacturing process. For design, the CCpk is based on the performance versus the variability of the components selected for all of the critical design specifications. For manufacturing, the CCpk is derived from the total number of operations necessary to completely manufacture the module. The module total Cpk (TCpk) is the combination of the design and manufacturing CCpk; it can be back-calculated by multiplying the yields or adding the DPUs of the design and manufacturing CCpk. It is a complete measure of the quality of design and manufacturing of the module.

The design and manufacturing CCpks allow for separating the quality of the modules into each functional area, and provide a good assessment of the overall system design. In a complex system, there are many CCpks, and the system designers need a quick method to assess the system quality. The Cpk tree, Figure 10.4, could be used as a representation of system quality. The CCpk is shown for each module; however, only the worst-case CCpk is carried over to the next level.

In a typical CCpk calculation, the worst-case CCpk, especially if it is much lower than the average CCpk, will dominate the rolled yield $Y_R$ and total defects TDPU for that module. Given that a complex system could have more than a hundred CCpk analyses, this focus on the worst-case CCpk is a good method to identify the element of the design or manufacturing cycle that needs attention, and therefore point to a reallocation of resources to reduce defects.

The assumption of normality of the manufacturing and design steps is an important issue in this methodology. Some suppliers preselect their components to match customer specifications, reducing the probability that their product characteristics are normally distributed.

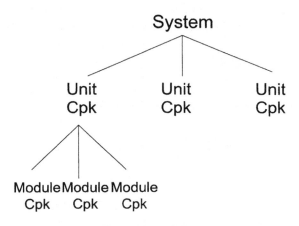

**Figure 10.4**  Cpk tree.

Some prototypes cannot be made in large enough quantities in order to test the characteristics of design or manufacturing steps for goodness of fit to the normal distribution. Whenever possible, design or manufacturing engineers should establish the statistical distribution of the key characteristics and components. The Cpk and defect calculations of processes and designs should reflect their statistical behavior, accordingly.

### 10.2.5  Selecting key characteristics for systems design analysis

In a typical system, there might be a multitude of functions that have to be specified with a nominal and a range. Pending the system architecture and partition of functions, the system design then has to be translated into many module specifications for the design engineers, as well as the manufacturing process requirements. A procedure is thus needed to select only the key system characteristics in order to monitor the quality level of the system.

Typical system parameters could be weight, size, power, cooling, speed, throughput, response time, system delay, range, resolution, accuracy, and repeatability. The process of selection of key system characteristics could be developed as follows:

1. List key systems functions. The list should include their relationship to the system attributes. Examples could be:
   - *Requirements*. What is the purpose of the system? At what level of accuracy, speed, repeatability, and reliability does the system need to perform?
   - *States (or modes) of operations*. At what levels should the system operate? Is there automatic as well as manual control?
   - *Customer desires*. What additional capabilities would the customer like to have? Does the system need to operate in a high-humidity environment?
   - *Trade-offs*. What are the relative merits of functionality versus cost trade-offs? A portable system could trade cost for weight, or a cost can be assigned for a unit of weight ($/lb.).
   - *Functional characteristics*. What makes the system work? What is the function of each subsystem?
   - *Relationships*. How does each function affect other characteristics or subsystems? Is this relationship well understood or mapped out?
   - *Benchmark*. What is currently available? Who makes the best system?

- *Targets.* How much power does the system use? What is the speed of performance?

2. Narrowing of characteristics. From the list of key functions, those characteristics that must be met in order for the system to basically function can be removed from the key list for Cpk analysis. These are prime characteristics, and therefore should be met without any variability in the design or manufacturing cycle of the product. Examples of these characteristics could be an emergency shutoff of the system or meeting minimum requirements of systems responses.

    For the remaining functions, those characteristics that can be variable are listed as key characteristics depending on component selection or manufacturing variability. In order to include them in the Cpk quality analysis, they should be measurable and controllable. The selection could be further augmented for these characteristics as follows:

    i. If they require decomposition. For example, when designing an amplifier with gain specification, other related specifications such as noise figure or sensitivity have to be selected as key characteristics as well. A composite CCpk can then be calculated for the related specifications.

    ii. If the specifications are difficult to meet, due to the current state of the technology or manufacturing capability, they should be added to the key list in order to focus on the most difficult ones.

    iii. If the specifications are in areas of high risk or unknown technology, they should be tested in the prototype phase prior to the detailed design stage. Adding them to the key list will ensure that the proper investigation is done prior to system design completion.

    iv. If the specification is deemed important to the customer, this will ensure customer satisfaction and the focus of the system on the customer needs.

    The objective of this phase is to select a list with a range of 3–10 key characteristics for each module or subsystem design. This list is added to the tasks performed by the system engineers, and is reported upon at each phase of the system design. Hence, the quality of the system design can be monitored along with other design, cost, delivery, and performance issues.

3. Reviewing the key characteristics list. Once the key specification list has been narrowed down, the selection process has to be audited to make sure that all of the system functions, key requirements,

**Table 10.2**  Cpk Design quality matrix selection for systems specifications and modes

| Operating Modes | System requirements | | | |
|---|---|---|---|---|
| | Detection error | | Multiple communication | |
| | Speed | Correction | Faults | Layer level |
| While network is on | X | | X | X |
| While network is off | | X | X | X |
| Internet messaging | X | | | X |
| Intranet messaging | X | X | | |

and operating modes are covered in the potential analysis. At least one of the system functions and one of the operating modes have to be included in the key characteristics list to ensure coverage of the Cpk analysis.

An example of such a process is in Table 10.2, where the quality analysis of a network communication fault detection and analysis system are shown. The X's in the matrix refer to characteristics that match the system's requirements with the customer requirements. Those that in boldface have been selected as the key characteristics. At least one of the system's specifications and one of the customer requirements should be selected for the cost and quality analysis.

### 10.2.6   Standardized procedures in design to determine the composite Cpk

The design CCpk is a measure of the design quality: how the design meets its intended specifications, regardless of the manufacturing steps necessary to produce the product or system. It is determined by the variability of the components specified in the design versus the overall system design performance to its specifications.

The application of the design CCpk is based on the selection of parts or components for the design. In a typical design consisting of multiple parts, each part's key characteristic must be characterized in terms of performance distribution. Part performances can be obtained from their suppliers. If this information is not available, then the design engineer can measure the performance from sample lots purchased for prototype runs. As a last resort, the design engineer can assume that parts are distributed normally, and part specifications are located at six sigma or any other appropriate sigma value from the mean. The module design can be analyzed or modeled in simulation to obtain a distribution of the module performance based on its components' distributions.

The methodology above can substitute for the worst-case analysis commonly used by engineers. By targeting a specific Cpk for the design, the designer can estimate the defect level due to the design, and be able to specify appropriate tolerance parts.

The composite Cpk design estimates can be made with the characteristic average of typical components as the design nominal and the components worst-case conditions as the specification limits. Components could be modeled as normal distributions of values between the specification limits for two-sided, or specification to limit for one-sided tolerances. Depending on the target Cpk for the design, the components distribution could be evaluated from the center to one side of the specification to measure either three sigma for Cpk = 1 or four sigma for Cpk = 1.33 or six sigma. The design is then evaluated for a composite Cpk on a statistical basis, as opposed to worst-case conditions.

This Cpk methodology can be applied to mechanical, electrical, and software module designs and manufacturing independently, then analyzed at the system level. They can be accumulated through the Cpk tree concept to present a complete design for quality analysis of the system. After the analysis is completed, the system specifications can be reviewed to determine which of the system modules are meeting their individual specifications easily and which ones are not. A reallocation of system specifications can then be made to more efficiently distribute the design tasks.

Another advantage of the Cpk system analysis is the determination of the number of defects generated and the yields at each phase of manufacturing. A detailed test strategy plan can then be developed to remove these defects in the most efficient manner, either by positioning testing and inspection operations at the proper stage of manufacturing, or by redesigning the modules so that that less testing can be performed depending on the Cpk level of the design.

### 10.2.7 Standardized procedures in manufacturing to determine the composite Cpk

For more complex designs or products, a simulation or modeling of the design is used to produce a distribution of the design characteristics, based on a random choice of the components' values. This module design distribution can be derived from the components' distributions according to the methodology outlined above, using discrete numbers from the components' distribution fed into the model or simulation of the design.

In determining the manufacturing CCpk, several steps have to be taken in order to evaluate the producibility of the system and the module's designs, and to ensure that the design and manufacturing engineers have taken steps necessary to reduce the defects produced when the system is in production:

1. Manufacturing has to determine the critical design parameters necessary to reduce defects in production. Called Cpk drivers, they are characteristics of the design specifications that are important in improving the CCpk value. Examples of these could be operator skills required, equipment setup, design geometry of the parts, tooling needs, and the selection of the equipment and processes based on appropriate capability. These drivers were discussed in Chapter 8.

2. Manufacturing has to outline the capability for every step of the manufacturing process, using standard statistical techniques. For each manufacturing process, the initial Cpk value based on capability and common specifications has to be evaluated as a baseline Cpk. Regularly, these Cpk values have to be reviewed and updated to reflect any improvements. The update period should be less that the one-half of the design time for new systems, and preferably done every quarter.

3. Sample sizes should be large enough (>30) for the baseline Cpk to ensure confidence levels of the average and $\sigma$ calculations above 90%, with a corresponding confidence. Quarterly checks should have a confidence level of 95% before changes are made. Changing the Cpk values too often could result in reduced credibility of the Cpk manufacturing values by the design engineers.

4. Since the Cpk requires specification limits as well as process average and variability, a preferred set of limits are given for every process. These will help guide the design engineers in selecting the appropriate manufacturing process for the design. For attribute processes, or those with defect data available, such as soldering or welding, the process Cpk can be back-calculated as was shown in Chapter 2.

   An example of such a capability study for a machining center is given in Table 10.3. In this example, the machining center is measuring the average and the standard deviation ($\sigma$) of each major operation performed at the center, as outlined in step 3 above. In addition, the center is also providing for a desired specification for each operation, in order to calculate the Cpk. Baseline as well as quarterly updates of the Cpk are provided to the design community. These Cpk and process data are used by the mechanical design engineers to calculate the manufacturing Cpk of their designs, in manner similar to the ones described in Chapter 8.

**Table 10.3** Example of a machining center Cpk status

| Process | Cpk baseline | Cpk this quarter | Specification status | limit (inches) |
|---|---|---|---|---|
| CNC mill | 1.42 | 1.61 | Recalculated | ± 0.005 |
| Bridgeport | 1.41 | 1.41 | Check OK | ± 0.005 |
| CNC lathes | 1.99 | 1.99 | Check OK | ± 0.005 |
| Manual lathes | 1.70 | 2.66 | Recalculated | ± 0.005 |
| CNC punch | 1.06 | 1.06 | Check OK | ± 0.005 |
| Brakes | 1.18 | 1.18 | Check OK | ± 0.005 |
| Paint | 1.06 | 1.06 | Check OK | ± 0.005 |
| Assembly | 1.72 | 1.72 | Check OK | Attribute |
| Welding | 1.70 | 1.70 | Check OK | Attribute |

5. Manufacturing has to calculate a CCpk value for every major element of the new product. A simple method of achieving this in a new design is to assign a critical process parameter as the quality driver for each step of the process. The number of critical process parameters for each step determines the CCpk of manufacturing the product.

6. Calculation of the total product Cpk. After completing the design and the manufacturing CCpk for the critical parameters of the design and manufacturing processes, the overall total or rolled product Cpk can be measured by adding the defect rates for design and manufacturing, then back-calculating the Cpk as shown in Chapter 6.

   In order to allow each functional area to work independently on resolving their quality issues, the defects caused by design and manufacturing are treated separately in their own CCpk terms. This is helpful in multidisciplinary designs, where software, electrical, and mechanical functions are required in the system. In some cases, system malfunction can be corrected by changing a component or a module, even when the removed element functions correctly to its individual specifications. This phenomena is sometimes referred to as "no trouble found" or NTF. Usually, this problem is caused by improperly matching system and module specifications. The design CCpk is a good indicator of potential NTF problems.

7. The design and manufacturing CCpks can also be used for test strategy development. They can help the design engineer estimate the amount of defects to be removed from the system and the nature of these defects, whether due to design issues or manufacturing variability. The system engineer can then develop the test strategy on where and when to perform functional and system tests at different stages of production to remove defects.

## 10.3 Conclusions

This design for quality Cpk methodology has been proposed to extend the use of well-established techniques for design and manufacture of individual parts and processes into guiding the design of large systems and products. It is intended for use in complex systems, comprised of many subsystems, modules, assemblies, and individual parts.

The chapter has outlined a methodology for the quality-based design of new systems and products through the use of the process capability index or Cpk. It presented several techniques and tools to achieve quality objectives, through a process of selecting key system characteristics and developing a composite Cpk (CCpk) value to measure the design and manufacturing steps. The resulting CCpk analysis has been used to guide and monitor system performance to specifications, component selection, and trade-offs in design and module specifications.

The proposed use of the CCpk is to improve system design by predicting design and manufacturing deficiencies and negotiate the required trade-offs in performance, cost, and manufacturing capability. In addition, a good manufacturing plan and test strategy could be developed to reduce system cost and production time. This methodology can encourage design engineers to quantify critical module specifications in terms of yield and to work closely with manufacturing for best process selection and improving the manufacturability of the system. It can also guide the manufacturing engineers to target specific processes for improvements.

This chapter also presented the management of the new product development cycle for electronic products, in terms of the product life cycle, the impact of technology, and development project tracking and control. The importance of the role of six sigma should be clarified in the business plan and identified in specific project goals. The method of implementing the six sigma in terms of methodology and tools was reviewed in the preceding chapters. Management should be specific about attaining the goals of six sigma in terms of setting design efficiency, product cost, reliability, and warranty targets for new products.

# Implementing Six Sigma in Electronics Design and Manufacturing

Product creation is the correct mix between technology advancement, market conditions, consumer trends, and competitive factors. Planning is the key to developing a coherent product introduction stream that anticipates the market mix. Otherwise, product creation becomes a reactive process, with the subsequent risk of developing a product too late to capture a significant presence in the market or lowering the existing product prices to protect market position.

The worst-case scenario is the scheduling a of new product with unrealistic expectations in quality, cost, or timeline. It results in undue strains on the organization in general and the development team in particular. In addition, marketing plans that are set in motion based on false schedules will be undermined for existing as well as new products.

Product creation should be a team-centered activity. The balance of the mix between technology and market input can be determined best by practitioners of both crafts. The inputs from marketing and sales organizations, the research laboratories, the advance development group, and the current development organization should be evaluated and a collective decision reached for the timing of the next product rollout. A risk–benefit analysis should be made of the trade-offs between rushing a new technology or idea to market versus properly investigating the development and manufacturing problems. Six sigma is good tool for reducing the uncertainty of using new technologies in products by allowing for an independent and systematic method of evaluating the readiness for adopting the technology in terms of product quality and cost.

## 11.1   Six Sigma Design Project Management Models

The six sigma project management model, discussed in the previous chapter, augments traditional project management by using six sigma analysis to plan and make decisions properly, instead of relying solely on past experiences.

Six sigma is best suited for team-directed project management. In this model, the emphasis is on collective decision making and communications assisted by the use of six sigma. There is a core project team that manages all project activities and is supported by subteams of the different functions involved. The team members are the individual engineers who are performing specified tasks. The project manager is the *driver* of the project, making sure the schedule is on time and the product within budget. There is a strong technical component to this project management, and the project manager makes decisions based on both technical and business evaluations of the project status. Six sigma based trade-offs are discussed collectively and decided upon to maintain the overall goals. These trade-offs include project scope, product performance, resources, and schedule.

### 11.1.1   Axioms for creating six sigma within the organization

In order to create a six sigma environment within the company, there should a move away from the process-based organization, with resource (matrix) management reinforcing the use of a standard model for efficient and specialized use of resources. This model is a good one for managing large and complex set of products and systems. In the current conditions of worldwide competition, the smaller, more efficient organizations are the ones that are nimble and fast reacting to the market. They are successful because of their focus on their business unit products, and can make fast decisions by micromanaging a smaller organization than by macromanaging a large business entity. The axioms for implementing a six sigma product creation process are as follows:

- Create a total quality culture within the organization
- Introduce a quality focus organization at the business unit level
- Emphasize the quality focus approach to project management

The implementation of these axioms at each functional level could be as outlined in the following three subsections.

**11.1.1.1 Create a total quality culture within the organization.** The implementation of total quality is a critical element of success for six sigma. It is the base from which all other ideas, procedures, methodologies, and tools of six sigma can be developed, nurtured, and successfully implemented.

Although the focus of a total quality culture is the control and enhancement of quality, it has evolved into highly successful methodologies for many different aspects of successful management and operation of companies. Total quality infuses the whole organization with a common set of terminology and procedures to perform the following important tasks:

1. Problem identification and resolution. The organization is trained to spot problems, in quality or otherwise, identify them promptly, and suggest methods for improvement. Alternatives are studied and weighed carefully, decisions properly made and adverse consequences evaluated. Management is kept informed and provides guidance, encouragement, and resources for successful completion of the tasks.

2. Team process. Total quality is synonymous with the team process. It encourages working in groups, helping team members reconcile individual versus group goals, set team objectives and expectations, make collective decisions and learn to operate with less management direction. All of these elements will be very important for the successful implementation of six sigma projects.

3. Continuous improvements. This is the idea of not being satisfied with the status quo, not doing things the same old way (SOW), and constantly seeking better performance from people, equipment, and processes. Part of continuous improvement is the challenge to realize that a limit has been reached with the current situation, and to seek other alternatives and original ideas for improvement. Expectations should be set correctly; improvements can be achieved in big steps only when using new technologies or methodologies. Total quality allows for small steps, which when accumulated over time, lead to large steps.

One of the inhibitors of total quality at the engineering level is the engineers' view that it is for manufacturing and less skilled personnel in the company—"fourth grade stuff." In addition, engineers by nature have been trained in universities to compete instead of collaborate:

grades and exams stress individual contribution rather than teamwork. These cultural inhibitors have to be dealt with by treating them as a procedural as well as training issues. Emphasizing quality and teamwork on performance evaluations sends a strong message that the company is serious about implementing a total quality program.

Total quality teams that provide for completion of successful projects at all the engineering and marketing functional levels are good precursors and training grounds for a thriving six sigma culture in the company.

### 11.1.1.2 Introduce a quality focus organization at the business unit level.
The company or the business units of a major corporations are comprised of a collection of functions required for managing a corporation, including the traditional three: marketing, development (or R&D), and manufacturing.

In the following text, each functional unit is analyzed and issues outlined for successful support of the six sigma process.

**Marketing/product management.** The *marketing* department is normally responsible for formulating plans and strategies to profitably penetrate existing markets and open new ones. The marketing department implements programs to support the field sales force. Market research and analysis is conducted to provide R&D with information on new market opportunities.

Marketing identifies trends and forecasts general business activities, providing long- and short-term forecasts. Detailed product sales forecasts are provided to establish production schedules.

The marketing department evaluates and reports on competition, providing information on trends, market share, and product features and positioning. It directs, in cooperation with the sales force, a continuing and coordinated program for sales and promotion including brochures, specification sheets, etc. It recommends new product pricing and performance levels and implements the introduction of new products into the field.

*The product management* department is the marketing representative on the new product project team. The product manager is the prime interface between the product and the market. The project manager should be the primary interface to the technology, according to the schedule time and cost goals.

The *development* department is responsible for the design of the product and, jointly with product management, must evaluates the design, making sure that it meets the market needs and justifies the project investments.

The formal processes for these important marketing functions should include the following:

Marketing research and analysis (market segmentation)
Product definition and positioning
New product business plans

**Business development.** A business development plan could consist of the following:

1. Statement of purpose of the plan
2. Specific market objectives to be achieved within the intermediate period (3–5 years)
3. Description of markets, including potential customers and channels of distribution
4. Description of the competition, their technical capabilities and potential, and their current product profile and emphasis
5. Description of products and services necessary for success in the next 3–5 years, and the plan for development or purchase of such products and service plans
6. Financial analysis of costs and returns
7. Risk assessment
8. Tactical implementation plan matching the tactical business horizon (12–18 months)

**System engineering.** The system engineering function is responsible for the architecture of the new product. Beginning with customer requirements, system engineers devise how the product structure will be divided among different modules and disciplines, each with its own set of specifications that come together to accomplish the product functions. They negotiate with the design engineers to best allocate the design activities, and actively pursue design trade-offs to accomplish the overall design functions.

Using six sigma design techniques, system engineers can augment the product realization process by changing the traditional role and culture of system engineering:

• Change from depending on past experiences for predicting design effort and specifications allocation risks to relying more on estimating the design function effort based on whether the design is difficult or easy to achieve, and how long will it take. In addition the design margin, which is the capability of the design to meet specifications, can be quantified using six sigma quality analysis.

• Six sigma analyses are more objective and can predict and focus future product quality and cost more accurately than traditional methods. System engineers can use six sigma to plan for investments in design and manufacturing capabilities, select and manage the supply chain, and budget for defect removal through integrated test strategies across different test systems and capabilities.

- System engineers can use six sigma analysis as the basis for design trade-offs and negotiations with design engineers.

**Development (design engineering).** The development department is responsible for the following functions:

1. A *communication function* with the corporate research laboratories or the general technology base of the company's business for initiating new product ideas and technologies.
2. An *advance development function* to transfer new technologies into product and process feasibility. It is beneficial to isolate this function from the tightly scheduled development effort.
3. A *development function* with cost and schedule control under project management and communications with system engineers to create new products.
4. A *technical support function* for the design systems that the engineers are using.

New product development projects are staffed by a team of multidisciplined engineers and managed by project managers that could be management appointed or team selected from either the development ranks or product management in marketing. The development department manager is responsible for the technical content and the technology to be used in the development project. He interfaces directly with the project manager and systems engineers on solving technical problems, as well as coordinating the project schedule and the assignments for tasks. Using six sigma design techniques, design engineers can augment the product realization process by changing the traditional role and culture of design engineering as follows.

- One of the important transitions in six sigma engineering is designing with statistical analysis versus worst-case design. This will result in a robust design rather than designs with unnecessary tight tolerance to control manufacturing variability or overdesigns that far exceed specifications.
- Six sigma design principles can help in the proper evaluation of design requirements based on design elements, component selection, and manufacturing process capability. This is important in making

sure that new designs will meet their specifications and the manufacturing cost and quality targets.

- Six sigma designs can quantify the quality of the design, so design engineers do not withhold margins. Design margin should be set to cover expected variability in manufacturing as determined by process capability analysis.
- Design engineers can evaluate the quality of newly designed products in manufacturing based on six sigma analysis. This would encourage them to seek out more information about manufacturability issues and how they can best meet DFM guidelines.

**Manufacturing.** The production department is responsible for the following functions for products and processes. Most of these functions can be enhanced and maintained through six sigma analysis. They include:

- *Current products support.* All operations, including scheduling, planning, documentation, assembly, and testing should be quantified using process capability analysis. In addition, process control should be in place either through control charting or defect analysis. Production should also be maintaining a continuous improvement posture to achieve higher quality, lower cost, and superior performance within the advertised specifications of the product and process.
- *New products introduction.* Manufacturing should work with design engineering to use six sigma to evaluate new designs by performing manufacturability assessment, prototype/production scheduling and layout, new material acquisition plans, product data transfer, and test development for in-line as well as final testing using a six sigma based test strategy.
- *Technical Support* for automation processes such as CAM and product data transfer
- *Process development* for assembly, test, and automation technology. Six sigma should be used as the target for process design and development. DoE should be used to optimize processes.
- *Communication* with the other parts of the organization: new product project management, logistics, materials, quality and development departments to ensure meeting of shipments and quality and cost targets.

The production department performs all the functions necessary to maintain a viable shipping profile. The interaction and communications with the logistics, materials, quality and other support functions, as well marketing and development should be open and bi directional.

**11.1.2.3 Emphasize the six sigma approach to project management.** The
team concept for project management has been shown to be very effective for new product development. In small projects, the product manager can double as the project manager. In this case, the original
product specifications can be preserved through development implementation, by having one person responsible for both technical and
project leadership.

In large projects, depending on the six sigma maturity of the organization, either an appointed or a de facto development project manager emerges, having the final authority to resolve, with other team
members, any technical problems that may arise. There is a core
team, composed of all the people who work on new development from
concept to production to field performance. Members of this core team
are the prime experts (or, initially, managers) of the different functions involved.

Reporting to this core team are the members of each activity, and
these members are identified by job function and contributions to the
project tasks. There are separate teams for electrical, mechanical,
software, production, logistics, and other functions, grouped by skill
sets, and they report the progress of the core team. In this model,
cross-functional communications occur directly from one team member to another, without having to pass through a management function. Cross-functional teams should be encouraged, as they shorten
communication loops.

The challenges to successful six sigma project management are as
follows:

1. Poor visibility of external and internal dependencies and their potential impact on the project schedule.
2. Focus mainly on the technical specification, with lack of coupling to
   financial, logistics, quality, production, and other functional expectations.
3. Lack of standard communication vehicles, both up to management
   and down to individual engineers.
4. Clear understanding of management goals and expectations.
5. Strong belief in the benefits of planning and executing project
   management guidelines and techniques.
6. Difficulty in allocating people in a dynamic environment, and
   defining the skill level necessary to successfully complete assigned
   tasks.
7. Unclear definition of project objectives and specifications, and reactive market decisions during the development phase.

The issues when implementing six sigma projects are:

- For management, it is the feeling that the project is on schedule, within the prescribed quality and cost, and with the defined functionality. This feeling *intensifies* as the project nears completion.
- For the development engineer, it is the clear understanding of the tasks required, the importance of the task in the overall project plan, and the amount of risks involved in meeting the technical objectives. The engineer must know when to report that a *schedule buster* problem has developed.
- For the project manager, it is how to operate between the rock and the hard place.

**Phased review and six sigma project management control.** One of the proven methods for alleviating these conditions is the use of the *phased review* technique. Specific phases are identified. Each phase can be viewed as a standalone entity with objectives, deliverables, product cost, quality, serviceability and manufacturability status, and project costs to date.

The phase review process brings the core project team and a selected management group together formally at the end of each phase (milestone) to review the status in terms of achieving objectives, analyze recommendations, make appropriate decisions, and commit to the next phase.

The spacing of these milestones is very important. They should be scheduled as required in groupings of like tasks, with more and closer milestones as the project nears completion. The functional team and top management should use the milestone meetings as an opportunity for detailed review of the project and its current direction and fit to the changing overall objectives.

Each functional area should have specific plans, measures, and goals as part of the overall project plans. These plans should be reviewed at the milestone meetings in addition to the technical review of the project. A sample of the plans could be:

- *Manufacturing* should have testability, quality, yield, and process capability goals, based on six sigma targets. Responsibilities include producibility feedback, test strategy and plans, production documentation, technical competence to handle the product after release to production, operator training, review of the manuals, and updates on production equipment installation and production process optimization.

- *Product management* should review the competition, market surveys, profitability, pricing, obsolescence, overall scheduling, and six sigma quality assessment at each milestone meeting with the general management.
- *Sales* should review the forecast, the product introduction plans, the promotional plans, and the feedback from customers and dealers.
- *Service* should review the mean time between failure and mean time to repair (MTBF and MTTR) and the service and repair plans, resources, training, and equipment.
- *Quality* should review the product quality specification and progress toward reaching six sigma goals according to the quality milestone plans.
- *Materials* should review part procurement and qualifications. Supplier status, especially overseas, should be also reviewed, and progress recorded toward achieving six sigma.
- *Controller department* should review expenditures to date and remaining funds and timing of major purchases. In addition, the costs and profit calculation should be redone if there are any changes.

A good project management plan will contain the schedule details, as well as input from all functions necessary for the overall success of the product. A set of goals for the plans should:

1. Provide concise project definitions with phases and milestones, and with specific deliverables at each phase.
2. Identify team members and their responsibilities, and keep management and team members informed of progress.
3. Force team members and managers to continuously evaluate and replan the project when problems occur.
4. Spot potential problems quickly, and help in taking preventative action in time.
5. Improve communications and delegation of duties and responsibilities.
6. Help the team focus on the activities at hand.

## 11.2 Cultural Issues with the Six Sigma Based System Design Process

The incorporation of a six sigma system design process is similar to earlier efforts to improve the product design and development process, such as concurrent engineering and design for manufacture. These new augmentations to the design process usually undergo several phases:

1. Management is the driver for the new six sigma effort, based on competitive factors or requests from customers. They set broad targets for quickly achieving the purported benefits from the new system. They will also set up support systems to help introduce the new concepts such as training programs and new tool champions. The efforts might be divided along skill sets such as electrical, mechanical, software, or multidisciplinary effort champions.

2. The engineers, including the system, design, and manufacturing engineers, react differently to the new six sigma requirements. Some see it as just as another "buzz word" with no redeeming benefits, and try to ignore it. Others see it as a burden that adds to the new product requirements with no additional resources or relaxation of the new product release schedule. Others resist it simply because they are comfortable with the current system, and they do not want to change by learning a new system. In addition, any new system changes the dynamics of the perceived skill set of engineers. An engineer who has been rated highly as a skilled and experienced designer with a reputation for making high-quality designs might feel threatened by a six sigma system that rates design quality independently. Such resistance to new ideas and systems has traditionally been known by the acronyms "SOW" (doing it the Same Old Way) or "NIH" (Not Invented Here).

3. A target product or system is designated as the beneficiary of the new six sigma system. A team is selected and the implementation of the new system is high on their priority. Three important requirements should be used to ensure the success of the team:

   i. The product being selected should be in the medium range of design effort and new technology content. It should not be a critical product to the company strategy, a new technology, or new market opportunity such as a "home run product." Nor should it be a simple redesign of or an add-on product to a product family. A target product to implement the new system could be a major redesign of a flagship product for lower cost or six sigma quality

   ii. The team selected to implement the new system should not be selected from the top performing engineers in the company. Such a "tiger team" might be unrepresentative of the engineering community of the company and difficult to emulate in future products.

    iii. A neutral facilitator or senior quality staff member familiar with the issues of statistics or familiar with six sigma from a different company could be available to act as advisor to the design team to make consensual decisions on issues raised by the new system.

4. The designated product team is very enthusiastic about the new system initially. Several problems begin to occur in terms of the specific implementation of the new system. This is a critical period; the team must pull together and resolve these problems. Management sensitivity at this point is very important in terms of providing additional resources or more time to resolve problems.

5. Gradually, champions will emerge who will use this new system to achieve unexpected benefits or exceed normal expectations in quality, cost, product design delivery, or customer satisfaction. These champions could come from the designated product group or from other projects that have leveraged the methods pioneered by the designated product team, or separately invented the methods.

6. The methods used in the six sigma system have to be transparent. They should be very easy to use and apply. Special tools are created or purchased to ease the application of these methods and techniques. A six sigma case study book should be created to document solutions of specific problems encountered in the new system, what assumptions were made and how they were resolved.

7. A general consensus will emerge that the new system is superior to the existing methodology and will gradually become the standard of new product design.

8. Some of the support staff or organizations for the new system will gradually drift away to other positions as the need for their services and knowledge in the new system decrease because of the general adoption and understanding of the new system issues by the engineering community. In many cases, the skills acquired in the support of the new system will be very valuable to those individuals, as they can share their expertise in new projects or other assignments.

### 11.3  Key Processes to Enhance the Six Sigma Product Creation Process

The following processes can be developed to operate in the suggested organizational structure in order to create an environment for continuos improvement and operational excellence through six sigma.

### 11.3.1  Six sigma phased review process

This should be the primary vehicle for project management. Specific phases are identified in the product development process, with each phase being a specific collection of task completions. Each phase can be viewed as a standalone entity with objectives, deliverables, product cost, quality, serviceability and manufacturability status, and project costs to date. Each phase should have a six sigma goal and an assessment of where the design is in meeting this goal.

The spider web chart is a method of visual presentation of the project meeting its separate goals in different phases through the project lifetime. A project on track will have concentric circles at each phase time period, represented for each phase in Figure 11.1. The scale of the spider web diagram is different for each parameter being measured. The center point of the diagram is the final goal of each project parameter being measured. The spider web chart represents a quick visual check of all project goals, and can be effective in spurring on different project groups to meet their goals concurrently with each other.

The project team should carefully plan each phase and milestone, with shorter time between the later milestones. This process should

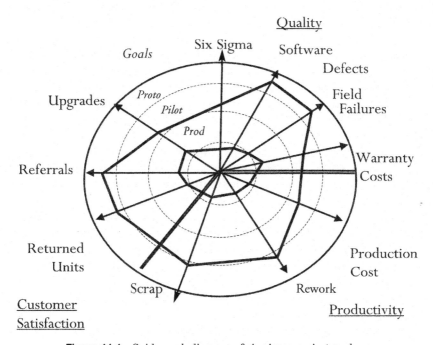

**Figure 11.1**  Spider web diagram of six sigma project goals.

be used as the primary vehicle to update management and project teams with the current status of the project. The phase review process brings the core project team and a selected management group together formally at the end of each phase (milestone) to review the status in terms of achieving objectives, analyze recommendations, make appropriate decisions, and commit to the next phase.

### 11.3.2   Six sigma quality advocacy and the quality systems review

This procedure is used to assure that the quality system is effective in achieving total quality and customer satisfaction. The historical focus on regulatory and product quality and reliability issues should be augmented by a quality advocacy at each functional level. The quality function at the company's highest management level should put sufficient emphasis on facilitating an organization-wide adoption of total quality methods (TQM) and six sigma across the total organization. It should have a process rather than a product focus. The role of the function is a consulting one assigned to assist other functions in integrating quality methods in their day-to-day operation.

The six sigma advocacy program could be initiated with some of these methodologies:

- Assign an organizational function to be the ultimate authority on six sigma. This function would make the decisions on six sigma policies and procedures. This function could reside in an existing organization such as the quality department, or in a committee made up of senior managers from different organizations in design and manufacturing.
- Enable a training program to educate the engineers and operators about six sigma. Establish a measure of the skills learned in six sigma, similar to the "belt" level in martial arts.
- Ensure that there is easy access to six sigma skills in the organizations. This could be accomplished by having very skilled personnel (six sigma "black belts") available in each department for consulting and encouragement on six sigma projects.
- Select a quality system goal such as six sigma or Cpk for a new product design or a manufacturing operation. Update all parts of the organization on the progress toward achieving the goal.
- Install a depository of six sigma project data documenting how six sigma was achieved and methods or tools used to successfully accomplish the project.

### 11.3.3    Six sigma manufacturability assessment and tactical plans in production

This process serves to evaluate new products for ease of manufacturing, to ensure a high level of quality, and to maintain lower production costs. The selection of the PCB components should be from those that are already in use in current products, if possible. The issues of production ease can be outlined in the following areas:

1. PCB assembly. This assessment ensures the selection of the proper six sigma measurement systems for the product. As discussed in Chapter 4, decisions have to be made in the selection of the metrics for six sigma quality design and control:

   - Which of the quality measurement systems are to be used as goals for new designs: six sigma (with or without 1.5 $\sigma$ average shift), Cpk, FTY, DPU (PPM), DPMO, or AQL levels?

   - What are the quality targets for PCB assembly prior to test, for current as well as new products, and in which measurement system?

   - What are the quality drivers for the six sigma design of new PCBs? PCB design guidelines should include the typical geometry requirements based on industry standards, as well as additional input from the manufacturing engineers as to what further selection should be made from available processes and suppliers of materials to ensure higher quality in production. For example, if a PCB is to be conformally coated, manufacturing should supply the performance specifications, quality level, and the cost of all available coating methodologies, with their recommended selection. If the process is not available locally, then they should recommend an approved supplier.

   - How should the opportunities for defects be standardized? By using components, terminations, or a combination of both, such as DPMO? In addition, is there a need to include defect data on certain component types, especially when the quality level is very high and the goal is in the part per billion range?

   - How should PCB processes be controlled and improved? Using sampling methods such as control charts ($\overline{X}$, $\overline{R}$, or $C$ charts), or using run charts and collecting individual defects for corrective action? The second choice is obvious for higher quality PCB assembly. In addition, decisions for data collection for PCB processes, PCB types, or a combination of both should be made. A manufacturing supplier or a PCB center serving multiple customers might collect data based on customers' PCBs as well as overall processes.

2. Testability assessment. This operation will ensure that all tests, including in-circuit, final, and systems tests, can be handled adequately by ensuring proper physical access to the PCBs and the product, using appropriate testing methodologies and integration of production and other service and self-test procedures and algorithms. The issues to be discussed and decisions to be made in test were discussed in Chapter 5 and are as follows:

- Identify factors that affect test effectiveness for new designs. These are given in Table 11.1 and should be reviewed by the test engineers in the design phase of new products.
- Identify the test goals in quality achieved, and the measurement systems for the quality.
- Identify the test effectiveness parameters such as test coverage, bad test effectiveness, and good test effectiveness on current PCBs. These are measures of a tester's ability to correctly distinguish between bad and good PCBs. The data should then be compared to the current quality output goals of the PCB operations, and decisions should be made as to which PCB test to improve first.
- Continuously examine the test plans for current products, and react swiftly when increases in quality can provide an opportunity for upgrading the test strategy, as shown in the example in Chapter 5.

3. Product assembly and supporting operations. This assessment ensures that proper current and future automation can be applied by reducing the number of distinct parts and assembly motions, simplifying part geometry and symmetry, and using other aides to en-

**Table 11.1**   Factors that affect test effectiveness

| Category | Subcategory | Examples |
|---|---|---|
| Technology | Circuit | Microwave circuits (require shielding) |
| | | Digital versus analog versus mixed |
| | Manufacturing | Through-hole versus SMT |
| | | Test pad size |
| | | Pitch size |
| | | Nodal access |
| | | Fixture fit |
| Business decisions | | Time and money budgeted for test and fixture development |
| | | Time allotted for in-line testing |
| Design for test (DFT) effort | | Existence of DFT effort |
| | | Use of built-in self-test (BIST) |
| | | Number of bytes of ROM dedicated to test |

hance automatic and robotic assembly. Process capability studies should be performed for all product assembly supporting operations such as the machine shop to supply design engineering with the data necessary to collect six sigma quality assessment of new designs.

## 11.4   Tools to Support Suggested Processes

A list of tools and strategies to support six sigma programs is shown below. Training on these tools should be provided to assist in achieving success in applying them.

| Topics/book chapter | Focus and tools |
| --- | --- |
| **Enterprise strategy** | |
| Chapter 2 | Goal setting at each level (six sigma, Cp, Cpk, FTY) |
| Chapter 11 | Six sigma advocacy through black belts |
| | |
| **Engineering design process** | |
| Chapter 10 | Project management/review including six sigma |
| Chapter 9 | New product lifecycle |
| Chapter 8 | Design quality assessment and statistical analysis |
| Chapter 7 | DoE use, especially in multi-disciplinary projects |
| Chapter 6 | Cost modeling and estimating tools |
| Chapter 5 | Establish process and gage capability baseline |
| Chapter 4 | Design manufacturing test map and strategy Manufacturing process control and improvement |
| Chapter 3 | Document and control the manufacturing process (TQM, control charts) |
| Chapter 1 | Improve manufacturing process (process mapping, QFD) |

# Index

**Sammy G. Shina,** P.E., is Professor of Mechanical Engineering at the University of Massachusetts, Lowell and has previously lectured at the University of Pennsylvania's EXMSE Program and at the University of California, Irvine. He is a past chairman of the Society of Manufacturing Engineers (SME) Robotics/FMS, a founding member of the Massachusetts Quality Award, and a member of the Mechanical Engineering advisory committee for NTU. He is the author of two best-selling books on concurrent engineering and has contributed two chapters and over 75 technical publications in his fields of research.

Dr. Shina is an international consultant, trainer and seminar provider on quality, six sigma, and DoE, as well as project management, technology supply chains, product design and development, and electronics manufacturing and automation. He worked for 22 years in high-technology companies developing new products and state of the art manufacturing technologies. He was a speaker for the HP Executive Seminars on Concurrent Product/Process Design, Mechanical CAD Design and Test, and the Motorola Six Sigma Institute. He received S.B. degrees in Electrical Engineering and Industrial Management from MIT, a S.M. degree in Computer Science from WPI, and a Sc.D. degree in Mechanical Engineering from Tufts University. He resides in Framingham, Massachusetts.

CPSIA information can be obtained at www.ICGtesting.com
Printed in the USA
BVOW06*1250010915

415631BV00002B/2/P